Sustainable Agricultural Marketing and Agribusiness Development: An African Perspective

_____

Sustainable Agricultural Marketing and Agribusiness Development: An African Perspective

---

Edited by

Brighton Nyagadza (CIM, Ph.D)

Marondera University of Agricultural Sciences and Technology (MUAST), Zimbabwe
Institute for the Future of Knowledge (IFK), University of Johannesburg, South Africa

Tanyaradzwa Rukasha (Ph.D)

Marondera University of Agricultural Sciences and Technology (MUAST), Zimbabwe

CABI

CABI is a trading name of CAB International

CABI
Nosworthy Way
Wallingford
Oxfordshire OX10 8DE
UK

CABI
200 Portland Street
Boston
MA 02114
USA

Tel: +44 (0)1491 832111
E-mail: info@cabi.org
Website: www.cabi.org

T: +1 (617)682-9015
E-mail: cabi-nao@cabi.org

A catalogue record for this book is available from the British Library, London, UK.

ISBN-13: 9781800622524   (hardback)
        9781800622531   (ePDF)
        9781800622548   (ePub)

DOI: 10.1079/9781800622548.0000

Commissioning Editor: Ward Cooper
Editorial Assistant: Lauren Davies
Production Editor: Rosie Hayden

Typeset by Exeter Premedia Services Pvt Ltd, Chennai, India

Dedication

To all emerging agricultural marketing and agribusiness enthusiasts.

The future is green!

Contents

Contributors

Abimbola Oluyemisi Adepoju, Department of Agricultural Economics, University of Ibadan, Ibadan, Nigeria; abimbola.adepoju@yahoo.com

Abolupe Oluyemi Ayanboye, Department of Fisheries and Aquatic Resources Management, Ejigbo Campus, Osun State University, Osogbo, Nigeria; oluyemi.ayanboye@uniosun.edu.ng

Adenike Omotola Ajibobare, Department of Agricultural Economics, University of Ibadan, Ibadan, Nigeria; aomotolami@gmail.com

Adulation Khayisano Ndlovu, Independent Researcher; adulationndlovu@gmail.com

Archeford Munyavhi, Department of Agribusiness and Management, Marondera University of Agricultural Sciences and Technology (MUAST), Zimbabwe; munyavhiarcheford@gmail.com; amunyavhi@muast.ac.zw

Basil Shumbanhete, Department of Agribusiness and Management, Marondera University of Agricultural Sciences and Technology (MUAST), Zimbabwe; bshumbanhete@gmail.com

Benard Chisiri, Department of Human Resource Management Manicaland State University of Applied Sciences (MSUAS), Mutare, Zimbabwe; benard.chisiri@staff.msuas.ac.zw

Benjamin Mudiwa, Zimbabwe Ezekiel Guti University (ZEGU), Bindura, Zimbabwe; b.mudiwa@gmail.com

Benson Ruzive, Faculty of Business & Economics, Modern College of Business & Science, Muscat, Sultanate of Oman; benson.ruzive@mcbs.edu.om

Brenda Guruwo, Department of Agricultural Extension and Technology, Marondera University of Agricultural Sciences and Technology (MUAST), Zimbabwe; guruwob@gmail.com

Brighton Nyagadza, Department of Marketing, Marondera University of Agricultural Sciences and Technology (MUAST), Zimbabwe and Institute for the Future of Knowledge (IFK), University of Johannesburg, South Africa; brightonnyagadza@gmail.com; bnyagadza@muast.ac.zw

Charles Tsikada, Department of Supply Chain Management, Marondera University of Agricultural Sciences and Technology (MUAST), Zimbabwe; tsikadac@gmail.com

Dumisani Rumbidzai Muzira, Department of Business Science, Africa University (AU), Mutare, Zimbabwe; rdmuzira@gmail.com

Emmanuel Ndhlovu, Vaal University of Technology (VUT), Vanderbijlpark, South Africa; manundhl@gmail.com

Esther Mafunda, Department of Business Management, Manicaland State University of Applied Sciences (MSUAS), Mutare, Zimbabwe; esther.mafunda@staff.msuas.ac.zw

Ernest Mugoni, Department of Supply Chain Management, Marondera University of Agricultural Sciences and Technology (MUAST), Zimbabwe; ernmugoni@gmail.com

Gibson Muridzi, Manicaland State University of Applied Science (MSUAS), Mutare, Zimbabwe; gibson.muridzi@staff.msuas.ac.zw

Gift Manhimanzi, Department of Economics, Zimbabwe Ezekiel Guti University (ZEGU); gcmanhie@gmail.com

Godwin Shumba, Department of Accounting, Manicaland State University of Applied Sciences (MSUAS), Mutare, Zimbabwe; godwin.shumba@staff.msuas.ac.zw

Jabulani Garwi, University of the Free State, Bloemfontein, South Africa; jabulanig400@gmail.com

Joseph P. Musara, Faculty of Natural and Agricultural Sciences, Department of Agricultural Economics, University of the Free State (UFS), South Africa; jpmusara@gmail.com

Julius Tapera, Lupane State University (LSU), Zimbabwe; juliuastapera@gmail.com

Luckmore Marodza, Department of Agribusiness and Management, Marondera University of Agricultural Sciences and Technology (MUAST), Zimbabwe; luckmaro@gmail.com

Maurice Kwembeya, Department of Psychology, Manicaland State University of Applied Sciences (MSUAS), Mutare, Zimbabwe; maurice.kwembeya@staff.msuas.ac.zw

Moses Jachi, Department of Accounting, Manicaland state University of Applied Sciences (MSUAS), Mutare, Zimbabwe; moses.jachi@staff.msuas.ac.zw

Noah Ariel Mutongoreni, Department of Human Resource Management, Manicaland State University of Applied Sciences, Mutare, Zimbabwe; noah.mutongoreni@staff.msuas.ac.zw

Nyasha Nyakuchena, Department of Agribusiness and Management, Marondera University of Agricultural Sciences and Technology (MUAST), Zimbabwe; gloriousnyakuchena@gmail.com

Iscacle Nyanhete, Department of Supply Chain, Insurance and Risk Sciences, Midlands State University (MSU) Zimbabwe; nyanheteiscacle74@gmail.com

Oluwakemi Adeola Obayelu, Department of Agricultural Economics, University of Ibadan, Ibadan, Nigeria; jkemmyade@yahoo.co.uk

Patrick Korera, Department of Accounting, Manicaland State University of Applied Sciences (MSUAS), Mutare, Zimbabwe; Patrick.korera@staff.msuas.ac.zw

Praise Zinhuku, Department of Performing and Visual Arts, Great Zimbabwe University, Masvingo, Zimbabwe; pzinhuku@gzu.ac.zw

Purity Hamunakwadi, Department of Building and Human Settlements Development, Faculty of Engineering, the Built Environment & Technology, Nelson Mandela University (NMU) South Africa; phamuna@gmail.com; s213442132@mandela.ac.za

Rahabhi Mashapure, Department of Agribusiness and Management, Marondera University of Agricultural Sciences and Technology (MUAST) Zimbabwe; cmashapure29@gmail.com

Reason Masengu, Department of Management Studies, Middle East College, Sultanate of Oman; masengu@mec.edu.om

Rumbidzai Pashapa, Department of Supply Chain Management, Marondera University of Agricultural Sciences and Technology (MUAST), Zimbabwe; rpashapa@muast.ac.zw

Savanhu Howard Manyere, Midlands State University (MSU), Gweru, Zimbabwe; savhomany@gmail.com

Tafadzwa Machaka, Department of Accounting, Manicaland State University of Applied Sciences (MSUAS), Mutare, Zimbabwe; tafadzwa.machaka@staff.msuas.ac.zw

Tafadzwa, Y. Chiwanza, Department of Agribusiness and Management, Marondera University of Agricultural Sciences and Technology (MUAST), Zimbabwe; tafyolanda@gmail.com

Tanyaradzwa Rukasha, Department of Development Sciences, Marondera University of Agricultural Sciences and Technology (MUAST), Zimbabwe; trukasha@muast.ac.zw

Thelma Lilian Munodawafah, Department of Supply Chain Management, Marondera University of Agricultural Sciences and Technology (MUAST), Zimbabwe; tmunodawafa@muast.ac.zw

Wellington Bandason, Department of Agribusiness Management, Faculty of Agricultural and Environmental Sciences, Women's University in Africa (WUA), Zimbabwe; wbandason@gmail.com

Preface

The vision and future of sustainable agricultural marketing and agribusiness development in Africa, as a continent with many potentially emerging economies, is set in the opening chapter by Nyagadza and Rukasha. The book is then divided into four sections.

The first section focuses on sustainable agriculture and includes chapters on sustainable agricultural practices that are applicable in the African context such as organic farming and urban agriculture. Garwi in Chapter 2 looks into sustainable agriculture development and rural transformation in Zimbabwe, by exploring the prospects and challenges of the smallholder farming sector. This is followed by Chapter 3 where Manyere constructs a systematic review of organic farming and organic foods in Southern Africa, towards organics 3.0. In Chapter 4, Oluwakemi *et al.* unpack the youths' participation in urban agriculture in Ibadan Metropolis, Nigeria. Further to this, human capital management - the conduit for unlocking agribusiness productivity in Zimbabwe issues - has been tackled in Chapter 5 by Mutongoreni *et al.*

The second section of the book focuses on agricultural marketing and entrepreneurship, and includes the challenges that are faced along the agricultural value chain, proposed procurement and supply chain strategies, as well as the role of women in agricultural entrepreneurship. In Chapter 6, Mudiwa and Ndlovu analyse smallholder farmers' enterprising tendencies and collective entrepreneurship towards increased incomes and poverty elimination with a perspective on Zimbabwe. This is positively connected to Chapter 7 by Mashapure *at al.* which discusses factors influencing rural female entrepreneurs in enhancing livelihoods from a global perspective. In Chapter 8, Ruzive and Masengu dig deep into sustainable agricultural supply chains on food security, with a systematic literature review methodology. In Chapter 9, Zinhuku explicates the eminence of music in agricultural marketing for sustainable development in Zimbabwe.

The book develops further in the third section by focusing on agricultural financing and investing. In Chapter 10 Muzira and Shumbanhete cover financial management for agricultural marketing and agribusiness development and, in Chapter 11, investment analysis in agribusiness and mitigation of capital risk. Chapter 12 by Munyavhi *et al.* looks into blockchain technology, sustainability and the future of public input distribution in Zimbabwe.

Lastly, the fourth section links the first three sections together by focusing on agricultural policy issues, and how the environment affects attainment of the goals set out in the previous sections. For sustainable agricultural marketing and sustainable agribusiness development to take place, there is need to engage in sustainable agriculture, have an in-depth understanding of the agricultural value chains and the players in the chain, have a financing and investing strategy and, lastly, have policies that enable these variables to be attainable. In Chapter 13, Mutongoreni *et al.* delve into re-envisaging agricultural public policy and governance in Zimbabwe. Resuscitation of neoliberalism in Zimbabwe by exploring implications for agriculture policy development is developed by Ndhlovu in Chapter 14. Nyakuchena *et al.* in Chapter 15 further analyse insights for sustainable rural agribusiness development policy. Chapter 16 by Tsikada *et al.* examines the procurement laws in agribusiness, which is closely connected to Nyanhete *et al.*'s Chapter 17 that focuses on agribusiness supply chain resilience. In Chapter 18 Tsikada *et al.* discuss sustainable supply chains in the agricultural sector. To conclude the book, in Chapter 19 Guruwo *et al.* assess the influence of village savings and loan associations on climate resilience and food security: a case of Domboshava, Zimbabwe.

By Brighton Nyagadza and Tanyaradzwa Rukasha

1 Introduction: Envisioning the Future of Sustainable Agricultural Marketing and Agribusiness Development

Brighton Nyagadza[1,2]* and Tanyaradzwa Rukasha[1]

[1]*Marondera University of Agricultural Sciences and Technology (MUAST), Zimbabwe; [2]Institute for the Future of Knowledge (IFK), University of Johannesburg, South Africa*

1.1 Introduction

This book comprises a collection of chapters relating to issues to do with sustainable agriculture, supply chain and procurement issues in the agricultural sector as well as policy implications. As this book offers practical insights into a wide range of topics on agricultural marketing and agribusiness management as well sustainability, it serves as a complement to the more specialized books that elaborate on each topic separately giving it a wide breadth of market and international appeal. Corporates in the world are going to be major players in the contribution to agricultural marketing and agribusiness management as well as sustainability, with readerships of millions. This publication is a significant addition. It provides companies, professionals, academics, and doctoral and graduate students in business with a comprehensive treatment of the nature of agricultural marketing and agribusiness management as well as sustainability transitions, and related practices in regions of the developing and the developed world. The text also serves as an invaluable resource for agricultural marketing practitioners requiring more than anecdotal evidence on the structure and operation of agricultural marketing and agribusiness management as well as sustainability in different organizations and geographical areas. It enables readers to compare and contrast the values of agricultural marketing and agribusiness management, and sustainability practices, covering different research methodological and settings. The book balances theory and experimentation by providing a thorough explication of the tools and techniques of agricultural marketing and agribusiness management, and sustainability relevant to the stakeholders' needs around the globe. This multi-faceted approach allows for a critical reflection on recent topical issues such as sustainability that have brought great paradigm shifts, based on a combination of academic insights and practical tactics.

The book is divided into four sections. The first section focuses on sustainable agriculture. This will include chapters on sustainable agricultural practices that are applicable in the African context, such as organic farming and urban agriculture. The second section of the book includes chapters focusing on agricultural marketing and entrepreneurship. This section will also include the challenges that are faced

*Corresponding author: brightonnyagadza@gmail.com

© CAB International 2023. *Sustainable Agricultural Marketing and Agribusiness Development: An African Perspective* (eds B. Nyagadza and T. Rukasha)
DOI: 10.1079/9781800622548.0001

along the agricultural value chain, proposed procurement and supply chain strategies as well as the role of women in agricultural entrepreneurship. The third section focuses on agricultural financing and investing. The fourth section links the first three sections by focusing on policy issues. The chapters in this section focus on how the environment affects attainment of the goals set out in the first three sections. For sustainable agricultural marketing and sustainable agribusiness development to take place, there is a need to engage in sustainable agriculture, have an in-depth understanding of the agricultural value chains and the players in the chain, have a financing and investing strategy and lastly have policies that enable these variables to be attainable.

1.2 Sustainable Agriculture

Agriculture is a major driver of economic growth in the greater part of the world. The sector is critical for food security, poverty reduction and industry linkage perspectives (Trabelsi *et al.*, 2016). The accelerating pace of climate change, population growth and changing dietary preferences, the global pandemic, and conflicts have threatened food security and the development of the agro-food sector. This has led to mounting pressure to shift the policy focus to the development of a more sustainable and resilient agro-food industry around the world (Velten *et al.*, 2015). Given the importance of the agricultural sector, emphasis is now being placed on sustainable agriculture. Sustainable development has been a guiding principle for all economics and politics in the past decade (Siebrecht, 2020). There is a general consensus between scholars and policy makers that sustainable agriculture plays a pivotal role in achieving global sustainable development (Trabelsi *et al.*, 2016). The main driving force for sustainable agriculture is to ensure that the current resource base is utilized in a manner that will not deplete the resources and to ensure that future generations can use these resources in continuity (Horrigan *et al.*, 2002). To that end, the urgency of developing sustainable agricultural production systems is widely acknowledged. This is even more important in the African context as the brunt of global climate change is most significant on the continent and if

care is not taken this will worsen food insecurity problems.

Sustainable agriculture as a concept involves ascertaining that food and fibre security is maintained for current use as well as future use (Gold, 2016). Apart from food and fibre security, sustainable agriculture also involves improving ecological conditions, improving soil fertility, encouraging biodiversity and supporting rural economic development among other goals (Gold, 2016). In order to achieve these goals a number of agricultural practices have been outlined. One of the most known practices is switching from chemical fertilizers to organic fertilizers or nitrogen-fixing plants, from aggressive pesticides to natural enemies, biodynamic and organic farming, and urban agriculture (Trabelsi *et al.*, 2016). These methods, especially in the African context, have proved to be more effective, unlike the 'Green Revolution' model. Africa was encouraged to adopt the green revolution, which emphasized using strains of crops that required agrochemical fertilizer, pesticides and irrigation, to increase yields. This method however proved to be unsustainable in Africa. This is largely because the continent imports 90% of its agrochemicals and because Africa is mainly characterized by small-scale farmers who cannot afford these chemicals. Sustainable agriculture will allow Africa to capitalize on its strengths: its land, local resources, indigenous plant varieties, indigenous knowledge, biologically diverse smallholder farms and limited use (to date) of agrochemicals. This will lead to an 'African Sustainable Green Revolution', which aims to increase agricultural productivity by using sustainable agricultural practices that minimize harm to the environment and build soil fertility. In the first section of the book, different researchers looked at the sustainable practices that are best suited for Africa including organic farming and urban agriculture.

1.3 Agriculture Marketing and Entrepreneurship

The second part of this book looks at the concept of agricultural marketing and entrepreneurship. Agricultural marketing and entrepreneurship in its simplest form involves the buying and selling of agricultural produce. Agricultural marketing is viewed as a process encompassing all the steps involved from the producers to the consumers including all pre- and postharvest operations. Agricultural entrepreneurship on the other hand is an ongoing process beginning with identification of values and ending with a strategic plan for addressing critical management functions (Timur *et al.*, 2020). Agricultural entrepreneurship can also be defined functionally as the process of applying the basic principles of entrepreneurship to agricultural and agrobased businesses or firms. Agri-entrepreneurship refers to the capacity of farmers to change, to abandon old models and to enter a new agricultural phase (Condor, 2020). The fundamental role of agricultural enterprises is to create a conducive environment as well as developing mechanisms for the sustainable development of the enterprise, providing employment and income for the rural population (Timur *et al.*, 2020). Agricultural marketing plays a fundamental role in achieving these goals.

Agricultural marketing in Africa is now playing an increasingly significant role especially after the launch of new economic policies and the consequent opening up of African markets to international markets. There has however been great concern in recent years with regards to the efficiency of marketing systems and marketing of agriculture produce in Africa. The general consensus from scholars is that there are poor linkages in marketing channels as well as poor marketing infrastructure and this is leading to high and fluctuating consumer prices. Secondly, this scenario is causing leakages and as such only a small proportion of the price paid by the consumer is reaching the farmers. There is also considerable wastage, deterioration in quality, and a considerable and frequent mismatch between demand and supply over time.

1.4   Finance and Investment in the Agriculture Sector

Agriculture provides livelihoods to many people in developing and emerging economies, especially those living in rural areas. Yet access to agricultural finance is often a hurdle. Access to finance is critical for the growth of the agriculture

sector. The shift from subsistence to commercial agricultural production requires funds. However, in developing countries, where agriculture is a source of livelihood for 86% of rural people (International Finance Corporation (IFC), 2013), financing for investments in agriculture is scarce, even for large investors. In Africa, less than 1% of commercial lending is destined to the agriculture sector (IFC, 2013). Lack of access to finance stops many farmers from adopting new technology and improving their efficiency. Responsible investment in agriculture and food systems is essential for enhancing food security and nutrition and supporting the progressive realization of the right to adequate food in the context of national food security (The Committee on World Food Security (CFS), 2014). Responsible investment makes a significant contribution to enhancing sustainable livelihoods, in particular for smallholders, and members of marginalized and vulnerable groups, creating decent work for all agricultural and food workers, eradicating poverty, fostering social and gender equality, eliminating the worst forms of child labour, promoting social participation and inclusiveness, increasing economic growth, and therefore achieving sustainable development.

Addressing the four dimensions of food security and nutrition – availability, access, stability and utilization – requires a significant increase in responsible investment in agriculture and food systems. Responsible investment in agriculture and food systems refers to the creation of productive assets and capital formation, which may comprise physical, human or intangible capital, oriented to support the realization of food security, nutrition and sustainable development, including increased production and productivity (Rezvanfar *et al.*, 2009). Responsible investment in agriculture and food systems requires respecting, protecting, and promoting human rights, including the progressive realization of the right to adequate food in the context of national food security, in line with the Universal Declaration of Human Rights and other relevant international human rights instruments. Responsible investment can be undertaken by a wide range of stakeholders.

Investing in agriculture and food systems can produce multiplier effects for complementary sectors, such as service or manufacturing industries, thus further contributing to food security and nutrition and overall economic development (Horrigan *et al.*, 2002). Without accompanying investment in public goods and services, such as infrastructure or a reinforced capacity for local government to deliver public services, many investments in agriculture and food systems would not be possible. However, the viability of investments in agriculture and food systems is also dependent on well-functioning ecosystems and sustainable use of natural resources. At the same time, the value of safety and health in generating productive agriculture and food systems is important and investing successfully means taking a holistic approach in terms of human, animal, environmental and overall public health (IFC, 2013). Responsible investment entails respect for gender equality, age and non-discrimination, and requires reliable, coherent and transparent law and regulations. Policies, laws and regulations must be well-designed and effectively implemented to ensure that such investments bring both economic and social benefits to the host country while guaranteeing a sustainable use of natural resources (Agula *et al.*, 2018).

The campaign in most countries to establish agriculture finance institutions as well as the drive to institute policies that promote lending to farmers emanates from the observation that there is a widespread shortage of finance, both long term and short term, in the sector compared to other sectors. This is perceived to be the cause of underdevelopment among the farming communities and low incomes for the farmers (Masiyandima *et al.*, 2011). Shortage of finance is also blamed for low technological advancement on farms, low mechanization, low fertilizer usage and the use of low yielding seed varieties, which all result in delayed growth and development of the sector. To date, commercial engagement in financing agriculture remains limited. The amounts mobilized from the private sector by official development finance going towards the agriculture sector averaged US$1.4 billion in 2019, which reflects 3% of the total amounts mobilized in that year (OECD, 2021).

1.5 Policy

Efficient agricultural policies are essential to meeting increasing demand for safe and nutritious food in a sustainable way (Gold, 2016). While growing demand for food, feed, fuel and fibre presents significant opportunities for agriculture, government policies must address challenges such as increasing productivity, enhancing environmental sustainability, reducing greenhouse gas emissions, and improving adaptation and resilience in the face of climate change and other unforeseen shocks (Horrigan *et al.*, 2002). Government support for agriculture has risen in recent years in response to global crises, but only a small amount has been directed at longer-term goals, such as climate change and other food systems challenges.

Estimates suggest that global food demand will increase by 70% by 2050 and at least $80 billion in annual investments throughout the value chains will be required in response (OECD, 2021). Most of this needs to come from the private sector due to limited public resources. Large-scale investments are needed for mechanization, climate-smart technologies, processing, and agrifood logistics. Smaller investments are also needed for farmers and agricultural micro-, small and medium-sized enterprises (MSMEs) to increase their productivity while reducing environmental impact and taking into account climate risks. Financial systems in most developing countries are ill-prepared to finance the shift to sustainable agriculture and agrifood industries. Banks, microfinance institutions and institutional investors have traditionally provided very limited resources for the sector. Agriculture loans and investments portfolios currently are disproportionately low compared to the agriculture sector's share of gross domestic product (GDP) (IFC, 2013). Important challenges for the financial markets include managing unique risks in agriculture, high transaction costs in dealing with large numbers of small farmers and MSMEs along the agriculture value chains, limited effective demand for finance, and lack of expertise of financial institutions in managing agricultural loan portfolios (Masiyandima *et al.*, 2011). On the other side, many countries have put in place inadequate or ineffective policies and instruments, which often limit the opportunities to mobilize private capital for the sector.

References

Agula, C., Akudugu, M.A. and Mabe, F.N. (2018) Promoting sustainable agriculture in Africa through ecosystem-based farm management practices: evidence from Ghana. *Agriculture & Food Security* 7(5). DOI: 10.1186/s40066-018-0157-5.

Condor, R. (2020) Entrepreneurship in agriculture: a literature review. *International Journal of Entrepreneurship and Small Business* 40(4), 516–562. DOI: 10.1504/IJESB.2020.109013.

Gold, M. (2016) Sustainable agriculture: definitions and terms. In: Gold, M. (ed.) *Sustainable Agriculture and Food Supply*. Apple Academic Press, Waretown, New Jersey, USA. DOI: 10.1201/b19837.

Horrigan, L., Lawrence, R.S. and Walker, P. (2002) How sustainable agriculture can address the environmental and human health harms of industrial agriculture. *Environmental Health Perspectives* 110(5), 445–456. DOI: 10.1289/ehp.02110445.

IFC (2013) *IFC and Agri-Finance: Creating Opportunity Where It's Needed Most*. International Finance Corporation.

Masiyandima, N., Chigumira, G. and Bara, A. (2011) Sustainable financing options for Zimbabwe. *ZEPARU Working Paper Series* 02(10).

OECD (2021) Amounts mobilised from the private sector for development. Available at: https://www.oecd.org/dac/financing-sustainable-development/development-finance-standards/mobilisation.htm (accessed 5 April 2023).

Rezvanfar, A., Samiee, A. and Faham, E. (2009) Analysis of factors affecting adoption of sustainable soil conservation practices among wheat growers. *World Applied Science Journal* 6(5), 644–651.

Siebrecht, N. (2020) Sustainable agriculture and its implementation gap—overcoming obstacles to implementation. *Sustainability* 12(9), 3853. DOI: 10.3390/su12093853.

The Committee on World Food Security (CFS) (2014) *Principles for Responsible Investment in Agriculture and Food Systems*. FAO.

Timur, N., Baltashev, J., Bayjanov, S. and Ismaylov, K. (2020) The importance of agricultural marketing services in the development of Agriculture of the Republic of Karakalpakstan. *Journal of Critical Reviews* 7, 2910–2914.

Trabelsi, M., Mandart, E., Le Grusse, P. and Bord, J.-P. (2016) How to measure the agroecological performance of farming in order to assist with the transition process. *Environmental Science and Pollution Research* 23(1), 139–156. DOI: 10.1007/s11356-015-5680-3.

Velten, S., Leventon, J., Jager, N. and Newig, J. (2015) What is sustainable agriculture? A systematic review. *Sustainability* 7(6), 7833–7865. DOI: 10.3390/su7067833.

2 Sustainable Agriculture Development and Rural Transformation in Zimbabwe: Exploring the Prospects and Challenges of the Smallholder Farming Sector

Jabulani Garwi[1]* and Reason Masengu[2]
[1]University of the Free State, Bloemfontein, South Africa; [2]Middle East College, Sultanate of Oman

Abstract

The possibility of achieving sustainable agriculture in sub-Saharan Africa has been a growing concern, particularly in Zimbabwe, where approximately 66% of the population rely on smallholder farming for their livelihoods. The smallholder farming sector in Zimbabwe is characterized by a focus on subsistence, extreme vulnerability to rainfall fluctuations, and poor productivity. The Zimbabwean government has been formulating and implementing policies and strategies to guide rural agricultural transformation, with a focus on sustainable smallholder agriculture. This chapter examines the question of whether sustainable agriculture is achievable in Zimbabwe by critiquing government agricultural policies and exploring the challenges confronting smallholder farmers. The authors utilized a systematic literature review methodology to investigate the attainment of sustainability in smallholder farming in Zimbabwe. The results indicate that, while the government has demonstrated a desire to transform peasant agriculture, numerous challenges threaten the sustainability of smallholder farming projects in Zimbabwe. These challenges include limited access to capital, insecure land tenure, inadequate infrastructure and technology, weak market institutions, and a lack of farmer involvement in production planning. To achieve productive and sustainable agriculture in Zimbabwe, the study recommends that significant investment should be made to address the challenges faced by smallholder farmers.

Keywords: Smallholder farmers, Sustainability, Agricultural policy, Poverty, Rural development

2.1 Introduction and Background to the Study

Agriculture is a critical sector in the economies of sub-Saharan Africa countries in terms of its contribution to food security, gross domestic product (GDP) (approximately 34%), job creation (65–75% of total workforce) and foreign exchange earnings (around 35%) (OECD-FAO, 2022). Improving agricultural sustainability is fundamental to poverty reduction, particularly in fulfilling elements of the Sustainable Development Goals (SDGs) by 2030. In Zimbabwe, the agricultural sector is regarded as a key driver of poverty reduction and a key enabler of economic growth (ZIMSTAT and World Bank, 2019). The sector offers employment and is a source of income for between 60 and 70% of the country's population;

*Corresponding author: jabulanig400@gmail.com

© CAB International 2023. *Sustainable Agricultural Marketing and Agribusiness Development: An African Perspective* (eds B. Nyagadza and T. Rukasha)
DOI: 10.1079/9781800622548.0002

it provides about 60% of the raw materials needed by the manufacturing industry and provides approximately 40% of total export earnings (Government of Zimbabwe, 2021a). The sector also contributes about 15–19% to the annual GDP, depending on the country's rainfall pattern (Government of Zimbabwe, 2021a). Despite the important role that agriculture plays in rural areas, concerns have been raised about the sustainability of smallholder agriculture in Zimbabwe (Gwara and Mazvimavi, 2016). The concerns emanate from the realization that there are many factors that continue to present insurmountable challenges to the sustainability of the smallholder farming sector.

The Zimbabwean government has been formulating and designing policies and strategies to guide overall national rural development, with particular focus on rural agricultural transformation since 1980. The government's agricultural development strategy placed emphasis on the commercialization of smallholder agriculture (Rubhara and Mudhara, 2019). Promoting sustainable agricultural practices is fundamental to the commercialization of the smallholder farming sector and has been one of the key pillars of the government's rural development strategy. Sustainable agriculture can be defined as farming practices that meet the needs of existing and future generations, while also ensuring profitability, environmental health and socio-economic equity (Ikerd, 1993; Ayoub, 2023). This chapter explores the agricultural policies adopted by the Zimbabwean government with the broad objective of evaluating the extent to which those policies embraced the three main components of sustainable agriculture identified by Ikerd, that is, economic viability, social profitability and environmentally sound practices. The chapter further explores the policies and strategies employed by the government to address the vagaries and variability of climate change among the smallholders.

2.1.1 Methodology

This study utilized a systematic literature review approach. This approach is part of a worldwide effort created by researchers to address the ongoing problem of insufficiently documented and transparent review methodologies that can be reported in published papers (Osobajo *et al.*,

2022). A systematic literature review involves a comprehensive and structured search of existing literature to identify relevant studies and data on a specific topic, followed by a rigorous analysis and synthesis of the findings (Todd, 2022). This method is considered a reliable way to identify and analyse the available evidence on a particular topic, and can help inform decision making and policy development (Kitchenham and Charters, 2021). The study used the following keywords to conduct the systematic literature review: smallholder farmers, sustainability, agricultural policy, poverty, and rural development. These keywords were used to identify relevant studies and data on the topic of interest, which in this case likely pertains to the challenges faced by smallholder farmers in developing countries, and the potential of sustainable agriculture and agricultural policy to alleviate poverty and promote rural development. By systematically reviewing the available literature on these topics, the study aimed to generate insights and recommendations for policy and practice that can contribute to more equitable and sustainable agricultural systems.

2.2 Literature Review

2.2.1 Critique of Zimbabwe agriculture policies

Since 1980, the Zimbabwean government has been implementing a series of agricultural policy reforms geared towards promoting sustainable smallholder agriculture. The policy reforms adopted by the government were largely shaped by the obtaining political and socio-economic environment. In the 1980–1990 period, the major dilemma confronting the Zimbabwean government was a high degree of rural poverty, emanating largely from the marginalization of the rural communities, who did not own land. Land is a key resource for any sustainable agricultural initiative (Rubhara and Mudhara, 2019). At independence, the Zimbabwe government inherited a dual economy characterized by skewed land distribution and ownership, which was biased in favour of a minority white community. Land was inaccessible to the black communal farmers, dating back to racist colonial policies instituted circa 1894 (Moyana, 2002; Sifile *et al.*, 2021). Given this state of affairs, the early policy reforms introduced by the newly independent Zimbabwe government focused more on land redistribution and provision of farming inputs as a rural transformation initiative.

2.2.2　The land reform policy and its implications on sustainable agriculture

Land possession is a key determining factor of smallholders' adaptive capacity in sustainable agriculture (Pickson and He, 2021; Kassa and Abdi, 2022). In Zimbabwe, the land redistribution exercise was carried out in three phases: 1980–1990 (first phase), 1998–1999 (second phase) and the post-2000 period (Fast Track Land Reform Programme – FTLRP). The land reform policy enhanced the growth of smallholder farming. Though the land reform programme successfully transferred the ownership of land to small-scale farmers, concern has been raised over concomitant environmental degradation and sustainability of resettlement schemes. Munongo (2014) carried out a study on the FTLRP in Masvingo and discovered that the majority of land beneficiaries who were coming from a communal farming background, brought with them poor animal husbandry practices that caused a lot of land degradation. The farmers' over-reliance on firewood as a source of energy also led to high rates of deforestation. The same findings were made by Moyo *et al.* (2009), who argue that uncontrolled deforestation in Mafungabusi Forest Reserve in Gokwe South District resulted in the loss of forest reserves and biodiversity. Findings from this study showed that the forested areas depleted at a rate of about 307 ha per annum (2.3%) during the FTLRP and at a rate of about 480 ha per year (5.1%) post-FTLRP (Moyo *et al.*, 2009). The absence of an environmental policy during that time made it hard to control the challenges arising from the land reform programme (Bevlyne *et al.*, 2003). The other criticism levelled against the land reform policy is that it has not been climate-sensitive and has not been able to resolve other intrinsic challenges threatening the viability of smallholder farming (Kudejira, 2014). For instance, agricultural productivity

in many smallholder farms has been undermined by changes in climatic conditions, which have affected farming seasons and have exposed smallholders to increased incidences of droughts and floods (Luhunga and Songoro, 2020; Mushore *et al.*, 2021).

2.2.3　The 1995–2020 Agricultural Policy Framework

As part of efforts to address agricultural production constraints faced by smallholder farmers, in 1994, the government came up with the 1995–2020 Agricultural Policy Framework. The policy framework advocated a free-market approach as the key answer to the agricultural productivity challenges (Government of Zimbabwe, 2012). The main objective of the policy framework, which was anticipated to guide agricultural development for two decades, was to implement land and agricultural reforms. It was anticipated that the policy would promote productive land use and increase smallholder farming output at a much faster rate than was obtaining during the period before the Economic Structural Adjustment Programme (ESAP), which was introduced from around 1991 to 1994 (Mutekwa and Mutekwa, 2009). According to Flaherty *et al.* (2011) and Mutisi *et al.* (1994), the specific objectives of the policy framework were to improve foreign currency earnings from agricultural exports, increase income for the smallholder farming communities as well as enhance national food security and ultimately contribute to regional food supplies. The goal was to transform smallholder farming into a fully-fledged commercial farming sector. Nevertheless, the 1995–2020 Agricultural Policy Framework did not achieve its intended objectives. Policy implementation was dogged by inadequate funding and limited political commitment, thereby undermining the government's efforts to resolve production challenges that were bedevilling the operations of smallholder farming projects. Worse still, the 1995–2020 Agricultural Policy Framework did little to cushion smallholders from the growing threat of climate change and variability in Zimbabwe (Kudejira, 2014).

2.2.4 Towards sustainable agriculture: the Comprehensive Agriculture Policy Framework (CAPF) (2012–2032)

The land policy reform pursued by the government, particularly the FTLRP, brought about fundamental changes in Zimbabwe's agrarian landscape. The policy reform resulted in an exponential growth in the number of smallholder farmers, which also brought with it several challenges. As noted by Kudejira (2014) the challenges brought about by the land reform policy made the 1995–2020 Agriculture Policy Framework obsolete. Concern was raised about the need to come up with better farming practices that addressed the growing threat of food insecurity and the need for smallholder farmers to embrace farming practices that protect the environment and that are sensitive to the new realities of climate change. This prompted the government to explore the possibility of coming up with a holistic policy framework that would cushion smallholders from the vagaries of climate change. By 2009, the government, through the then Ministry of Agriculture, Mechanization and Irrigation Development (MAMID), now Ministry of Lands, Agriculture, Water, Fisheries and Rural Development (MLAWFRD), together with other stakeholders, had already begun the process of updating the 1994 framework, which gave birth to the Comprehensive Agriculture Policy Framework (CAPF) 2012–2032 (Government of Zimbabwe, 2012). The new framework's vision was anchored on ensuring 'a prosperous, diverse and competitive agriculture sector, ensuring food and nutrition security significantly contributing to national development' (Government of Zimbabwe, 2012). Given the vulnerability of the smallholder farming sector to the changing climatic conditions, the Comprehensive Agriculture Policy Framework placed emphasis on climate change coping mechanisms.

2.3 Climate Smart Agriculture Policy (CSAP)

The CAPF (2012–2032) prioritized the use of climate-smart technologies for sustainable agriculture, including the adoption of conservation agriculture, crop diversification, irrigation development, investment into climate change research, farmer capacity-building programmes and improved agricultural extension support services.

2.3.1 Conservation agriculture

Conservation agriculture is one of the climate-smart agriculture (CSA) activities widely practised in Zimbabwe. The farming system has been practised as far back as the 1980s in different parts of the country (Rubhara and Mudhara, 2019). Conservation agriculture is a form of sustainable farming practice that 'minimizes soil disturbance, applies more precise timing of planting, and utilizes crop residue to retain moisture and enrich the soil' (Bolwig *et al.*, 2009; Mazwi *et al.*, 2018). In 2018, the government scaled up and strengthened conservation agriculture through the adoption of what has come to be known as *Pfumvudza/ Intwasa* programme (Government of Zimbabwe, 2021b). *Pfumvudza/Intwasa* is a crop cultivation system that embraces intensive farming methods to encourage efficient use of resources such as farming inputs and labour on a smaller piece of land to optimize its management. The Food and Agriculture Organization (FAO) and Foundations for Farming (FfF) have been offering training and support services to government extension workers to help them implement the *Pfumvudza* concept (FAO, 2022). When the *Pvumvudza/Intwasa* programme was launched in 2018, the government's target was to train 1.8 million farmers in conservation agriculture by October 2020 (Government of Zimbabwe, 2021b).

2.3.2 Crop and livelihood diversification

Crop diversification is a key component of Zimbabwe's CSAP. Crop diversification has been defined as 'the cultivation of more than one variety of crops belonging to the same or different species in a given area in the form of rotations and or inter-cropping' (Dorward and Kydd, 2002). It is regarded as one of the most cost-effective and eco-friendly farming practices. Several farming communities in Zimbabwe, particularly those located in the drier parts of the country, have embraced CSA through crop diversification, including adopting a variety of drought-resistant crops such as sorghum, rapoko and millet. Several non-governmental organizations (NGOs) including CARE International, Oxfam, Concern Worldwide and ActionFaim have been instrumental in promoting crop diversification through the introduction of a variety of drought-resistant crops (Kang'ethe and Serima, 2018). Research carried out by Nyashadzashe et al. (2014) and Zembe et al. (2014) revealed that sorghum and finger millet have become crops of choice for farmers in the drought-prone rural communities of Gutu District in Masvingo Province, owing to their drought tolerance and insensitivity to temperature fluctuations in comparison to maize. This finding is in sync with the observation of Gwara and Mazvimavi (2016) that the cultivation of drought-resistant crops, such as finger millet, has cushioned farmers from the risks of climate change-induced droughts that have hard-hit several parts of Bikita for the past years.

2.3.3 Irrigation development

Irrigation development has long been recognized as a key aspect of smallholder farming in Zimbabwe. The zeal to promote irrigation agriculture has been demonstrated over the years through several policy implementation drives. For instance, in September 2016, the government approved the Smallholder Irrigation Revitalization Programme (SIRP) with an overall objective of improving agricultural productivity of the smallholder farming sector and cushioning smallholders from the growing threat of climate change (Government of Zimbabwe, 2021b). Recently, the government introduced the Accelerated Irrigation Rehabilitation and Development Plan (AIRDP): 2021–2025 as its new policy framework to guide the implementation of irrigation projects in Zimbabwe. The specific objectives of the AIRDP are to improve governance and management of irrigation schemes, promote efficiency in the use of water for irrigated farms, and improve smallholders' accessibility to funding, markets and farming inputs (Government of Zimbabwe, 2021b). Studies have, however, shown that inappropriate management models, land tenure insecurity and water scarcity in some parts of the country have resulted in poor performance of many irrigation schemes (Oxfam SARPN, 2021).

2.3.4 Climate change research programme

In so far as research is concerned, the government established the Department of Research and Specialist Services (DRSS) to take a lead in CSA research programmes. The department has been working with the Crop Breeding Institute (CBI) to train farmers in crop breeding courses (de Villiers, 2003). A variety of seeds have been produced through the company's research programmes, including the drought-tolerant open-pollinated varieties, which have been distributed by agro-based firms such as the Agriseeds, Agricultural and Rural Development Authority (ARDA) and Champion (FAO, 2022). Institutions of higher learning, namely the University of Zimbabwe and Chinhoyi University of Technology are also carrying out CSA-related research programmes, particularly on soil fertility management issues. The other players participating in CSA research programmes include such centres as the International Maize and Wheat Improvement Center (CIMMYT) whose focus has been mainly on developing drought-tolerant maize varieties (Jahedi, 2016). CIMMYT has also been collaborating with international organizations such as the FAO in supporting government programmes aimed at building key policies and an enabling environment for CSA activities. Civil society organizations (CSOs) have also been very active in CSA-related activities. For instance, World Vision, Oxfam and Practical Action have been carrying out advocacy work through the Climate Change Working Group. The primary focus of these CSOs has been on adaptation strategies owing to the fact that they work mainly with the most vulnerable communities residing in the arid regions of the country (FAO, 2022).

2.3.5 Agricultural extension support services

The Agricultural Advisory Services (formerly AGRITEX) has the primary responsibility of providing agricultural extension support services to smallholder farmers in Zimbabwe. The department's key objective is to increase farm productivity while simultaneously maintaining the sustainability of the agricultural production base (Chiremba and Masters, 2003). It does this through the provision of general extension services, training of smallholders in the use of new agricultural technologies such as CSA activities, identifying farmer groups for proper extension efforts, carrying out environmental awareness campaigns among smallholders, and equipping smallholders in resolving their own challenges and promoting agricultural development.

2.4   Challenges to Sustaining Smallholder Agriculture in Zimbabwe

Despite the various efforts by government to promote sustainable agriculture, there are a myriad of factors undermining the productive capacity of smallholder farmers in Zimbabwe. The challenges confronting smallholders are multifaceted in nature and these include, among others, weak agricultural support services, poor infrastructure, lack of adequate inputs, low human capacity and insecure land tenure.

2.4.1   Insecure land tenure

Following the implementation of the FTLRP, the issue of land tenure has become a central concern in Zimbabwe's agrarian studies. Unlike the pre-2000 land redistribution programme, which offered title deeds to land beneficiaries, beneficiaries of the FTLRP were not accorded land ownership rights; they were allowed to settle and work on the farms on account of 99 year leases though the government held the exclusive right to cancel those leases at any time, with three-months' notice (Charinda *et al.*, 2012). As noted by de Villiers (2003), the absence of land ownership rights in the FTLRP plunged smallholders into a high degree of uncertainty about the security of their tenures. Nyashadzashe *et al.* (2014) contend that this uncertainty and insecurity of land tenure has been a major demotivating factor for farmers who want to make long-term investments on their farms. This observation resonates well with the World Bank's (2016) conclusion that lack of tenure system is the major cause of low agricultural productivity in most communal farms of the Least Developed Countries (LDCs).

Critics of the FTLRP argue that due to the lack of title to the land, experienced farmers are reluctant to invest in communal areas.

2.4.2   Weak agricultural support services

The FTLRP coincided with a decline in key agricultural services, particularly technical and financial support services. Experience elsewhere has shown that provision of adequate financial support services and technical expertise is pivotal in enhancing sustainability of smallholder farming (Chambers and Conway, 2014). Armed with this understanding, soon after attaining independence in 1980, the government made a lot of investments aimed at capacitating AGRITEX, the department that is mandated with providing agricultural extension services in Zimbabwe. According to Moyo *et al.* (2009), while AGRITEX did very well in equipping smallholder farmers in the pre-2000 period, the situation changed after 2000, especially with the implementation of the FTLRP. The authors elaborated that there was a marked decline in agricultural support services following the FTLRP as extension officers failed to cope with the growing number of smallholder farmers in resettlement schemes. To date, smallholder farmers are still confronted with challenges of restricted extension, technical and research support services resulting from a prolonged period of economic decline spanning over a period of two decades (Zimbabwe Vulnerability Assessment Committee, 2021).

2.4.3   Low human capacities to drive farming

The government's failure to provide adequate technical and extension support services to the resettlement schemes translated into a rise in the number of unskilled smallholder farmers (Scoones *et al.*, 2018). Moreover, since the FTLRP was more of a political rather than economic initiative, most of the beneficiaries who received the land did not have the requisite farming skills (Kang'ethe and Serima, 2018). As observed by Kang'ethe and Serima (2018), some of the FTLRP beneficiaries received land only for prestigious and patronage reasons, and because they were members of the ruling party, without the capacity to do farming. This has resulted in several farms lying idle. Reports have also been made of land beneficiaries vandalizing the already built infrastructure that they inherited from the former white commercial farm owners.

2.4.4   Inadequate access to farming inputs

Adequate and timely supply of farming inputs is central to the attainment of sustainable agriculture. Supply lethargy and deficiency has acted as a key impediment to the productivity of the smallholder farming sector in Zimbabwe (FAO, 2022). Mazwi et al. (2018) pinpointed that the majority of resettled farms have remained largely unsustainable and have not contributed to expectations around poverty reduction and food security due to poorly designed farming input supply schemes. The authors elaborated that most land reform beneficiaries were unemployed families and the rural poor who could not and still cannot afford to buy farming inputs on their own. The plight of these smallholders has worsened due to economic hardships experienced in the country for the past two decades (Mazwi et al., 2018).

2.4.5   Poor infrastructure

Another key factor threatening the sustainability of smallholder farming in Zimbabwe is the lack of adequate and suitable infrastructure. Concern has been raised over the inaccessibility of many farming areas owing to dilapidated conveyance infrastructure. Research carried out by Scoones et al. (2018) found that poor road networks were affecting smallholder farming operations in Masvingo, particularly the FTLRP beneficiaries. The poor road network is affecting delivery of farming inputs and marketing of farming produce. Similar concerns were raised by Hahlani and Garwi (2014) who decried the state of roads in Mayfield Dairy Scheme in Manicaland Province. The authors reported that the roads in Mayfield rural community were in a very poor state and were last serviced some six to

7 years ago. The whole road network servicing dairy farmers had deteriorated into dangerous gullies. The poor state of roads affected extension officers who relied on the use of motorcycles when visiting dairy farmers.

2.5   Conclusions and Recommendations

The preceding discussion has revealed significant deficiencies in Zimbabwe's agricultural policy reform process. Despite the government's apparent interest in promoting sustainable smallholder agriculture, achieving this goal has remained elusive. Zimbabwe's agricultural policy transformation has been primarily driven by political considerations rather than sustainable and profitable farming practices, with the land reform policy serving as the main anchor. Consequently, a range of socio-economic difficulties that threaten the sustainability of smallholder farming have not been adequately addressed. Although the agrarian policy reform process has successfully distributed land to the rural poor, the majority of land recipients still struggle with issues such as inadequate farming supplies, insufficient infrastructure, inadequate technical and extension support services, insecure land tenure, weak support structures, and a lack of human capacity to support smallholder farming. These challenges have been further exacerbated by ongoing shifts in climatic conditions, which have resulted in increased poverty and vulnerability among smallholder farmers to natural disasters.

A range of policy recommendations can be proffered to enhance the government's efforts towards promoting sustainable agriculture in Zimbabwe:

- Due to the limited farming skills of most smallholders, it may be necessary to prioritize and enhance training programmes for land beneficiaries to ensure the sustainability and overall progress of smallholder farming. Successful farming requires comprehensive knowledge, and farmers can only increase productivity with adequate training.
- An intensive training program for agricultural extension workers is also necessary to equip them with the relevant knowledge of

sustainable agriculture and coping mechanisms for climate change and variability.

- The insecurity of land tenure creates a high degree of uncertainty, which hinders farmers who want to make long-term investments in their farms. Urgent action is needed to address this issue.
- The National Climate Change Research Strategy (NCCRS) mandated by the government to lead climate change research is currently underfunded and operating below capacity. To improve the operation of the NCCRS, it is important to provide funding and other key resources.
- Improving the farming input delivery system is also necessary to support farmers in their businesses.
- Many resettlement schemes have neglected roads, which is a major issue. The government should provide free, all-weather major roads and upgrade existing road networks to address this problem.

References

Ayoub, M. (2023) One size does not fit all: the plurality of knowledge sources for transition to sustainable farming. *Journal of Rural Studies* 97, 243–254. DOI: 10.1016/j.jrurstud.2022.12.007.

Bevlyne, S., Bruce, C., Dale, D. and Witness, K. (2003) Narratives on land: state-peasant relations over fast track land reform in Zimbabwe. *African Studies Quarterly* 7, 81–95.

Bolwig, S., Gibbon, P. and Jones, S. (2009) The economics of smallholder organic contract farming in Tropical Africa. *World Development* 37(6), 1094–1104. DOI: 10.1016/j.worlddev.2008.09.012.

Chambers, R. and Conway, G. (2014) *Sustainable Rural Livelihoods: Practical Concepts for the 21st Century*. IDS Discussion Paper.

Charinda, L., Bello, H.M. and Thomas, B. (2012) Sustainable solutions to the resuscitation of agricultural cooperatives in Zimbabwe: a case for Manicaland Province. MSc Thesis, University of Namibia.

Chiremba, S. and Masters, W. (2003) The experience of resettled farmers in Zimbabwe. *African Studies Quarterly* 7, 97–117.

de Villiers, B. (2003) *Land reform: Issues and Challenges: A Comparative Overview of Experiences in Zimbabwe, South Africa and Australia*. Konrad Adenauer Foundation, Namibia.

Dorward, A. and Kydd, J. (2002) Locked in and locked out: smallholder farmers and the new economy in low income countries. *Presented at the 13th International Farm Management Congress*, July 7–12, 2002. Available at: https://citeseerx.ist.psu.edu/document?repid=rep1&type=pdf&doi=1b90a40c23 37117647e173deae6c0157dfd4680b (accessed 18 July 2023).

FAO (2022) *FAO Publications Catalogue 2022*.

Flaherty, K., Chipunza, P. and Nyamukapa, A. (2011) Key trends since 2002. Available at: www.asti. cgiar.org/zimbabwe (accessed 14 February 2023).

Government of Zimbabwe (2012) Comprehensive Agricultural Policy Framework Executive Summary.

Government of Zimbabwe (2021a) Twenty-third post-cabinet press briefing-Ministry of information, publicity & Broadcasting services. Cabinet press briefing. Available at: https://www.infomin.org.zw/twenty-third-post-cabinet-press-briefing-2/ (accessed 14 February 2023).

Government of Zimbabwe (2021b) Post Cabinet Press Briefing Statement 3 August (3).

Gwara, S. and Mazvimavi, K. (2016) The adoption of a portfolio of sustainable agricultural practices by smallholder farmers in Zimbabwe. In: *5th International Conference of the African Association of Agricultural Economists*, Addis Ababa, Ethiopia, September 23–26, 2016. Available at: https://www.researchgate.net/publication/330846261

Hahlani, C.D. and Garwi, J. (2014) Operational challenges to smallholder dairy farming: the case of Mayfield Dairy Settlement Scheme in Chipinge District of Zimbabwe. *IOSR Journal of Humanities and Social Science* 19(1), 87–94. DOI: 10.9790/0837-19148794.

Ikerd, J.E. (1993) The need for a system approach to sustainable agriculture. *Agriculture, Ecosystems & Environment* 46(1–4), 147–160. DOI: 10.1016/0167-8809(93)90020-P.

Jahedi, R. (2016) *Creating a Business Conducive Environment to Attract Foreign Direct Investment in Bangladesh*. GRIN Verlag.

Kang'ethe, S.M. and Serima, J. (2018) Exploring challenges and opportunities embedded in small-scale farming in Zimbabwe. *Kamla Raj Enterprises* 46(2), 177–185. DOI: 10.1080/09709274.2014.11906718.

Kassa, B.A. and Abdi, A.T. (2022) Factors influencing the adoption of climate-smart agricultural practice by small-scale farming households in Wondo Genet, Southern Ethiopia. *SAGE Open* 12, 3. DOI: 10.1177/21582440221121604.

Kitchenham, B. and Charters, S.M. (2021) Guidelines for performing systematic literature reviews in software engineering. EBSE Technical Report.

Kudejira, D. (2014) *Early Career Fellowship Programme: An Integrated Approach Towards Moderating the Effects of Climate Change on Agriculture: A Policy Perspective for Zimbabwe*. Future Agricultures.

Luhunga, P.M. and Songoro, A.E. (2020) Analysis of climate change and extreme climatic events in the Lake Victoria Region of Tanzania. *Frontiers in Climate* 2. DOI: 10.3389/fclim.2020.559584.

Mazwi, F., Chambati, W. and Mutodi, K. (2018) *Contract Farming Arrangement and Poor Resourced Farmers in Zimbabwe*. SMAIAS.

Moyo, S., Chambati, W., Murisa, T., Siziba, D., Dangwa, C, *et al.* (2009) *Fast Track Land Reform Baseline Survey in Zimbabwe: Trends and Tendencies, 2005/06*. African Institute for Agrarian Studies.

Moyana, H. (2002) *The Political Economy of Land in Zimbabwe*. Mambo Press, Gweru, Zimbabwe.

Munongo, S. (2014) Welfare impact of private sector interventions on rural livelihoods: the case of Masvingo and Chiredzi smallholder farmers. *Russian Journal of Agricultural and Socio-Economic Sciences* 10, 3–9. DOI: 10.18551/rjoas.2012-10.01.

Mushore, T.D., Mhizha, T., Manjowe, M., Mashawi, L., Matandirotya, E. *et al.* (2021) Climate change adaptation and mitigation strategies for small holder farmers: a case of Nyanga District in Zimbabwe. *Frontiers in Climate* 3, 82. DOI: 10.3389/fclim.2021.676495.

Mutekwa, T. and Mutekwa, V.T. (2009) Climate change impacts and adaptation in the agricultural sector: the case of smallholder farmers in Zimbabwe. *Journal of Sustainable Development in Africa* 11, 2.

Mutisi, C., Gomez, M., Madsen, J. and Hvelplund, T. (1994) *Proceedings of the Workshop on Integrated Livestock/Crop Production Systems in The Small Scale and Communal Farming Sectors in Zimbabwe*. University of Zimbabwe Press, Harare, Zimbabwe.

Nyashadzashe, Z., Edmore, M., Fungai, H.M. and Elizabeth, C. (2014) An assessment of the impact of the fast track land reform programme on the environment: the case of Eastdale Farm in Gutu District, Masvingo. *Journal of Geography and Regional Planning* 7(8), 160–175. DOI: 10.5897/JGRP2013.0417.

OECD-FAO (2022) Part I. Agriculture in Sub-Saharan Africa: prospects and challenges for the next decade. In: *Agricultural Outlook 2022–2031*. OECD Publishing, Paris.

Osobajo, O.A., Oke, A., Omotayo, T. and Obi, L.I. (2022) A systematic review of circular economy research in the construction industry. *Smart and Sustainable Built Environment* 11(1), 39–64. DOI: 10.1108/SASBE-04-2020-0034.

Oxfam SARPN. (2021) SARPN – Zimbabwe. Available at: https://sarpn.org/documents/d0001707/index.php (accessed 14 February 2023).

Pickson, R.B. and He, G. (2021) Smallholder farmers' perceptions, adaptation constraints, and determinants of adaptive capacity to climate change in Chengdu. *SAGE Open* 11(3). DOI: 10.1177/21582440211032638.

Rubhara, T. and Mudhara, M. (2019) Commercialization and its determinants among smallholder farmers in Zimbabwe. A case of Shamva District, Mashonaland Central Province. *African Journal of Science, Technology, Innovation and Development* 11(6), 711–718. DOI: 10.1080/20421338.2019.1571150.

Scoones, I., Marongwe, N., Mavedzenge, B., Murimbarimba, F., Mahenehene, J. *et al.* (2018) *Zimbabwe's Land Reform: A Summary of Findings*. IDS, Brighton.

Sifile, J., Chiweshe, M.K. and Mutopo, P. (2021) Political economy of resettlement planning and beneficiary selection in A1 and A2 settlement models in Zimbabwe Post 2000. *OALib* 08(08), 1–17. DOI: 10.4236/oalib.1107758.

Todd, P. (2022) Subject guides: systematic review: PRISMA. Available at: https://guides.lib.monash.edu/systematic-review/manuals-and-prisma (accessed 19 August 2022).

World Bank (2016) The Growth Report: strategies for sustained growth and inclusive development. Commission on Growth and Development, World Bank.

Zembe, N., Mbokochena, E., Mudzengerere, F.H. and Chikwiri, E. (2014) An assessment of the impact of the fast track land reform programme on the environment: the case of Eastdale Farm in Gutu District, Masvingo. *Journal of Geography and Regional Planning* 7(8), 160–175. DOI: 10.5897/JGRP2013.0417.

Zimbabwe Vulnerability Assessment Committee (2021) 2021 Rural Livelihood Assessment Report. Food Security Cluster. Available at: https://fscluster.org/zimbabwe/document/2021-rural-livelihood-assessment-report (accessed 14 February 2023).

ZIMSTAT and World Bank (2019) Analysis of spatial patterns of settlement, internal migration, and welfare inequality in Zimbabwe 1 Joint Report.

3 A Systematic Review of Organic Farming and Organic Foods in Southern Africa: Towards Organics 3.0

Savanhu Howard Manyere*

Midlands State University, Gweru, Zimbabwe

Abstract

Purchase decisions or consumption represent the 'trigger' of the demand side of sustainable consumption and production (SCP) while production decisions represent the supply side. Organic food systems have the ability to address challenges of massive deforestation, water scarcity, soil depletion, erosion of biodiversity and human health hazards in line with the Sustainable Development Goals. Such a desirable direction is only possible through investments in agronomic/technical, consumer and institutional research. However, in Southern Africa, there is a paucity of literature and research on organic farming and organic foods. This systematic review and bibliometric analysis is the first attempt to shed light on organic foods and organic farming in Southern Africa from farmers', consumers', agronomic and institutional perspectives. The search was performed in Lens Database in May 2022 to provide an analysis of topics covered in Southern African literature so as to inform future studies thereby filling this gap. The results show that South Africa is the most active country with publications totalling 24 out of 26. There is a huge research gap in organic farming and foods in other Southern African countries particularly Zimbabwe and Zambia, which each have one publication. Angola, Botswana, Lesotho, Malawi, Mozambique, Namibia and Swaziland have no single study specifically dealing with organic farming and organic foods. For the inclusion of small-scale farmers, it is recommended that agronomic, consumer and institutional research and support are reinforced through sound organic farming policy in order to achieve organics 3.0.

Keywords: Sustainable consumption and production, Organic farming, Organic foods, Organics 3.0

3.1 Introduction

Massive deforestation, water scarcity and deteriorating natural resources have accompanied world food systems (FAO, 2021), hence the unsustainability of food systems. Also, the COVID-19 pandemic has left the fragilities of national agrifood systems widely exposed (FAO, 2021). According to Geiger *et al.* (2017), organic agriculture has the potential to transform global agriculture into sustainable food systems. Purchase decisions or consumption represent the 'trigger' of the demand side of sustainable consumption and production (SCP) while production decisions represent the supply side. Organic food systems can address challenges of massive deforestation, water scarcity, soil depletion, erosion of biodiversity and human health hazards in line with the Sustainable Development Goals (SDGs). Such a desirable direction is only possible through investments in agronomic/technical, consumer

*savhomany@gmail.com

© CAB International 2023. *Sustainable Agricultural Marketing and Agribusiness Development: An African Perspective* (eds B. Nyagadza and T. Rukasha)
DOI: 10.1079/9781800622548.0003

and institutional research. There is a need for research on consumer preferences, farmer perceptions and their adoption processes, and institutional and policy frameworks of organic foods.

3.2   Sustainable Consumption and Production

SDG 12: Sustainable Consumption and Production, focuses on the role of both consumers and producers in achieving sustainable consumption and production. On the one hand, consumers must be encouraged to shift to nutritious and safe diets with a lower environmental footprint. On the other hand, there is a need for producers to grow more food while reducing the deleterious environmental consequences so as to feed the world sustainably (FAO, 2015). Oslo Symposium on Sustainable Consumption (1994) coined SCP as: 'the use of services and related products which respond to basic needs and bring a better quality of life while minimising the use of natural resources and toxic materials as well as the emissions of waste and pollutants over the life cycle of the services or products so as not to jeopardize or compromise the needs of the future generations'.

It has been stated that in order to achieve commitment and public support for sustainable consumption and production, it is imperative for people to understand the key issues related to SCP, identify key priority areas and generate region-specific knowledge (United Nations Environment Programme, 2011). Organic agriculture (OA) focuses on food production by avoiding the use of chemosynthetic fertilizers, pesticides and growth regulators among other principles. According to the World Health Organization (2015), there is a need to scrutinize dietary patterns for their impact on human health, the environment and climate change. The World Health Organization (2015) also notes that about thirty global hazards caused a total of 600 million foodborne illnesses and 420,000 deaths in 2010. However, in Southern Africa, there is a paucity of literature and research on organic farming and organic foods.

3.3   Organic Agriculture

OA concerns itself with maintaining and improving the overall health of the individual farm's soil microbe–plant–animal system which has a heavy bearing on present and future yields (FAO, 1998). According to IFOAM-Organics International (2005), OA is a production system that sustains the health of soils, ecosystems and people by relying on ecological processes, biodiversity and cycles adapted to local conditions, rather than the use of inputs with adverse effects.

3.3.1   Global organic agriculture production and marketing

The organic market has witnessed a rise from €15.1 billion in 2000 to €120.6 billion in 2022. In terms of growth in organic farmland, there has been an increase of 3 million hectares, which translates to 41% in 2020. Although Africa is the second largest region/continent in the world, Table 3.1 shows that it lags behind in terms of global shares of organic agricultural land.

3.3.2   The need for organics 3.0

Organics 1.0 involved visionaries laying out how people can healthily nourish themselves while protecting the environment (IFOAM-Organics International, 2017). Organics 2.0 involved the formation of the movement, certification schemes, codification of standards and enforced rules that have established organic agriculture in 82 countries with a market value of over US$72 billion per year (Otaiku, 2020). Organic 3.0 is the future of OA as an inclusive sustainable innovation, a dimension that Organics 2.0 has failed to achieve (Otaiku, 2020). The Organic 3.0 concept seeks to change this by positioning OA as a modern, innovative system that emphasizes the results and impacts of farming in the foreground. Organics 3.0 is a revised understanding of the role of the organic movement, which positions OA as a contributor to the achievements of the widely acknowledged SDGs (IFOAM-Organics International, 2017).

Table 3.1. World organic agricultural land (including in-conversion areas) and regions' share of the global organic agricultural land in 2020. From FiBL Survey (Willer *et al.*, 2022).

Region	Organic agricultural land (ha)	Region's share of the global organic agricultural land
Africa	2,086,859	2.8%
Asia	6,146,235	8.2%
Europe	17,098,134	22.8%
Latin America	9,949,461	13.3%
Northern America	3,744,163	5.0%
Oceania	35,908,876	47.9%
World	74,926,006	100.00%

Table 3.2. Organic farming and consumption and marketing in Southern Africa. Author's construction using FiBL Survey (Willer *et al.*, 2022).

Country	Organic area (ha)	Share of organic area to total agricultural land	Producers	Processors	Importers	Exporters
Angola	–	–	–	–	–	–
Botswana	–	–	1	1	–	1
Lesotho	–	–	1	1	–	1
Swaziland	1156	0.1%	2	1	–	1
South Africa	40,954	0.04%	–	35	–	–
Malawi	232	–	21	1	1	1
Mozambique	14,438	0.03	134	–	1	–
Zambia	691	0.003	10,100	9	–	8
Zimbabwe	1043	0.01	963	–	–	10
Namibia	–	–	–	7	–	–
World	74,926,006	1.6%	3,368,254	112,911	7,703	9,963

Therefore, Organics 3.0 hinges on the living relationship between consumers, producers and the environment. Organics 3.0 positions itself in SCP by emphasizing the link between organic producers, organic consumers and the environment through transparency and integrity.

3.3.3 The organic agriculture sector in southern africa

There were more than two million hectares of certified organic agricultural land in Africa in 2020, 149,000 hectares more than in 2019 (Willer *et al.*, 2022). However, Table 3.2 shows that countries in Southern Africa are yet to mainstream OA in their agricultural practices.

In terms of regulations, only South Africa is at the drafting stage (Willer *et al.*, 2022). In other parts of the world, for instance in Asia, many countries are now in partial implementation while in Latin America a number of countries have fully implemented, and in Europe almost all countries have fully implemented it. Auerbach (2020) argues that contrary to the common assumption that agriculture in Africa is 'organic' by default rather than 'resource constrained', the real organic production system is knowledge-intensive and cannot be treated as a naive practice because it uses natural materials. Therefore, one of the reasons for the stunted growth of OA in Southern Africa might be the lack of proper research and development systems. This search was performed in Lens Database in May 2022 to provide an analysis of topics covered in Southern African literature so as to inform future studies thereby filling this gap. It is an attempt to produce a comprehensive analysis of organic foods and organic agriculture research conducted in Southern Africa. The purpose of this study is to investigate the development of research related to organic farming and organic foods in Southern Africa in terms of the distribution of bibliometric maps and research/publication trends on the Lens Database using VOSviewer software, version 1.6.18. According to Nandiyanto et al. (2020), bibliometric analysis is effective for giving a data set that can be used by policy makers, researchers and other stakeholders to improve the quality of research.

3.4   Materials and Methods

The analysis is a systematic review of all documents indexed in the Lens Database following the PRISMA guidelines (Preferred Reporting Items for Systematic Reviews and Meta-Analyses) (Moher et al., 2009). The search was conducted on 25 May 2022. It was specified as 'Title-Abs'-Key search query: ({organic foods} AND {organic farming} OR {organic agriculture} AND ('Southern Africa' OR 'Angola' OR 'Botswana' OR 'Lesotho' OR 'Malawi' OR 'Mozambique' OR 'Namibia' OR 'South Africa' OR 'Swaziland' OR 'Zambia' OR 'Zimbabwe'). Three inclusion criteria used are geographical area (the document deals with one or more countries in Southern Africa), document type (only journal articles, book chapters or conference papers were selected) and thematic focus (the main topic is organic food). There is no time period used in this sampling because the objective is to cover all research conducted in Southern Africa. Fig. 3.1 shows the inclusion criteria used to come up with the documents that were reviewed.

Sample articles were downloaded in *.csv format. VOSviewer was used to visualize and analyse trends in the form of bibliometric maps. VOSviewer can make publication maps based on co-authorship, co-occurrence, citation,

bibliographic coupling and co-citation (van Eck and Waltman, 2022).

3.5   Results and Discussion

3.5.1   Visualization of topic area using VOSviewer

A map based on bibliographic data was constructed using VOSviewer. The research settled for 'co-occurrence' and 'all keywords' as the type of analysis and unit of analysis respectively. The minimum occurrences of a keyword are set at default (1). After being analysed using Vosviewer, there were two clusters (red and green), which showed the relationship between one topic and another. VOSviewer was used to display bibliometric mapping in three different visualizations, namely network visualization (Fig. 3.2), overlay visualization (Fig. 3.3), and density visualization (Fig. 3.4). Thirty-two keywords were selected and analysed through the three visualizations. Keywords were labelled with coloured circles with the size of the circle being positively correlated with the appearance of keywords in the titles and abstracts. Therefore, the size of letters and circles was determined by the frequency of occurrences. In VOSviewer, the more often a keyword appears, the greater the size of the letters and circles.

Fig. 3.2 shows clusters in each of the topic areas studied. It can be seen that the keywords 'consumer behaviour', 'food safety', 'perceptions', 'purchase intentions', 'female', 'middle-aged', 'adult', 'fruit', 'ethnicity', 'cross-sectional studies', 'multiple regression model' and 'costs and cost analysis' are in the same cluster (red area), hence there is a close relationship between them. Other keywords such as 'vegetable/microbiology', 'plant leaves/microbiology', 'hygiene', 'enterobacteriaceae', 'bacterial load', 'health knowledge' and 'attitudes' are in the same cluster together with humans and South Africa (green area). The latter cluster has more to do with the health concepts of organic foods. Fig. 3.3 shows the trend from year to year related to organic foods in Southern Africa. Research on organic foods and organic farming in Southern Africa increased in 2014, with much research activity between 2016 and 2020. Fig. 3.4 shows the depth of research; the

Systematic review phases   Flow of information

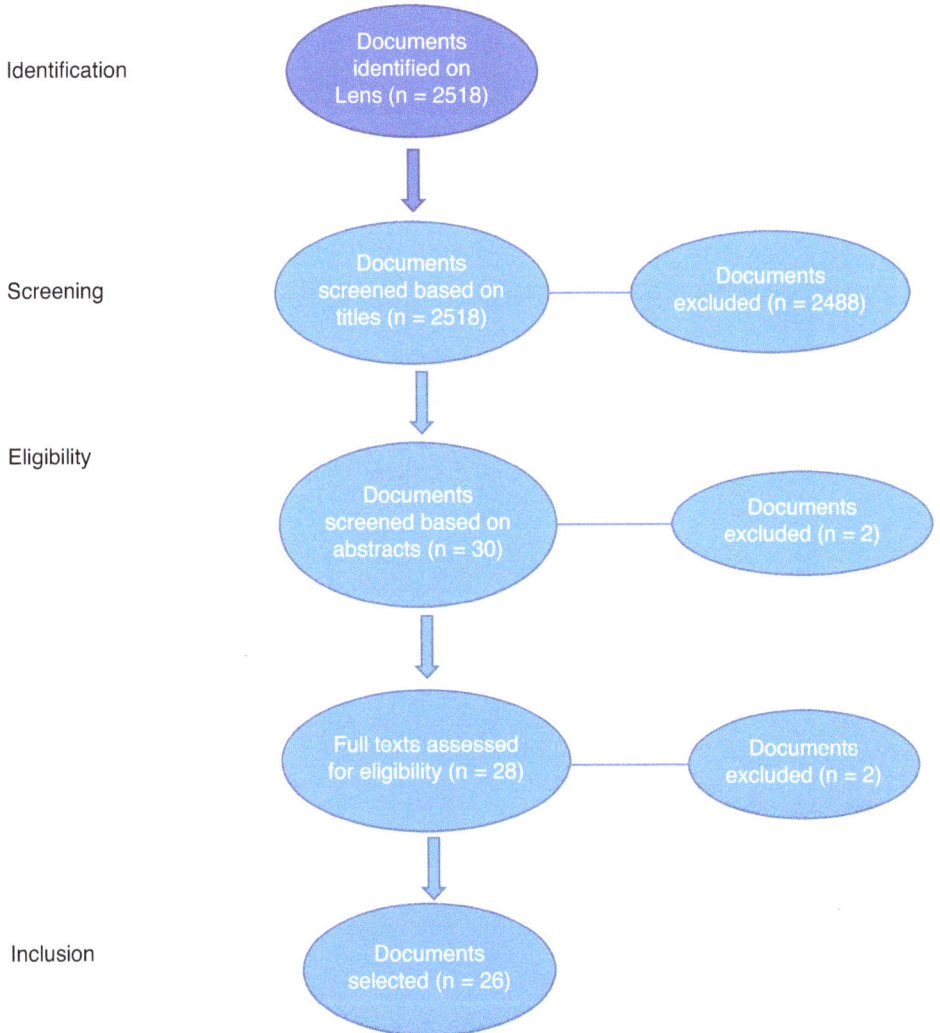

Identification

Documents identified on Lens (n = 2518)

Screening

Documents screened based on titles (n = 2518)

Documents excluded (n = 2488)

Eligibility

Documents screened based on abstracts (n = 30)

Documents excluded (n = 2)

Full texts assessed for eligibility (n = 28)

Documents excluded (n = 2)

Inclusion

Documents selected (n = 26)

Fig. 3.1. Inclusion criteria for documents analysed. Author's own construction.

more concentrated the colours appear, the greater the number of research studies.

Figs 3.2–3.4, show that the keywords that often appeared were 'South Africa' and 'humans' while others such as 'perceptions' and 'purchase intentions' among others appeared equally (only once). This active stance of South Africa should not be used to reflect the involvement of the whole Southern Africa region. This data shows that novel organic food research can be conducted, for instance, research related to the use of technology in tracing the original source of organic foods, which can ultimately increase trust and hence increase demand or consumption of organic foods. This can be used to assess the potential of organic foods to develop markets for small-scale farmers.

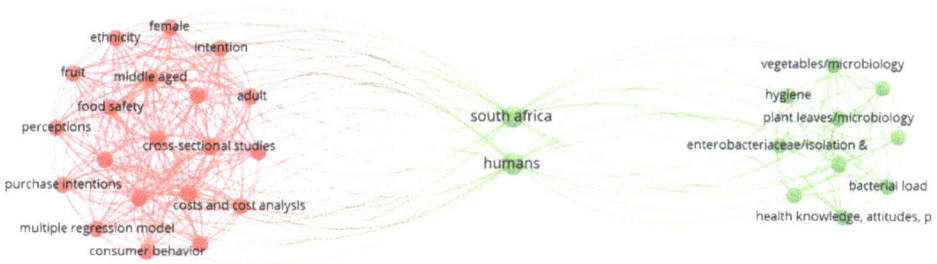

Fig. 3.2. Visualization with network visualization.

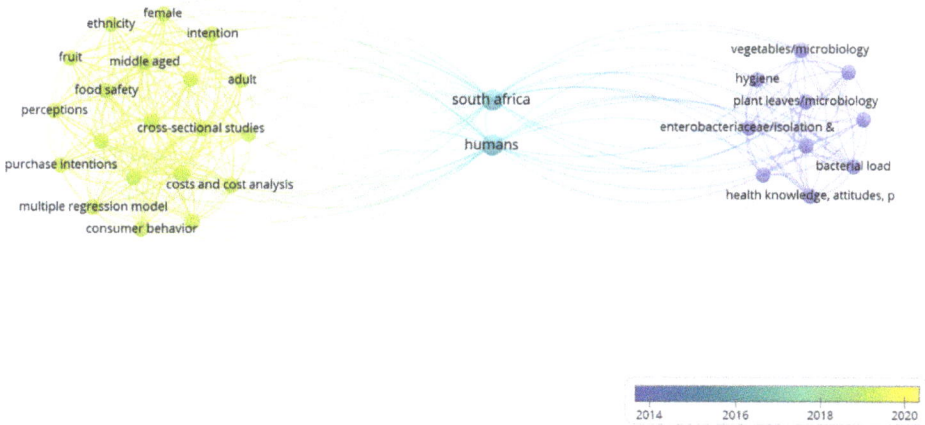

Fig. 3.3. Visualization with overlay visualization.

The analysis of the selected documents indicates that research on organic food and farming is not a new phenomenon in Southern Africa. The first document that specifically deals with organic foods dates back to 2003. The period 2010–2022 has a total of 20 research studies with an average of two documents per year. These range from one per year (2010, 2011) to a maximum of 13 in 2013. The peak of the number of publications in 2016–2018 also saw Zambia and Zimbabwe taking part in a single research study each. The period witnessed three documents in 2016, four documents in 2017 and four documents in 2018. However, the period 2019–2021 has seen a decrease in the number of research studies to three.

The bibliometric analysis shows that the most prominent authors are Hendricks, S.L (five articles), Thamaaga-Chitja, J. (three), Auerbach, R.M. (two), Kunene-Ngubane, P. (two), Muposhi,

A. (two) and Lim Tung, O.J. (two). Analysis of the geography of research in the region shows that research on organic foods and organic farming is mainly conducted in South Africa (24 out of 26 selected documents). The reasons for this might be two-fold. First, South Africa is the largest and most populous country in Southern Africa. Bilali (2020) notes that it is essential to take into account the countries' sizes, which are often associated with their research systems (for example, the number of scientific articles per million inhabitants is used to assess a country's research performance). Secondly, a large percentage of South Africans are white, whose concerns for health and environmental issues are high. The results show that South Africa is the most active country with publications totalling 24 out of 26. Angola, Botswana, Lesotho, Malawi, Mozambique, Namibia and Swaziland have no single article dealing with

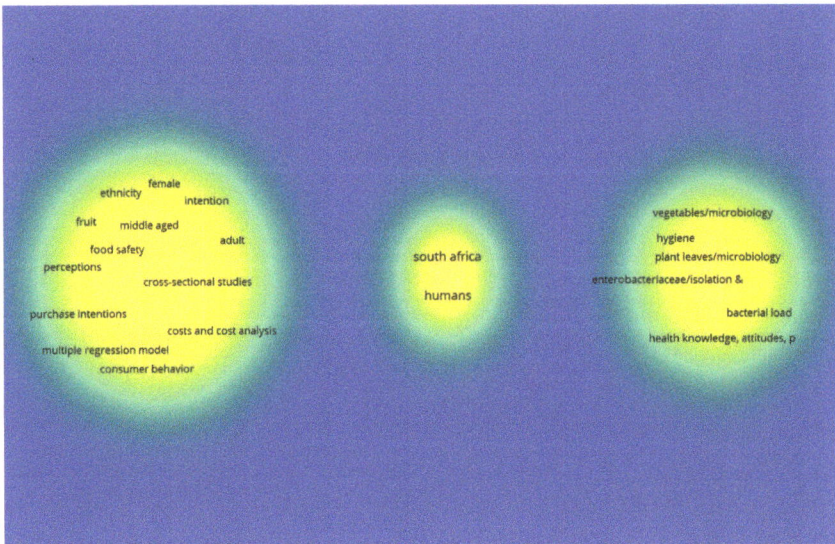

Fig. 3.4. Visualization with density visualization.

organic foods and organic farming showing a lack of research activity in these countries. Furthermore, there are no cross-country studies on any aspect dealing with organic foods and organic farming in Southern Africa. In terms of funding, only three research studies have received funding. These are: 'Exploring the Potential of Developmental Work Research and Change Laboratory to Support Sustainability Transformations: A Case Study of Organic Agriculture in Zimbabwe', which received funding from the International Social Science Council; 'Perceptions on organic farming and selected organic fertilizers by subsistence farmers in Ga-Rankuwa, Pretoria, South Africa', funded bySefako Makgatho Health Sciences University; and 'Organic farming promotes bee abundance in vineyards in Italy but not in South Africa' funded by DAAD, EU.

3.6   Analysis of Topics

3.6.1   Organic versus conventional farming

Mashele and Auerbach (2016) carried out a long-term controlled comparative experiment in South Africa to assess the differences between yield and nutritional qualities of crops grown under conventional and organic conditions. Meyer *et al.* (2015) analysed the effects of organic vs conventional floor management practices in an apple orchard.

3.6.2   Production

The analysis of topics of research shows that most of the research concentrating on the production side dealt with small-scale farmers. These ranged from perceptions of small-scale farmers on organic farming and selected organic fertilizers (Muguvhisa *et al.*, 2017), emerging issues in smallholder production (Thamaga-Chitja and Hendriks, 2008) to hygiene in organic production (Mdluli *et al.*, 2013). For crops grown, the analysis shows leafy vegetables (Mdluli *et al.*, 2013) and organic potatoes (Katundu *et al.*, 2010). Only research by Kunene-Ngubane *et al.* (2018) explored the possibility of organic beef production by small-scale farmers. While such initiatives should be applauded, research in the field is scarce to inform policymaking and drive adoption. Gadzikwa *et al.* (2006) examined the effects of the collective action of a smallholder farming group in reducing production costs and transactional costs in the supply chain.

3.6.3   Conversion to organic farming

Farmers, being rational, can only convert to organic agriculture after a cost–benefit analysis, which can be informed by research. Hence it is critical to understand the drivers of conversion from the perspectives of farmers. Only one study by Niemeyer and Lombard (2003) examined the socio-demographic issues, motivations and challenges of farmers in converting from conventional to organic farming practices.

3.6.4   Nutrient management

A few studies examined nutrient management in order to enhance productivity in organic cropping systems (Mafongoya and Jiri, 2016; Ajibade *et al.*, 2022) in South Africa and Zambia respectively. There is a dearth of knowledge on the nutrient management systems in the Southern Africa region.

3.6.5   Commercialization

Commercialization of organic farming helps to achieve sustainable rural development. Only one research study by Hendriks and Msaki (2009) analysed the possibility of commercialization of organic production by small-scale farmers by first considering their food security issues through food consumption patterns, dietary diversity and consumption elasticities.

3.6.6   Certification of organic foods

Organic foods are credence goods, hence physical attributes such as freshness alone are not enough to earn the trust of the consumers. Thus, external cues such as certification labels and their information may generate expectations about exceptionally high eating quality (Brunso *et al.*, 2002). Tung (2017) assessed

whether there is a need for a mechanism for private sector certification of organic foods.

3.6.7   Consumer perceptions, intentions and willingness to pay

Research on consumers not only shapes market development but also reflects the level of profitability and sustainability of organic farming. Baiyegunhi *et al.* (2018) evaluated consumer willingness to pay for organic fruits and vegetables. Muposhi and Dhurup (2016) and Wekeza and Sibanda (2019) examined the selection attributes of organic foods and consumer purchase intentions respectively.

3.6.8   Marketing

While production is an important process in organic agriculture, marketing is equally important because it is the channel by which the products reach the end user. Marketing of organic food products has been addressed by Thamaga-Chitja and Hendriks (2008) and Muposhi and Dhurup (2017).

3.6.9   Organic product standards

Tung (2018) analysed the need to inspire the development of harmonized organic product standards in the whole African region. Such research focusing on prospects and limitations drives the development of organic agriculture not only in Southern Africa but in Africa as a continent. Table 3.3 lists the number of publications conducted on organic agriculture.

3.7   Recommendations for the Development of Organic Agriculture in Southern Africa

The analysis of organic farming has shown that much should be done if organic agriculture is to be wholly embraced in Southern Africa. First, there is not enough research to advise and enhance the production, productivity and

Table 3.3. List of selected documents found in Lens Database.

Year	Document number	References	Countries
2022	1	(Ajibade *et al.*, 2022)	South Africa
2021	1	Eyinade *et al.* (2021)	South Africa
2019	2	Wekeza and Sibanda (2019)	South Africa
		Auerbach (2019)	South Africa
2018	4	Baiyegunhi *et al.* (2018)	South Africa
		Mukute *et al.* (2018)	Zimbabwe
		Tung (2018)	South Africa
		Kunene-Ngubane *et al.* (2018)	South Africa
2017	4	Muguvhisa *et al.* (2017)	South Africa
		Tung (2017)	South Africa
		Kehinde *et al.* (2018)	South Africa
		Muposhi and Dhurup (2017)	South Africa
2016	3	Mashele and Auerbach (2016)	South Africa
		Muposhi and Dhurup (2016)	South Africa
		Mafongoya and Jiri (2016)	Zambia
2015	1	Meyer *et al.* (2015)	South Africa
2014	3	Kunene-Ngubane *et al.* (2018)	South Africa
		Mduli *et al.* (2014)	South Africa
		Meyer *et al.* (2014)	South Africa
2013	1	Mdluli *et al.* (2013)	South Africa
2010	1	Katundu *et al.* (2010)	South Africa
2009	2	Hendriks and Msaki (2009)	South Africa
		Hendriks (2009)	South Africa
2008	1	Thamaga-Chitja and Hendriks (2008)	South Africa
2006	2	Gadzikwa *et al.* (2006)	South Africa
		Darroch and Mushayanyama (2006)	South Africa
2003	1	Niemeyer and Lombard (2003)	South Africa

consumption of organic foods. It is important to note that only 3 out of 26 documents focused on consumers. If organic agriculture development is to be accelerated, there is a need to synchronize consumption and production. In this regard, education for both consumers and producers should be the building block for adoption by farmers and consumers alike. Education is likely to also positively impact the market development of organic foods. Vehapi and Dolicanin (2016) argue that the deciding influence on the demand for organic food is knowledge of the concept of organic food. They state that more information on the organic food market leads to greater consumer knowledge of organic food and has a positive effect on consumers' attitudes towards these products.

There is a need to design research that examines the willingness to pay for organic foods because the price has been cited as the main barrier to the consumption of organic foods (Hughner *et al.*, 2007; Shafie and Rennie, 2012). Oluwoye (2017) asserts that one area that has been extensively researched in other regions is the willingness to pay. However, this is not the case in Southern Africa. Certification and labelling of organic foods have also been known to increase trust from consumers (Mohamed *et al.*, 2012). Not much research has been expended towards third-party certification. Also, according to Tutunjian, 2008, the organic food industry is steadily moving from niche markets such as small speciality shops to mainstream markets such as large supermarket chains. Not much has been done in Southern Africa in terms of organic food channel distribution.

The research also recommends that the government and other stakeholders extend funding for organic farming and organic foods research. Agricultural policies in organic foods and organic farming should also be aligned. Such trends have been observed, for instance, in West Africa (Emeana *et al.*, 2019). Nicolay (2019) propounds that organic agriculture hinges on socio-economic and political networks, hence fully developed institutions are an important determinant of the development of organic farming. In West Africa, it has been found that non-governmental organizations (NGOs) have been playing a prominent role in the development of organic farming (Glin *et al.*, 2012). As such, promoting alliances with many organizations is pivotal in ensuring Organics 3.0 and the development of organic agriculture in Southern Africa.

3.8 Conclusion

The bibliometric and systematic review analysed the development and direction of organic foods and organic farming in Southern Africa. While a synthesis of systematic and non-systematic analyses shows that organic farming dates back to 2001, it is the speed of its evolution in Southern Africa that is worrisome. Studies have concentrated on small-scale organic vegetable production, nutrient management, consumer perceptions, intentions and willingness to pay, analysing organic vs conventional farming systems, and marketing. Research in organic foods and organic farming is concentrated in South Africa, with Zimbabwe, Zambia and Malawi contributing little. Angola, Botswana, Lesotho, Mozambique, Namibia and Swaziland have not been involved in organic food and farming research. While regional studies have been featured, there are no cross-country studies in Southern Africa. The study recommends the urgent need to invest funds in research on production, supply chain, consumption and institutional levels to foster the development of organic agriculture. Many countries have sound agricultural policies, however, organic agriculture is either not included or only mentioned in passing. Developing and aligning policies to boost organic agriculture is also premised on the desire of the government to work with private players and NGOs in certification, raising awareness and production.

References

Ajibade, S., Mupambwa, H.A., Manyevere, A. and Mnkeni, P.N.S. (2022) Vermicompost amended with rock phosphate as a climate smart technology for production of organic swiss chard (Beta vulgaris subsp. vulgaris). *Frontiers in Sustainable Food Systems* 6, 757792. DOI: 10.3389/fsufs.2022.757792.

Auerbach, R. (2019) Sustainable food systems for Africa. *Economics Agro-Alimentare* 20(3), 301–320.

Auerbach, R. (2020) Organic food systems: meeting the needs of Southern Africa. In: *Organic Food Systems: Meeting the Needs of Southern Africa*. CAB International, UK. DOI: 10.1079/9781786399601.0000.

Baiyegunhi, L.J.S., Mashabane, S.E. and Sambo, N.C. (2018) Influence of socio-psychological factors on consumer willingness to pay (WTP) for organic food products. *Journal of Economics and Behavioral Studies* 10(5), 208–219. DOI: 10.22610/jebs.v10i5(J).2510.

Bilali, H. (2020) Organic food and farming in West Africa: a systematic review. *Journal of Sustainable Organic Agriculture Systems* 70(2), 94–102.

Brunso, K., Fjord, T.A. and Grunert, K.G. (2002) *Consumers' Food Choice and Quality Perception*. The Aarhus School of Business, Working Paper No. 77.

Darroch, M.A.G. and Mushayanyama, T. (2006) Improving working relationships for small holder farmers in formal organic supply chains: Evidence from KwaZulu-Natal, South Africa. *Agrekon* 45(2), 339–360.

Emeana, E., Trenchard, L., Dehnen-Schmutz, K. and Shaikh, S. (2019) Evaluating the role of public agricultural extension and advisory services in promoting agro-ecology transition in Southeast Nigeria. *Agroecology and Sustainable Food Systems* 43(2), 123–144. DOI: 10.1080/21683565.2018.1509410.

Eyinade, G.A., Mushunje, A. and Yusuf, S.F.G. (2021) The willingness to consume organic food: a review. *Food and Agricultural Immunology* 32(1), 78–104.

FAO (1998) *Evaluating the Potential Contribution of Organic Agriculture to Sustainability Goals*. FAO, Mar del Plata, Argentina.

FAO (2015) FAO and the 17 Sustainable Development Goals. Available at: www.fao.org/post-2015-mdg (accessed 4 May 2020).

FAO (2021) *The State of Food and Agriculture: Making Agrifood Systems More Resilient To Shocks and Stresses*. FAO, Rome.

Gadzikwa, I., Lyne, M.C and Hendriks, S.L (2006) Collective action in smallholder organic farming: a study of the Ezemvelo farmers' organization in Kwazulu-Natal. *The South African Journal of Economics* 74(2), 344–358. DOI: 10.1111/j.1813-6982.2006.00070.x.

Geiger, S., Fischer, D. and Schrader, U. (2017) Measuring what matters in sustainable consumption: an integrative framework for the selection of relevant behaviors. *Sustainable Development* 26(1), 18–33. DOI: 10.1002/sd.1688.

Glin, L., Mol, A.P., Oosterveer, P. and Vodouhê, S.D. (2012) Governing the transnational organic cotton network from Benin. *Global Networks* 12(3), 333–354. DOI: 10.1111/j.1471-0374.2011.00340.x.

Hendriks, S. (2009) Sociological perspectives on organic agriculture: from pioneer to policy. *Development Southern Africa* 26(3), 517–518.

Hendriks, S. and Msaki, M.M. (2009) The impact of smallholder commercialisation of organic crops on food consumption patterns, dietary diversity and consumption elasticities. *Agrekon* 48(2), 184–199. DOI: 10.1080/03031853.2009.9523823.

Hughner, R.S., Prothero, A., McDonagh, P., Shultz, C.J. and Stanton, J. (2007) Who are organic food consumers? A compilation and review of why people purchase organic food. *Journal of Consumer Behaviour* 6(2–3), 1–17. DOI: 10.1002/cb.210.

IFOAM-Organics International (2005) General Assembly of IFOAM-Organics International. IFOAM-Organics International, Adelaide.

IFOAM-Organics International (2017) ORGANIC 3.0 for truly sustainable farming & consumption. IFOAM-Organics International, New Delhi.

Katundu, M., Hendriks, S.L, Bower, J. and Siwela, M. (2010) Can sequential harvesting help small holder organic farmers meet consumer expectations for organic potatoes? *Food Quality and Preference* 21(4), 379–384. DOI: 10.1016/j.foodqual.2009.09.003.

Kehinde, T., von Wehrden, H. and Briattain, C. (2018) Organic farming in Italy but not in South Africa. *Journal of Insect Conservation* 22, 61–67.

Kunene-Ngubane, P., Chimonyo, M. and Kolanisi, U. (2018) Possibility of organic beef production on South African communal farms. *South African Journal of Agricultural Extension* 46(1), 1–13. DOI: 10.17159/2413-3221/2018/v46n1a390.

Mafongoya, P. and Jiri, O. (2016) Nutrient dynamics in Wetland organic vegetable production systems in Eastern Zambia. *Sustainable Agriculture Research* 5(1), 78–85. DOI: 10.5539/sar.v5n1p78.

Mashele, N. and Auerbach, R.M. (2016) Evaluating crop yields, crop quality and soil fertility from organic and conventional farming systems in South Africa's southern Cape. *South African Journal of Geology* 119(1), 25–32. DOI: 10.2113/gssajg.119.1.25.

Mdluli, F., Thamaga-Chitja, J. and Schmidt, S. (2013) Appraisal of hygiene indicators and farming practices in the production of leafy vegetables by organic small-scale farmers in uMbumbulu (rural KwaZulu-Natal, South Africa). *International Journal of Environmental Research and Public Health* 10(9), 4323–4338. DOI: 10.3390/ijerph10094323.

Mduli, F., Thamaga-Chitja, J., Schmidt, S. and Shimelis, H. (2014) Production hygiene and training influences on rural small scale organic farmer practices: South Africa. *Tydskrif vir Gesinsekologie en Verbruikerswetenskappe* 42(1), 17–23.

Meyer, H.A., Woolridge, J. and Dame, J. (2014) Relationship between soil alteration index three (AI3), soil organic matter and tree performance in a 'Cripps Pink/M7' apple orchard. *South African Journal of Plant and Soil* 31(3), 173–175.

Meyer, A., Wooldridge, J. and Dames, J.F. (2015) Effect of conventional and organic orchard floor management practices on enzyme activities and microbial counts in a 'Cripp's Pink'/M7 apple orchard. *South African Journal of Plant and Soil* 32(2), 105–112. DOI: 10.1080/02571862.2015.1006274.

Mohamed, M., Chymis, A. and Shelaby, A.A. (2012) Determinants of organic food consumption in Egypt. *International Journal of Business Modeling and Economics* 3(3), 183–191.

Moher, D., Liberati, A., Tetzlaff, J., and Altman, D.G. (2009) Preferred reporting items for systematic reviews and meta-analyses: the PRISMA statement. *PLoS Medicine* 6(6), e1000097. DOI: 10.1371/journal.pmed.1000097.

Muguvhisa, L., Olowoyo, J.O. and Mzimba, D. (2017) Perceptions on organic farming and selected organic fertilizers by subsistence farmers in Ga-Rankuwa, Pretoria, South Africa. *African Journal of Science, Technology, Innovation and Development* 9(1), 85–91. DOI: 10.1080/20421338.2016.1269459.

Mukute, M., Mudokwani, K., McAllister, G. and Nyikahadzoi, K. (2018) Exploring the potential of developmental work research and change laboratory to support sustainability transformations: a case study of organic agriculture in Zimbabwe. *Mind, Culture and Activity* 25(3), 229–246.

Muposhi, A. and Dhurup, M. (2016) A qualitative inquiry of generation Y consumers' selection attributes in the case of organic products. *International Business & Economics Research Journal* 15(1), 1–14. DOI: 10.19030/iber.v15i1.9571.

Muposhi, A. and Dhurup, M. (2017) The influence of green marketing tools on green eating efficacy and green eating behaviour. *Journal of Economics and Behavioral Studies* 9(2), 76–87. DOI: 10.22610/jebs.v9i2(J).1651.

Nandiyanto, A., Biddinika, M.K. and Triawan, F. (2020) How bibliographic dataset portrays decreasing number of scientific publication from Indonesia. *Indonesian Journal of Science and Technology* 5(1), 154–175. DOI: 10.17509/ijost.v5i1.22265.

Nicolay, G. (2019) Understanding and changing farming, food & fiber systems. The organic cotton case in Mali and West Africa. *Open Agriculture* 4(1), 86–97. DOI: 10.1515/opag-2019-0008.

Niemeyer, K. and Lombard, J. (2003) Identifying Problems and Potential of the Conversion to Organic Farming in South Africa. In: *the 41st Annual Conference of the Agricultural Economic Association of South Africa (AEASA)*, Pretoria, South Africa, October 2–3, 2003.

Oluwoye, J. (2017) The association between consumers' socio economic factors and knowledge of organic food products in Huntsville, Alabama: a pilot study. *International Journal of Agricultural Research, Sustainability and Food Sufficiency* 4(4), 202–210.

Oslo Symposium on Sustainable Consumption (1994) In: Oslo Roundtable on Sustainable Consumption and Production, Nowergian Ministry of the Environment, Oslo.

Otaiku, A. (2020) Uganda Agriculture human capacity development: organic 3.0 agriculture gap and need analysis. *World Journal of Agriculture and Soil Science* (7), 1–20.

Shafie, F.A. and Rennie, D. (2012) Consumer perceptions towards organic food. *Procedia - Social and Behavioral Sciences* 49, 360–367. DOI: 10.1016/j.sbspro.2012.07.034.

Thamaga-Chitja, J. and Hendriks, S.L. (2008) Emerging issues in smallholder organic production and marketing in South Africa. *Development Southern Africa* 26(3), 317–326. DOI: 10.1080/03768350802212113.

Tung, O. (2017) Organic food certification in South Africa: a private sector mechanism in need of state regulation? *Potchefstroom Electronic Law Journal* 19(1), 1–48. DOI: 10.17159/1727-3781/2016/v19i0a584.

Tung, O. (2018) African organic product standards for the African continent? Prospects and limitations. *Potchefstroom Electronic Law Journal* 21, 1–38. DOI: 10.17159/1727-3781/2018/v21i0a4308.

Tutunjian, J. (2008) Market survey 2007. *Canadian Grocer* 122(1), 26–34.

United Nations Environment Programme (2011) *Sustainable Consumption and Production in Africa.* UNEP, Nairobi.

van Eck, N.J. and Waltman, L. (2022) Manual for VOSviewer version 1.6.18.

Vehapi, S. and Dolicanin, E. (2016) Consumers behavior on organic food: evidence from the Republic of Serbia. *Economics of Agriculture* 63(3), 871–889. DOI: 10.5937/ekoPolj1603871V.

Wekeza, S.V. and Sibanda, M. (2019) Factors influencing consumer purchase intentions of organically grown products in Shelly Centre, Port Shepstone, South Africa. *International Journal of Environmental Research and Public Health* 16(6), 956. DOI: 10.3390/ijerph16060956.

Willer, H., Travnick, J., Meir, C. and Schlatter, B. (2022) *The World of Organic Agriculture Statistics and Emerging Trends 2022.* Research Institute ofOrganic AgricultureFiBL, IFOAM-Organics International.

World Health Organization (2015) Global burden of foodborne diseases. Available at: https://apps.who.int/iris/bitstream/handle/10665/199350/9789241565165_eng.pdf (accessed 5 May 2020).

4 Youths' Participation in Urban Agriculture in Ibadan Metropolis, Nigeria

Oluwakemi Adeola Obayelu[1]*, Adenike Omotola Ajibobare[1],
Abimbola Oluyemisi Adepoju[1] and Abolupe Oluyemi Ayanboye[2]

[1]University of Ibadan, Ibadan, Nigeria; [2]Ejigbo Campus, Osun State University, Osogbo, Nigeria

Abstract

Urban agricultural production is a key policy thrust to improving food security, urban employment and income generation in developing countries. Increasing youth population in urban areas exacts supply pressures on the thinning industrial labour markets, hence the call for youths' participation in urban agriculture as an opportunity to expand the agricultural sector. This study therefore assessed the perceptions of youths about urban agriculture and identified factors influencing youths' participation in urban agriculture. Data obtained from youths (aged 18 to 35 years) in Ibadan metropolis, Nigeria were analysed using descriptive statistics, Likert analysis and logit regression. Results showed that 70.1% of the youths were in the range of 21–30 years and 79.1% of them were single. About 84.3% of the youths were literate and attained one form of formal education or other with mean annual income of ₦1,321,852. Furthermore, 38.8% of them participated in urban agriculture with 75.0 % of the youth farmers having less than six years of farming experience. The youths disagreed that agriculture is a low status, dirty, unprofitable livelihood activity meant for the poor and uneducated individuals. Having access to urban farmland and being a married male and older youth increased the likelihood of participating in urban agriculture, while longer market distance, attaining a higher level of formal education, and poor access to credit and land reduced it.

Keywords: Urban farming, Youths, Perception, Socio-economic characteristics

4.1 Introduction

Urban agriculture is defined as the growing, processing and distribution of crops and seldom raising livestock in the cities for feeding urban populations growing and distributing fruits (Hendrickson and Porth, 2012). The connection between urban agriculture and food security has been acknowledged for many years (Rezai *et al.*, 2016). Studies have shown that urban agriculture contributes to enhanced urban food security, recycling of nutrients, community development, job opportunities that generate income, poverty alleviation, and the maintenance of green spaces (Stewart *et al.*, 2013; Badami and Ramankutty, 2015). However, poverty and food insecurity are some of the challenges facing Nigeria's population and governments at various levels today. These problems are more serious in urban areas than in rural areas because of high share on food expenditure of the total income of low-income households in

*Corresponding author: jkemmyade@yahoo.co.uk

urban areas (Lozano-Gracia and Young, 2014). Consequently, the urban poor still need to pay for other services and basic necessities of life that the income received cannot take care of. Successive governments in the country have launched various programmes that were meant to curb food insecurity and alleviate poverty. However, most of the programmes could not stand the test of time while some failed totally and this aggravated poverty and food insecurity. Urban agriculture is therefore pivotal to solving these problems.

Although agriculture provides over 36% of total employment opportunities in Nigeria, it remains unattractive to youths (World Bank, 2019; Obayelu and Fadele, 2019). Agriculture in the country is mostly done by the ageing rural farming population owing to rural–urban migration of youths. However, the urban areas cannot generate enough jobs consistent with escalating population growth rate, leading to high levels of youth unemployment and restiveness in urban areas (Gelan and Seifu, 2016). Youths have been noted for their unique capabilities and they could constitute a formidable force in agricultural production activities in any nation. Pan (2014) argues that the contribution of agriculture to farmers' income and urban agriculture depends on the active participation of youth who are the potential labour force. They are characterized by innovative behaviour, minimal risk aversion, less fear of failure, less conservativeness, greater physical strength and greater knowledge acquisition propensity. However, most of the youths perceive urban agriculture as a part-time job and not as a profession (Nnadi and Akwiwu, 2008).

The Nigerian population was estimated at over 193 million in 2016 with an annual population growth rate of 3.2% and youths within the age range of 20–34 represented about a quarter of the population (NBS, 2018). The burgeoning youth population not only imposes supply pressures on labour markets, but also suggests that a rising share of the total population is made up of working-age population. It is expected that youths will use their potential to contribute meaningfully to economic development, especially in food production, and youths' participation in urban agriculture presents the nation with an opportunity to expand the agricultural sector. Lack of education means youth cannot gain formal employment, but migration to the cities to partake in informal small-scale enterprises remains preferable to farming (Haruna et al., 2019). Though youths have desirable qualities and capabilities that can promote agricultural production activities, most of them do not want to participate in it (Twumasi et al., 2019). This has increased youths' participation in violence and other social vices. These include youths' restiveness in the Niger Delta, Boko Haram insurgence in the northern part of Nigeria and kidnapping activities across Nigeria.

Despite the substantial investment by the government, as well as development partners, in providing funds and capacity-building support to youth groups, young people in Nigeria have not embraced the opportunities to engage in farming for employment creation and food security (Kising'U, 2016). The participation of youths in urban agriculture is necessary to ensure sustainability of agricultural production and realization of the food and nutrition security aspect of the Sustainable Development Goals (SDGs). There is substantive literature on youths' participation in agriculture in rural Nigeria (Edeoghon and Okoedo-Okojie, 2015; Nwaogwugwu and Obele, 2017; Nsikak-Abasi and Udoh, 2018). However, there is limited empirical literature on youths' participation in urban agriculture in Nigeria. This study is therefore important to the local and national government policy makers, as well as development partners and youth agribusiness financiers, in designing urban food security and youth development policies and programmes, especially in developing countries. The study therefore assessed the perceptions of the youth on urban agriculture and identified the determinants of youths' participation in urban agriculture in Ibadan metropolis.

4.2 Methodology

The study was conducted in Ibadan, the largest city in West Africa. Data was obtained from youths in Ibadan North Local Government Area (LGA) in February 2019. Four out of the twelve wards in the LGA were randomly selected. A total of 134 youths aged 18 to 35 years were randomly selected from these wards proportionate to the sizes of the wards. Data collected included age of the respondents, sex, marital status, level of education, household size, primary occupation, secondary occupation, access to credit, access to land, distance from homestead to the market, participation in urban agriculture, nature of agricultural activities engaged in and whether they were considering agriculture as a source of livelihood. Data were analysed using descriptive statistics, Likert scale and logit regression.

The five-point Likert scale allows individual respondents to express how much they agree or disagree with a particular statement, with the neutral point being neither agree, disagree or undecided. The mean score (MS), by the summation of the product of rating points and observation divided by the total number of sample size, is obtained as expressed in the formula below:

$$\text{Weighted score} = \sum \text{frequency of scores} \quad (4.1)$$

$$\text{Weighted mean score} = \frac{\sum \text{Weighted scores}}{\text{Total number of respondents}} \quad (4.2)$$

$$\text{Contribution} = \frac{\text{Weighted mean score}}{\text{Total weighted mean scores}} \times 100 \quad (4.3)$$

Logit regression is a qualitative response model used widely to investigate factors affecting an individual's choice from among two or more alternatives (Faralu, 2013; Victor, 2013).

$$\begin{aligned} \text{Log } Y &= \log\left(\frac{P\,(Y=1)}{1-P\,(Y=1)}\right) \\ &= \beta_0 + \beta_1 X_1 + \beta_2 X_2 + \dots + \beta_n X_n + \varepsilon \end{aligned} \quad (4.4)$$

where,

Y_i = dependent variable (engaged in urban agriculture; 0 if otherwise)

β_0 = intercept

β = regression coefficient

ε = error term

n = nth variable.

Let $P(Y=1)$ be the probability that a youth is engaged in urban agriculture while the probability that a student is not engaged in urban agriculture is $1 - P(Y=1) = P(Y=0)$. $\frac{P\,(Y=1)}{1-P\,(Y=1)}$ is the odd-ratio. Apriori expectations of the explanatory variables are presented in Appendix I.

4.3   Results and Discussion

4.3.1   Description of respondents

A larger percentage (61.2%) of the youths was engaged in urban agriculture, with about 65.4% of the participants being male (Fig. 4.1). This implied that urban agriculture was not attractive to the female youths. This could be largely due to the fact that males are more likely to cope with the tedious nature of farming activities than females (Akpan, 2010). This result is contrary to the view of Chikezie *et al.* (2012) that gender is no barrier to active involvement in agricultural livelihood activities. However, the result is consistent with the findings of Adenegan *et al.* (2016) that the majority of urban farmers in Ibadan, Nigeria, were male.

The majority (79.8%) of urban agriculture participants were unmarried (Fig. 4.2). This could be due to the fact that married youths were more attracted to non-farm livelihood activities with higher returns to meet the marital responsibilities of feeding additional members (Nnadi *et al.*, 2021). In addition, the highest proportion of urban agricultural participants were between the ages of 21 and 30 years (78.8%) (Table 4.1). Their relatively young age may make them receptive to new innovations unlike the older ones who may be resistant to change (Nwaogwugwu and Obele, 2017).

Most (84.3%) of the youths had tertiary education (Table 4.1). This high level of literacy is expected to have a positive influence in urban agriculture. However, it is obvious that most respondents had a considerable level of literacy among the youths in the study area. The implication of this result is that the respondents stand a better chance of accessing agricultural information than otherwise. Education is a form of human capital for agricultural development as it enhances adoption of agricultural innovations by youth, thereby sustaining a strong farming population (Akpan, 2010). The

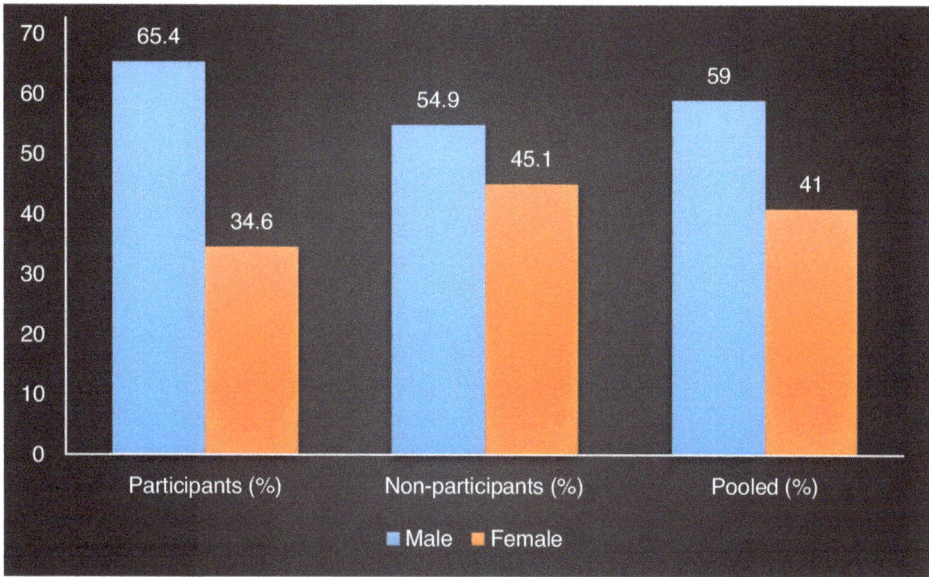

Fig. 4.1. Youths' urban agriculture participation profile by sex.

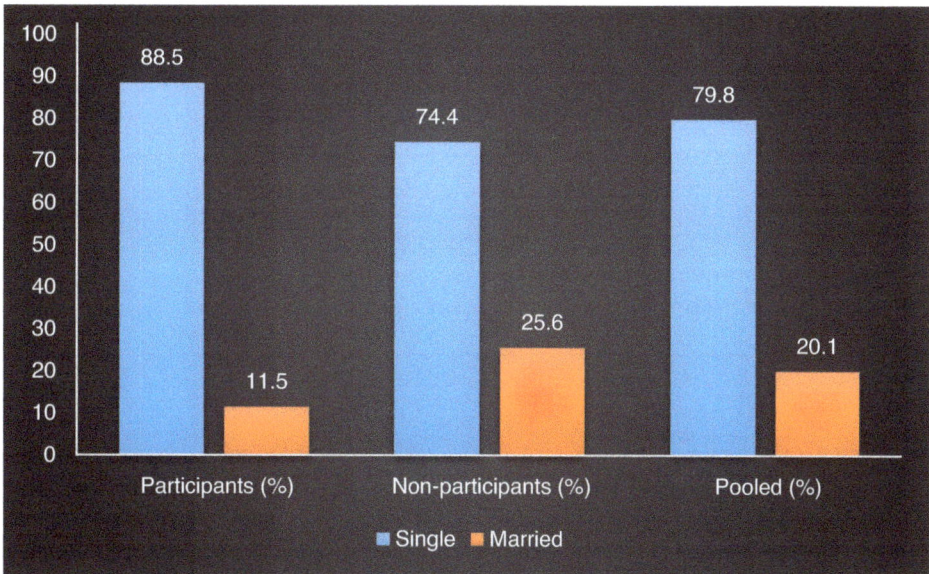

Fig. 4.2. Youths' urban agriculture participation profile by marital status.

highest proportion (67.9%) of the youths' had households that did not exceed the average size of 5.35 ± 1.814 members implying that household size may not explain their participation in urban agriculture. Youths with less than six household members (57.7%) also participated more in agricultural production than those with larger household sizes, implying that

Table 4.1. Respondents' profile by participation and non-participation in urban agriculture.

Characteristics	Participants (N = 52)	Non-participants (N = 82)	Pooled (N = 134)
Age			
< 20	13.5	24.4	20.1
21–30	78.8	64.6	70.1
> 31	7.7	11.0	9.7
Level of education			
No formal education	1.9	2.4	2.2
Secondary education	9.6	15.9	13.4
Tertiary education	88.5	81.7	84.3
Household size			
1–5	57.7	74.4	67.9
6–10	40.4	24.4	30.6
> 10	1.9	1.2	1.5
Access to credit			
Yes	23.1	23.2	
No	76.9	76.8	76.9
Access to land			
Yes	48.1	30.5	37.3
No	51.9	69.5	62.7
Market distance (km)			
< 2	53.8	47.6	50.0
2–4	42.3	45.1	44.0
> 4	3.8	7.3	6.0
Farm distance from homestead (km)			
< 2	94.2	93.9	94.0
2–4	1.9	1.2	1.5
> 4	3.8	4.9	4.5

household size might limit participation in urban agriculture.

A larger percentage of the youths had no access to credit (76.9%) and farmland (51.9%). This implies that there was little or no farmland in the city that could accommodate urban agricultural practice. Thus, lack of access to credit and land would not encourage youths' participation in urban agriculture. The result is consistent with the findings of Adenegan *et al.* (2016) that most urban farmers had no access to credit and almost none of them owned land. Similarly, findings by Akpan *et al.* (2016),

Osabohien *et al.* (2021) and Haruna *et al.* (2019) showed that access to credit significantly influenced participation in urban agriculture, while larger farm sizes have often been associated with higher output and profitability. This therefore, could be an attraction to participation in urban agriculture among the respondents. More youths (53.8%) covered a market distance of 2 km from farmstead, while others went through a long walk to the market, which could discourage them from practising agriculture. The majority (94.2%) of the youths travelled less than 2 km from homestead to farm, which could

Table 4.2. Distribution of youths by human capital and financial assets.

Variables	Frequency	Percentage
Training attendance		
0	30	22.4
1–3	92	68.7
4–6	10	7.5
> 6	2	1.4
Farming experience		
1–5	39	75.0
6–10	10	19.2
> 10	3	5.8
Annual revenue from agriculture (Naira)		
< 50,000	20	38.5
50,000–100,000	1	1.9
> 1,00,000	31	59.6

encourage the youths to farm as they would not be too exhausted before getting to the nearest tarred road with their goods.

4.3.2 Human capital and financial assets

The result showed that the largest percentage (68.7%) had attended two trainings, while 1.4% of them had attended 8 and 10 trainings (Table 4.2). This implies that the majority of the youths do not attend training at all. Table 4.2 shows that almost (75.0%) of the respondents had 1–5 years of farming experience. Farming experience is accumulated over years and experienced farmers are expected to have adequate knowledge in agriculture. The average year of poultry farming experience was approximately 5 ± 4 years. This implies that the majority of the youths had farming experience, which in turn would help good production. Additionally, 59.6% make profit of above ₦150,000. The mean annual revenue from agriculture was ₦1,321,852.89 ± 2,41,6860.72, while the median was ₦615,000.00. This implies that

urban agricultural enterprises were profitable to the youths.

4.3.3 Agricultural training attended

About a third (33.6%) of the urban youth farmers had attended training on livestock management, while less than 1% (0.7%) of them had attended post-planting and livestock management and post-planting, storage and processing (Table 4.3). However, 22.4% of the urban youth farmers had never attended any agricultural training. This implies that an average of the urban youths had attended training on livestock management only. The majority (76.9%) of the urban youth farmers that had attended a training in agricultural production found the training relevant and adequate, while 2.9% of them posited that the training topics did not meet their needs at all (Table 4.3).

4.3.4 Perception of youths on urban agriculture

Likert scale was used to analyse the frequency of responses of youth's perceptions about urban agriculture. The youths disagreed that agriculture is a low status, dirty, unprofitable livelihood activity meant for the poor and uneducated individuals (Table 4.4). They also disagreed that their parents would not hinder them from practising urban agriculture. However, they agreed that they needed tertiary education in agriculture to practise urban agriculture but were undecided whether agriculture is profitable or not. The total mean for the perception of youth participation was 2.2 indicating a low perception of youth towards urban agriculture.

4.3.5 Constraints to youth participation in urban agriculture

The main constraints to practising urban agriculture among the youths were scarcity of land for farming in the city, use of crude agricultural implements, and inadequate financial capital (Table 4.5). However they disagreed

Table 4.3. Distribution of youths by type of agricultural training attended.

Training attended	Frequency	Percentage
Types of training attended		
None	30	22.4
Pre-planting	14	10.4
Post-planting	9	6.7
Storage	5	3.7
Processing	18	13.4
Livestock management	45	33.6
Post-planting and livestock management	1	0.7
Pre-planting, post-planting, storage, processing, livestock management	2	1.5
Pre-planting, post-planting, livestock management	2	1.5
Postharvest, storage, processing	1	0.7
Pre-planting and post-planting	5	3.7
Processing and livestock management	2	1.5
Total	134	100.0
Adequacy of training attended		
Very great extent	13	12.5
Great extent	67	64.4
Moderate extent	16	15.4
Less extent	5	4.8
Not at all	3	2.9
Total	104	100.0

that veterinary services were expensive, that there were inadequate infrastructural facilities to support farming activities, and that their non-farm activities could constrain them from engaging in urban agriculture.

4.3.6 Factors influencing the participation of youths in urban agriculture

The significant Chi-square value of 104.12 indicated that all the explanatory variables jointly explain youths' participation in urban agriculture. Results showed that gender, age, sex, marital status, educational status, primary occupation, access to credit, access to land, market distance and crop revenue significantly explained participation of youths in urban agriculture (Table 4.6). Being male increased the

likelihood of participation in urban agriculture among the youths as the results revealed that more males participate in urban agriculture than their female counterparts. This affirms the finding of Akpan (2010) that males are often more energetic and could be readily available for energy-demanding jobs/activities. Age had a positive effect on participation of youths in urban agriculture – being a year older increased their likelihood of participation in urban agriculture by 0.23 units. Thus, when the youths are younger their energies could be harnessed for productive ventures. Their relatively young age may make them receptive to new innovations unlike the older ones who may be resistant to change (Nwaogwugwu and Obele, 2017).

Furthermore, a unit increase in crop revenue increased the log-likelihood of youths' participation in urban agriculture by 2.43 units. This implies that a unit increase in crop

Table 4.4. Perception of youths about urban agriculture by respondents.

Perception attributes	Weighted score (WS)	Mean weighted score	%	Rank order	Remark
Agriculture is meant for the uneducated	232	1.73	9.9	5th	Disagreed
Agriculture is tedious	441	3.34	19.1	2nd	Undecided
Agriculture is a dirty job	281	2.10	12.0	3rd	Disagreed
Agriculture is not a profitable business	197	1.47	8.4	8th	Disagreed
Agriculture is a low status profession	213	1.59	9.1	7th	Disagreed
Urban agriculture is for poor individuals	241	1.80	10.3	4th	Disagreed
Agriculture needs higher education	511	3.81	21.8	1st	Agreed
My parents will not allow me to practise agriculture	222	1.66	9.5	6th	Disagreed

revenue increased the likelihood of participation in urban agriculture among the youths. Similarly, access to credit and land facilities increased the likelihood of participation in urban agriculture among the youths. The result of this study is in line with previous studies that access to credit (Twumasi *et al.*, 2019) and land (Njeru *et al.*, 2015) facilitates youths' participation in urban agriculture. Farm size has been found to be a critical factor in production and adoption of technologies in agriculture, implying that a youth with larger farm size was expected to obtain higher income from their productive activities. This suggests that youths with larger farm size have greater chance of access to farm credit, invest more in agriculture and consequently earn more income (Yunusa and Giroh, 2017).

Conversely, results showed that unmarried youths were more likely to participate in urban agriculture than their married counterparts. Marriage comes with more responsibility to feed additional mouths, increased per capita food expenditure and food insecurity, which might propel married urban youths to engage in non-agricultural livelihoods with higher returns in order to reduce their household food expenditure (Gamhewage *et al.*, 2015; Obayelu and Oyekola, 2018; Obayelu and Osho, 2020). Similarly, years

of formal education had a negative influence on the log-likelihood of youths' participation in agriculture implying that an additional year of formal education among the youths reduces the likelihood of their participation in urban agriculture by 1.78 units, possibly owing to higher returns to education (Agwu *et al.*, 2012). Thus, highly educated youths were less willing to participate in urban agriculture. Moreover, an additional kilometre in market distance from homestead reduced the log-likelihood of youths' participation in urban agriculture by 2.99 units. This suggests that proximity to the market plays an important role in youths' participation in urban agriculture.

4.4   Conclusion and Recommendations

The study found that having access to urban farmland and being a married male and older youth increased the likelihood of participating in urban agriculture, while years of formal education, market distance, poor access to credit and land significantly reduced it. Urban agriculture should therefore be included in urban planning by the planners so that the youths could be

Table 4.5. Constraints to youths' participation in urban agriculture.

Constraints attributes	Weighted score (WS)	Mean weighted score	%	Rank	Remark
Inadequate capital	524	3.91	5.3	3rd	Agreed
Urban agriculture generates a lot of pollution (soil, water, air)	384	2.87	3.9	17th	Agreed
Inadequate information on urban agriculture	450	3.35	4.6	15th	Agreed
Expensive veterinary services	305	2.28	3.1	20th	Disagreed
Urban agriculture cannot be practised on a large scale	457	3.41	4.7	14th	Agreed
Urban agriculture cannot be easily managed	365	2.72	3.7	18th	Agreed
Inadequate farmland in the city	565	4.22	5.8	1st	Agreed
My work does not allow me to practise agriculture	316	2.36	3.2	18th	Disagreed
My parents will not allow me to practise agriculture	241	1.80	2.5	21st	Disagreed
Industrialization (presence of industries)	479	3.57	4.9	13th	Agreed
Poor health status	423	3.16	4.3	16th	Agreed
Unfavourable land tenure system	465	3.61	4.9	12th	Agreed
Inadequate access to farmlands	509	3.80	5.2	7th	Agreed
Poor extension services	486	3.63	4.1	11th	Agreed
Inadequate infrastructural facilities to support farming activities	223	1.66	2.3	22nd	Disagreed
Inadequate processing and storage facilities	520	3.88	5.3	5th	Agreed
Poor marketing system	504	3.76	5.1	9th	Agreed
Short supply of farm labour	497	3.71	5.2	10th	Agreed
Climate change	508	3.79	5.2	8th	Agreed
Lack of agricultural implements and equipment	527	3.93	5.4	2nd	Agreed
Environmental hazards like erosion, flooding, drought	517	3.89	5.3	4th	Agreed
Inadequacy in the supply of improved farm inputs, like fertilizers, seeds, cuttings	511	3.81	5.2	6th	Agreed

Table 4.6. Determinants of youths' participation in urban agriculture.

Explanatory variables	Coefficient	Marginal effect
Gender	2.6398**	0.1808**
	(1.0363)	(0.0918)
Age of youths	0.0674***	0.2554***
	(0.0356)	(0.0965)
Marital status	−1.3514**	−0.2684**
	(0.7064)	(0.1123)
Education level	−2.4688*	−0.1259*
	(1.0512)	(0.0706)
Household size	0.1153	0.0268
	(0.1177)	(0.0274)
Primary occupation	0.6983	0.1705
	(0.8275)	(0.2051)
Access to credit	0.5436	0.1789**
	(0.2558)	(0.0805)
Access to land	0.8879**	0.2093**
	(0.4678)	(0.1094)
Market distance	−1.6328***	−0.3512***
	(0.8065)	(0.1174)
Road distance	−0.0135	−0.0031
	(0.0186)	(0.0043)
Number of agricultural trainings	0.0518	0.0120
	(0.1392)	(0.0324)
Agricultural contributions	−0.0001	−0.0000
	(0.0000)	(0.0000)
Number of agricultural meetings attended	0.0005	0.0108
	(0.2153)	(0.0501)
Crop revenue	6.2865**	0.3344**
	(1.7095)	(0.1373)
Livestock revenue	0.0086	0.0020
	(0.0325)	(0.0076)
Constant	−3.2018	
	(2.1287)	
Number of observations	134	
LR chi^2 (15)	104.12	
Prob > chi^2	0.0000	
Pseudo R^2	0.1040	
Log likelihood	−80.185914	

*Significance at 10% level
**Significance at 5% level
***Significance at 1% level

provided with production resources, especially land. This is because if the youths have greater control over land, the incentive for them to increase production may be greater. Female youths should also be encouraged to participate actively in urban agriculture by giving them adequate recognition and attention so that they can have that sense of belonging to participate. Furthermore, agricultural credit facilities should be made available to young people who want to engage in urban agriculture, especially crop farming, which yields quicker returns. The banking scheme as well as other microcredit agencies should be encouraged to accommodate young people that are interested in urban agriculture. The study further recommends that local and state governments should establish a development and training programme to enlighten the youths and develop their positive perception about agriculture.

References

Adenegan, K.O., Balogun, O.L. and Yusuf, T.O. (2016) Initial household assets and profitability of urban farming. *International Journal of Vegetable Science* 22(2), 153–160. DOI: 10.1080/19315260.2014.974793.

Agwu, N.M., Nwankwo, E.E. and Anyanwu, C.I. (2012) Determinants of agricultural labour participation among youths in Abia state, Nigeria. *International Journal of Food and Agricultural Economics* 2, 157–164.

Akpan, S.B. (2010) Encouraging youth's involvement in agricultural production and processing. International Food Policy Research Institute, Abuja, Nigeria.

Akpan, S.B., Inimfon, V.P. and Udoka, S.J. (2016) Youth involvement in agricultural production: Factors that influence credit accessibility among rural youths in Akwa Ibom State. In: *Proceedings of the 50th Annual Conference of Agricultural Society of Nigeria, held at the Library Complex*, National Root Crops Research Institute (NRCRI), UmudikeAbia State, October 3–7, 2016.

Badami, M.G. and Ramankutty, N. (2015) Urban agriculture and food security: a critique based on an assessment of urban land constraints. *Global Food Security* 4, 8–15. DOI: 10.1016/j.gfs.2014.10.003.

Chikezie, N.P., Chikaire, J., Osuagwu, C.O., Ihenacho, R.A., Ejiogu-Okereke, N. *et al.* (2012) Factors constraining rural youth involvement in cassava production in Onu-IMO local government area of IMO state, Nigeria. *Global Advanced Research Journal of Agricultural Science* 1, 223–232.

Edeoghon, C.O. and Okoedo-Okojie, D.U. (2015) Information needs of youths involved in urban agriculture as strategy for checking unemployment in Epe LGA of Lagos State, Nigeria. *Journal of Applied Sciences and Environmental Management* 19(1), 37–42. DOI: 10.4314/jasem.v19i1.5.

Gamhewage, M.I., Sivashankar, P. and Mahaliyana, R.P. (2015) Women participation in urban agriculture and its influence on family economy – Sri Lankan experience. *The Journal of Agricultural Sciences* 10(3), 192–206. DOI: 10.4038/jas.v10i3.8072.

Gelan, D.T. and Seifu, G. (2016) Determinates of employment generation through urban agriculture. *International Journal of African and Asian Studies* 26, 49–55.

Haruna, O.I., Asogwa, V.C. and Ezhim, I.A. (2019) Challenges and enhancement of youth participation in agricultural education for sustainable food security. *African Educational Research Journal* 7(4), 174–182. DOI: 10.30918/AERJ.74.19.028.

Hendrickson, M.K. and Porth, M. (2012) Urban Agriculture – best practices and possibilities. University of Missouri Division of Applied Social Sciences. Available at: http://extension.missouri.edu/foodsystems /urbanagriculture.aspx (accessed 10 November 2020).

Kising'U, J.M. (2016) Factors influencing youth participation in agricultural value chain projects in Kenya: A case of Kathiani sub-county, Machakos county, Kenya. Unpublished MSc Thesis, University of Nairobi, Nairobi.

Lozano-Gracia, N. and Young, C. (2014) Housing consumption and urbanization. Policy Research Working Paper; No. 7112. World Bank Group, Washington, DC. Available at: https://openknowledge.worldbank .org/handle/10986/20653 (accessed 8 November 2020).

NBS (National Bureau of Statistics) (2018) 2017 Demographic Statistics Bulletin. Abuja, Nigeria.

Njeru, L., Gichimu, B., Lopokoiyit, M. and Mwangi, J. (2015) Influence of Kenyan Youth's Perception towards agriculture and necessary interventions: a review. *Asian Journal of Agricultural Extension, Economics & Sociology* 5(1), 40–45. DOI: 10.9734/AJAEES/2015/15178.

Nnadi, F.N. and Akwiwu, C.D. (2008) Determinants of youths` participation in rural agriculture in Imo State, Nigeria. *Journal of Applied Sciences* 8(2), 328–333. DOI: 10.3923/jas.2008.328.333.

Nnadi, G.S., Madu, I.A., Ossai, O.G. and Ihinegbu, C. (2021) Effects of non-farm activities on the economy of rural communities in Enugu State, Nigeria. *Journal of Human Behavior in the Social Environment* 31(5), 642–660. DOI: 10.1080/10911359.2020.1803173.

Nsikak-Abasi, A. E. and Udoh, E.J. (2018) Willingness of youths to participate in agricultural activities: implication for poverty reduction. *American Journal of Social Sciences* 6, 1–5.

Nwaogwugwu O.N. and Obele, K.N. (2017) Factors limiting youth participation in agriculture-based livelihoods in Eleme Local Government Area of the Niger Delta, Nigeria. *Scientia Agriculturae* 17, 105–111.

Obayelu, O.A. and Oyekola, T. (2018) Food insecurity in urban slums: evidence from Ibadan metropolis, Southwest Nigeria. *Journal for the Advancement of Developing Economies* 7, 1–17.

Obayelu, O.A. and Fadele, I.O. (2019) Choosing a career path in agriculture: a tough calling for youths in Ibadan metropolis, Nigeria. *Agricultura Tropica et Subtropica* 52(1), 27–37. DOI: 10.2478/ats-2019-0004.

Obayelu, O.A. and Osho, F.R. (2020) How diverse are the diets of low-income urban households in Nigeria? *Journal of Agriculture and Food Research* 2, 100018. DOI: 10.1016/j.jafr.2019.100018.

Osabohien, R., Wiredu, A.N., Nguezet, P.M.D., Mignouna, D.B., Abdoulaye, T. *et al.* (2021) Youth Participation in Agriculture and Poverty Reduction in Nigeria. *Sustainability* 13(14), 7795. DOI: 10.3390/su13147795.

Pan, Y.D. (2014) Rural-urban migration among youths in Nigeria: the impacts on agriculture rural development. *IOSR Journal of Humanities and Social Science* 19(3), 120–123. DOI: 10.9790/0837-1932120123.

Rezai, G., Shamsudin, M.N. and Mohamed, Z. (2016) Urban agriculture: a way forward to food and nutrition security in Malaysia. *Procedia - Social and Behavioral Sciences* 216, 39–45. DOI: 10.1016/j.sbspro.2015.12.006.

Stewart, R., Korth, M., Langer, L., Rafferty, S., Da Silva, N.R. *et al.* (2013) What are the impacts of urban agriculture programs on food security in low and middle-income countries? *Journal of the Collaboration for Environmental Evidence* 2(1), 7. DOI: 10.1186/2047-2382-2-7.

Twumasi, M.A, Jiang, Y. and Acheampong, M.O. (2019) Determinants of agriculture participation among tertiary institution youths in Ghana Martinson Ankrah. *Journal of Agricultural Extension and Rural Development* 11(3), 56–66. DOI: 10.5897/JAERD2018.1011.

World Bank (2019) Employment in agriculture (% of total employment), (modeled ILO estimate). Available at: https://data.worldbank.org/indicator/sl.agr.empl.zs (accessed 7 June 2020).

Yunusa, P.M. and Giroh, D.Y. (2017) Determinants of youth participation in food crops production in song local government area of Adamawa State, Nigeria. Scientific Papers Series Management, Economic Engineering in Agriculture and Rural Development 17, 427–434.

Appendix

Table A1. Description and *a priori* expectations of explanatory variables.

Variable	Description	Literature
Sex	Male = 1; 0 if otherwise	Chikezie *et al.*, 2012
Age	Age of youth (in years)	Akpan (2010); Chikezie *et al.* (2012); Yunusa and Giroh (2017)
Marital status	Single = 1; 0 if otherwise	Nwaogwugwu and Obele (2017)
Education status	Years of formal education (in years)	Akpan (2010); Yunusa and Giroh (2017)
Household size	Number of household members	Nwaogwugwu and Obele (2017); Yunusa and Giroh (2017)
Primary occupation	Farming = 1; 0 if otherwise	Yunusa and Giroh (2017)
Access to credit	1 = Have access to credit; 0 if otherwise	Kising'U (2016); Twumasi *et al.* (2019)
Access to land	1 = Have access to land; 0 if otherwise	Nwaogwugwu and Obele (2017)
Market distance	Market distance from homestead in kilometres	Kising'U (2016)
Road distance	In kilometres	Kising'U (2016)
Number of agricultural training attended	Number of trainings attended	Kising'U (2016)
Agricultural contributions	Contributions to agricultural livelihood in Naira	Kising'U (2016)
Extension contact	Number of extension contact	Kising'U (2016)
Crop revenue	In Naira	Nwaogwugwu and Obele (2017)
Livestock revenue	In Naira	Nwaogwugwu and Obele (2017)

5 Human Capital Management: The Conduit for Unlocking Agribusiness Productivity in Zimbabwe

Noah Ariel Mutongoreni*, Benard Chisiri, Maurice Kwembeya, Moses Jachi, Tafadzwa Machaka and Esther Mafunda
Manicaland State University of Applied Sciences, Mutare, Zimbabwe

Abstract

In a similar vein to the world over, Zimbabwe is engulfed by an unprecedented level of food shortages and general agricultural output shrinkage (FAO, 2015). This is despite the government's notable expenditure towards agricultural financing, which now averages 5.4% of the county's gross domestic product, the highest in the world . The situation has been exacerbated by national internal strife and instability, natural disasters, harsh economic environment, climatic change, the Covid-19 pandemic and lately the Ukraine–Russian war. Zimbabwe strives to arrest this food crisis through a handful of initiatives. These efforts however, exclude a holistic human capital management strategy. The side-lining and relegation of the human capital management strategy in the investment equation exists against the backdrop of the current realization that investment in human capital leads to superior agribusiness performance, productivity and competitive advantage. This chapter seeks to uphold the non-substitutability of human capital in the agribusiness success matrix in Zimbabwe. The research was underpinned by the Human Capital Theory, which opines that expenditure towards training and development of the employee is a boost on the capital nature of labour, and therefore is not a cost but an investment. The conceptual framework brings a loud holistic human capital strategy and positive agribusiness success nexus. The study adopts a qualitative approach. Results from the study reveal that agribusiness success is anchored on investment in human capital. Without a vibrant human capital strategy, success of the agribusiness industry will remain a mirage.

Keywords: Human capital, Agribusiness, Employee performance, Productivity, Sustainable livelihoods

5.1 Introduction

The global community is bedevilled with challenges related to hunger and starvation, decline in agricultural production, climate change (FAO, 2021; OECD, 2022), pandemics, internal strife and instability, natural disasters and disruption of agricultural supply chains due to the Covid-19 pandemic policies (FAO, 2021; Aminetzah *et al.*, 2022; OECD, 2022). Zimbabwe, like any other African country, was depending on Russia, Ukraine, China, South Africa and Brazil among others for its agribusiness supply chain (OECD, 2022). Zimbabwe has favourable conditions for agriculture and a thriving agribusiness (Sneller and Schelling, 2014). However, due to natural disasters and human conflict, the nexus between Zimbabwe and its agricultural supply partners

*Corresponding author: noah.mutongoreni@staff.msuas.ac.zw

© CAB International 2023. *Sustainable Agricultural Marketing and Agribusiness Development: An African Perspective* (eds B. Nyagadza and T. Rukasha)
DOI: 10.1079/9781800622548.0005

has undergone a sudden lull (OECD, 2022). This sudden stoppage in the supply chain is coming at a time when the Government of Zimbabwe under the second Republic is on overdrive for a paradigm shift in agricultural production at communal and commercial levels. The desire for transformation is coming against a background of deep economic challenges characterized by rolling power shortages and massive human capital flight with about 3 million people having migrated to other countries (Chitiyo *et al.*, 2019). There is also unprecedented emphasis on industrialization, value addition and beneficiation. The grand question relates to whether this is indeed an innovation on strategic thrust or a reinvention of the agricultural wheel. The quandary before the government is that while uttermost energy and support is being channelled towards agricultural industry, there is not much evidence that justifies the resource and effort being expended. In other words, there is a negative correlation between expenditure and output/productivity (Mutambara, 2016). What is most intriguing and evident in the agribusiness ecosystem, which should point to the failure of this sector, is the serious neglect and non-consideration of the human element as a game changer towards the productivity agenda. Despite all the vibrant policies and practice, without a sound human capital strategy, all effort is in vain.

The agribusiness ecosystem thrives in an environment where all subunits, elements or parts are fully functional and breed a healthy interdependence framework. The failure of the agribusiness ecosystem is attributed to a dysfunctional human capital strategy. There is scant attention paid to investment in human capital in the agricultural industry (Davis *et al.*, 2021). All businesses should transition from the current production perspective, which treats the human component peripherally, to the adoption of a new paradigm that handles employees as critical assets in line with the purview of the knowledge economy (McCracken *et al.*, 2017). Against this realization, agricultural business is not an exception. Salary and conditions of service in the agribusiness sector, for instance, are in a dire state.

In Zimbabwe, conditions of service are determined by the respective National Employment Councils (NEC) in terms of the Labour Act, Chapter 28:01. The power matrix inherent in the tripartite negotiations, which result in the design of conditions of service, is tilted in favour of the employer. This exacerbation of the employee quagmire is born of the reality that the farmer is either the government or the employer or both. Government, which is supposed to be the epitome of fairness and justice, is thus conflicted at the detriment of employees' conditions of service. The bulk of the Fast Track Land Reform Programme beneficiaries are government elites and officials who also play a critical role in the determination of wages and conditions for agribusiness employees. This has resulted in house maids and gardeners having highly favourable conditions of service compared with agribusiness employees. This has created an unfortunate paradox where hygiene has become more important than food in Zimbabwe. While excessive labour costs discourage businesses (LEDRIZ, 2020), the serious non-consideration and disenfranchisement of the agribusiness value chain employees has caused serious disengagement, misidentification and demoralization, which could have severely affected agricultural productivity. Pursuant to the foregoing, it is imperative to regard the employee as a bigger piece of the productivity puzzle and as the missing element in the current Zimbabwean agribusiness high production agenda. Production is human oriented. Pro-employee management approaches have a human capital inclination, which is an essential condition for high organization productivity and a positive employee citizenry environment. This chapter presents human capital management as the key to unlocking business in general and in Zimbabwean agribusiness productivity in particular.

5.2 Theoretical Framework: Human Capital Theory as Underpinning Human Capital Strategy Towards Agribusiness Development

The Human Capital Strategy as a means to unlock business potential is informed by the Human Capital Theory. Fix (2021) posits that famous figures in the progeny of this theory included Mincer (1958), Schultz (1961) and Becker

(1962). This theory regards human capital as being not comparable to any other assets (physical, social or natural), which are perishable and can be destroyed (Davis *et al.*, 2021). Swanson and Holton (2009, in Ju, 2019) indicates that the Human Capital Theory has microeconomic and macroeconomic implications. Its applicability cuts across all businesses including agribusiness in all its various forms (Davis *et al.*, 2021). The theory envisages that the employee's value is incomparable to the values of any other assets in the organization. The idea is that individuals can gain skills (human capital) through training and development, which makes them more productive in the present and in the future (Fix, 2021). In order to maintain this unparalleled and unsurpassed value, regular upskilling of capabilities and competencies through training and development is quite critical (Mutambara, 2016; Kabonga, 2020; GoZ, 2022; Davis *et al.*, 2021). Human Capital Theory's central tenet is that education, training and development makes people more productive, increasing their marginal output in comparison to less-educated workers (Merwe, 2010).

The human asset entails possession of education, skills, competencies, knowledge, work habits and attitudes, which are vital for a successful agribusiness undertaking. These components that make up the human capital suffer obsolescence if they are not regularly revisited and titivated through training and development (Merwe, 2010; Hassanzoy, 2019). The era of selective breeding and maintenance of fit (Fix, 2021) has long gone. Human Capital Theory is now the new-found 'Eugenic light' that can position the organization competitively through training of employees. Theodore Schultz (1960) in Holden and Biddle (2017), noting the importance of training, posited the following assertion: 'I propose to treat education as an investment in man and to treat its consequences as a form of capita' (Holden and Biddle, 2017, p. 2). What this insinuated was also buttressed by Ju (2019) who opined that the Human Capital Theory is all about knowledge, expertise and skills development.

5.3 Human Capital and Dynamism

The whole intention of adopting the human capital tenets is for organizations to be able to adapt, co-evolve, change and suit the changing environment (Schultz, 1961; Nelson and Phelps, 1966; Farr and Faber, 2019). Nelson and Phelps (1966) argued that organizations that do not incorporate human capital development strategy suffer from backwardness, stagnation and all forms of inertia. Ishikawa and Ryan (2002, in Fleischhauer, 2011) suggest that it is the stock of human capital that predominantly determines the earnings of individuals. As a result, to maintain relevance in an organization's payment and reward matrix, individuals should continuously strive to improve themselves if organizations neglect them. The requirements of organizations are constantly changing each day owing to turbulence in the current operating environment (Wangenge-Ouma and Kupe, 2020). The situation has been compounded by the emergence of the Fourth Industrial Revolution, which is characterized by high levels of digitalization (Schwab, 2016). This technological paradigm shift has witnessed the emergence of smart farming. The use of drones in scouting pests, animal surveillance and employee supervision among others calls upon the need for technological retooling of employees. Rfid technology, vertical cropping, weather tracking, soil and water sensing, precision agriculture, satellite imaging, pervasive automation and robotics among others (Manida and Ganeshan, 2021) are some of the ubiquitous technologies that justify the requirement for dynamic capabilities. In light of this, co-evolution and adaption and possession of dynamic capabilities are the new critical success factors that represent the core of modern and future organizations (Panizzon and Barcellos, 2020). The agribusiness sector is guaranteed a proactive and multi-skilled pool if a human capital thrust is followed. Multiskilling has an advantage of cutting costs of engaging many employees (Mutongoreni, 2015). The underlining principle suggested in this research is agricultural institutions, big or small, should fight outdatedness and obsolescence of skills through training and development as one element of the productivity game-changing toolkit. No matter how hard-earned employee skills are (Ngara, 2017), continuous reskilling is an investment organizations cannot do without given the constantly changing internal and external agricultural business environment.

5.4 Methodology

A qualitative research methodology was adopted and it hence relied on content analysis. As a result, a thorough review of government policy documents and related literature was undertaken.

5.5 Role of Human Capital in a Transitioning Agribusiness Context in Zimbabwe

The history of agribusiness in Zimbabwe can be traced to the ninety years of British settler rule in Zimbabwe stretching from 1890 to 1980. The agribusiness sector during this period was predominately under the minority white settlers (Rutherford, 2001; Waeterloos and Rutherford, 2004). The black farm employees acted as the inferior frontiers of the white farmers. It is the land, which is a key cog in the agribusiness value chain, that prompted the bloody fourteen years of the liberation struggle (Moore, 2012). Although agriculture was thriving under the minority white settlers, the relationship between capital and labour was skewed in favour of the former (Masaka, 2012). It was not only dehumanizing (Masaka, 2012), but lacked the critical tenets of human capital management. On attainment of majority rule in 1980, the Mugabe-led administration was keen to transform the agribusiness landscape to accommodate the majority of the black population. Strategies adopted included the 'Growth with Equity' policy and the Land Resettlement Programme. The former opened access and support to black agribusiness actors while the latter led to resettlement of blacks on land accessed through the willing buyer and willing seller arrangement. The second wave in the acquisition of land as a strategy to widen access of black farmers and correct land ownership imbalances was the Fast Track Land Reform Programme (Mlambo and Gwekwerere, 2019). This ad hoc and violent programme witnessed the mass entry of the hitherto marginalized black majority into the agribusiness sector. The management of human capital did not witness a positive transformation as evidenced by the downward spiral of productivity.

Efforts were directed towards the acquisition of more land at the expense of automation, which drives modern smart agribusiness. As a result, labour-intensive agrarian approaches of production endured to become the status quo. This status quo is characterized by labour-intensive agribusiness practices, low technical and conceptual skills and low output per employee, which culminates in a depressed agribusiness system. To transition from this state, human capital intervention strategies are required. These people-oriented mechanisms entail e-resourcing, performance management, training and development and competitive compensation. If these human capital management practices are comprehensively executed the ideal agribusiness outcomes are attained, ceteris paribus. The ideal human capital agribusiness context is characterized by high productivity per employee, high agribusiness performance, smart agriculture practices, labour retention, corporate citizenship behaviour, more mental effort and less physical effort. Fostering a human capital management working culture therefore has a positive natural outcome on agribusiness performance, which is the most desired situation. Fig. 5.1 shows the state of the agribusiness environment in Zimbabwe compared with the desirable and ideal scenario. A healthy human capital and agribusiness nexus is an essential condition for a successful agribusiness environment.

5.6 Transitioning Agribusiness From the Current State to Ideal Context

In line with the human capital management perspectives, a glance at the state of human capital practices in the agribusiness sector in Zimbabwe show a dearth of the application of the merit principle in resourcing. More specifically, the resourcing of senior government officials mandated with the design and execution of agribusiness policy is replete with corruption, nepotism and partisanship allegiances (Chigudu, 2015). In fact, corruption in Zimbabwe is endemic (Makumbe, 1994; Chigudu, 2015; Van Fleet, 2016; Zinyama, 2021). These malpractices are also evident in the appointments of senior staff in public

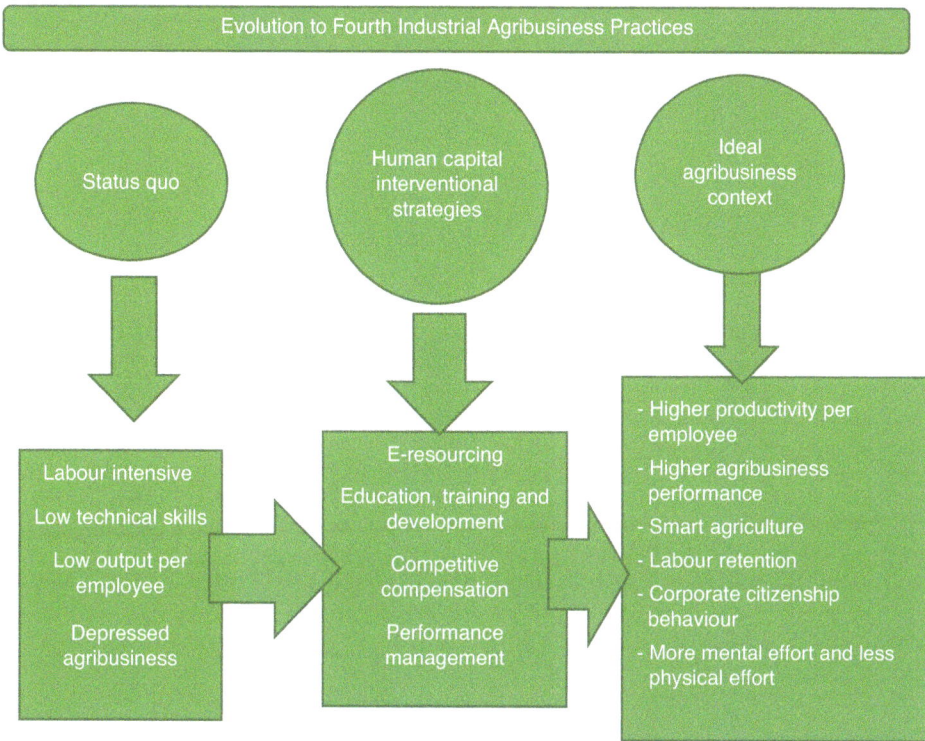

Fig. 5.1. Agribusiness evolution to the Fourth Industrial Revolution (4IR) and the necessary conditions. Researcher's conceptualization, 2022.

entities in agribusiness value chains such as the Tobacco Research Board (TMB), Agricultural and Rural Development Authority (ARDA), Cold Storage Company (CSC), AFC, Agricultural Research Council (ARC) and Tobacco Industry and Marketing Board (TIMB) (GoZ, 2022). Other public entities where resourcing takes a partisan thrust include higher education institutions under the Ministry of Lands, Agriculture, Water, Fisheries, Climate and Rural Resettlement and those under the Ministry of Higher Education and Tertiary Education, Innovation, Science and Technology Development. Staff appointed in these entities are appointed on partisan grounds as opposed to meritocracy (Chigudu, 2015). As a strategy to unlock the full potential of the agribusiness system, the government must retrace its meritocratic pathway in line with best practices. It should ensure that all its arms are functional to curb management override as well as political interference in the recruitment and

selection process of staff. The adoption of the merit principle is the low-hanging fruit for the attainment of candidates with the right agribusiness attributes, acumen, knowledge, abilities and competencies. These will breed a fertile ground for a vibrant agribusiness environment.

A glance at the private sector agribusiness resourcing practices in Zimbabwe shows a strong resemblance to the public sector domain. The merit principle is also not respected. Modern recruitment and selection practices such as e-recruitment and psychometrics are rarely used. E-recruitment widens the candidate net and has the propensity to attract agribusiness techno-savvies who are relevant in modern-day agricultural assignments. Meritocratic resourcing often characterized by implementation of examinations either by the Civil Service Commission or by any agribusiness entity has been seen as an effective strategy for diminishing corruption (USAID, 2017). Honesty tests could also be used as a pre-screening tool for potential hires in an endeavour to reduce corrupt practices and augment integrity levels in the workforce. The human capital practices should speak loudly to the technological demands of the Fourth Industrial Revolution, which is characterized by Internet of Things, robotics, artificial intelligence and 3D printing among others (Schwab, 2016). The outcome of a robust recruitment and selection process should be to acquire a versatile and agile agribusiness contributor. The use of psychometric tests entails a move towards adoption of science in the selection process. This helps the agribusiness entities to acquire employees with the best skills, traits and cognitive abilities to drive them towards acquiring competitive advantage. Psychometric selection technologies choose the best from the rest. An organization that embraces a full package of contemporary recruitment practices survives in a difficult environment.

As soon as individuals are absorbed into the agribusiness environment, their performance should be managed. The probationary period should be marked by regular performance reviews. In line with best practice, performance targets should be clear and unambiguous. The situation in Zimbabwe's agribusiness sector is the antithesis of established and best practices. Those entrusted with the design and execution of agribusiness policies and practices shun performance management. The untold story is that these employees entered the system through either partisan or unorthodox pathways. To them, performance management is there to expose them of their inadequacies and hence should not succeed. Those who come first and those who occupy higher positions in institutions technically should be more knowledgeable than others. While the public sector has adopted the result-based management system as the performance management framework, there is nothing to show in terms of improved performance. A senior manager at ARDA farm Bubi Lupane irrigation project was fired for alleged incompetence by those responsible on 21 March 2021 (The Zimbabwean Chronicle, 2021). In dismissing the manager, the minister remarked: 'You are still in the past ARDA must wake up to the realities...of using agriculture as the vehicle for attaining vision 2030'.

Although the government has introduced performance contracts for senior public sector employees, effective 2021, it is something that is still in its infancy. On the ground there is nothing to show in terms of improved performance. However, on paper everything appears to be ideal. The agribusiness private sector in Zimbabwe on the other hand is generally subdued. The thesis that is postulated is that some of the critical value chain players, especially government, are not performing to standard. A disturbance of one subunit of a system has a higher likelihood of affecting all other subunits (Dunlop, 1958; Katz and Kahn, 1966). An agribusiness under this perspective is a system built by energetic output where the energy coming from the subunit reactivates and sustains the system. The demise of the private agribusiness operators cannot be solely attributed to their failures but to the entirety of the whole agribusiness chain (World Bank, 2022). Going into the future both public sector and private sector agribusiness should fully embrace digitalization, automation and mechanization for timely identification and correction of deficiencies. Performance should also be tied to rewards and this should be actualized.

Deficiencies that are noted at performance management stage inform an accurate training and development intervention. Training is multifaceted and it can be formal and informal. Of late it has become virtual (e-training). The tragedy in the agribusiness sector and all other sectors is that training and development is conceived as a cost, notwithstanding the benefits accrued to the organization. In Zimbabwe, prior to the adoption of the Fast Track Land Reform Programme in the year 2000, large-scale agribusiness entities tended to prioritize training and development for their employees. They had well-developed training programmes intended to create a highly competent and stable workforce (Narman, 1991). However, the post Fast Track Land Reform Programme large-scale commercial farmers perceive training as a cost. They do not have programmes for training and development of employees. Consequently, the performance of their agribusiness is subdued (Zikhali, 2017). Training and development could go a long way in allowing for the transition to climate-smart agricultural technologies, which are vital for boosting productivity and competitiveness (Ngara, 2017). The Agricultural Extension Services (Agritex) in Zimbabwe is the government's arm mandated to offer training on agricultural-related matters. Its prime mandate entails provision of extension services to farmers. Its Mission Statement states as follows: 'To generate, provide and promote agricultural technologies, information, linkages, support and regulatory service, through high quality research, extension co-ordination and liaison, thus enhance economically viable agriculture on sustainable basis.' (GoZ, 2022).

The provision of this mandate has however been below expectation (Zikhali, 2017). As a consequence, agriculture production in Zimbabwe from 2000 to date is on a downward spiral (Mutambara, 2016; Zikhali, 2017). According to the World Bank (2004, in Zikhali, 2017), expenditure on extension services worldwide dropped from 18% in 1979 to 3.5% by 2007. This is notwithstanding the view that agribusiness holds the key to reducing poverty and increasing sustainability for the majority of the population (Collett and Gale, 2009). Training of farmers has thus decreased and this has impacted on productivity of agribusiness entities. Related to this is the fact that the Agriculture Extension Services staff are not receiving adequate training and development. The Zimbabwean government is subdued in terms of resources and hence unable to fully invest in training and development so that it acts as a springboard for enhancing productivity (Zikhali, 2017). Consequently, Zimbabwe has lost its flagship status as the country with the best agricultural extension services in Africa (Raikes, 1988). In fact, Zimbabwe was the pride of Africa in agricultural extension services prior to the 1990s' Economic Structural Adjustment Programme and the Fast Track Land Reform Programme post-2000 (Narman, 1991; Zikhali, 2017). Those who received training, rarely disseminate their expertise to farmers and this has had a negative impact on productivity. There are therefore deficiencies in skills, knowledge and ability among extension workers in Zimbabwe (Zikhali, 2017). This dearth of skills among the extension service workers is also manifesting itself on the farmers themselves. This in turn results in depressed productivity and derailment of the production value chain among the agribusiness players.

The idea of investing in farmers is called agriculture human capital (Davis *et al.*, 2021). Agriculture human capital is essential to discovering creative solutions for more resilient and climate-smart agriculture as well as tackling difficulties facing global agri-food systems, such as feeding the world's expanding population in a sustainable manner with food that is safe, healthy and nourishing. It is equally as crucial as making investments in physical capital such as infrastructure (Davis *et al.*, 2021). Agriculture human capital is the missing link in the productivity nexus of the large-scale commercial farmers who benefited from the Fast Track Land Reform programme of the post-2000 period.

Pursuant to this and in line with best practice, the Agricultural Extension Services should be fully capacitated. The agribusiness sector is a fast-paced field where technology has not shied. With the rise in climate change and the rapid development of technology beginning at the advent of the new millennium, it behoves government and other agribusiness actors to consider aligning its staff with the latest technologies. This should start with the extension workers who will then cascade it to individual farmers. The main duties and responsibilities of agricultural services employees should be to transmit new techniques to farmers (Narman, 1991). This could be done through non-formal training on the application of new production techniques, the economic benefits and financial returns that can be yielded when new techniques are used, among others. The farmers should not shy away from new technologies such as smart agriculture as these have been proved to augment productivity and are climate friendly. There should always be an existing healthy communication between farmers and extension services employees. Such a development can only exist if the extension services employees are trained and developed in the new and emerging technologies.

Human capital development should be a must for all progressive organizations, and agribusiness organizations cannot be an exception. It is through human capital training and development that organizations may be well positioned to attain competitive advantage through their human capital. This training could be formal and informal. It can be delivered virtually or blended.

Apart from training and development, there is need to address the reward and compensation regime in the agribusiness sector. Conditions of service in the agriculture sector are governed by the National Employment Council (NEC) for the agriculture sector. The salaries and benefits for farm workers are so low that they discourage entry of skilled personnel. The situation in Zimbabwe has been worsened by the fact that some of the players in the NEC are playing the role of both employers and regulators. Most government officials, including judges, are beneficiaries of the Fast Track Land Reform Programme. Their partiality on issues concerning reward levels of farm employees is thus questionable. Compensation in agribusiness tends to improve in the agribusiness sectors involved in manufacturing, processing and retailing. Employees working for the government are also underpaid, which creates a breeding ground for corruption and bureaucratic bottlenecks. In order to decrease the incentive for corruption, an effective bureaucratic system must be made up of well-paid officers (Van Rijckeghem and Weder di Mauro, 1997). Apart from corruption, a poorly rewarded workforce is more apt to resort to absenteeism, pilferage and presenteeism all of which are an antithesis of productivity in the agribusiness environment.

5.7 Recommendations

The agribusiness sector in Zimbabwe should take a 'people first' perspective and consider making meaningful investment in human capital. It is the strength of human capital that gives agribusiness organizations a firm competitive advantage. Appointments should be premised on merit. Training and development programmes for Fast Track Land Reform Programme beneficiaries must be developed by government through the Agritex Department and be made mandatory for agribusiness enterprises as a means for disseminating new skills, methods, technologies and practices in agribusiness. The strength of this category of farmers would boost the agribusiness value chain. Resourcing must be premised on meritocracy and in line with modern technological practices. The performance management system should be digitalized and be bespoke to the reward management strategy. The industrial relations climate should be made conducive, and perceived structural and extra-legal parties should be denied a part in the employment relationship. Lastly, there should be regular interaction between agribusiness training institutions and the agribusiness sector to ensure that training and development programmes are offered formally and informally. Training and development should also be emphasized on virtual platforms and non-formal platforms to improve trainer–trainee interactions as well widen the trainee base.

5.8 Conclusion

Sound human capital management strategies are an essential condition for unlocking agribusiness potential, sustainability and productivity in Zimbabwe. The agribusiness sector in Zimbabwe is currently depressed and hence requires a 'reviving moment'. Such a 'reviving moment' can only be born out of a sound human capital strategy. It is through human capital management that the agribusiness sector in Zimbabwe can unlock its potential and be competitive in the midst of a highly competitive and pervasive global environment.

References

Aminetzah, D., Baroyan, A., Kravchenko, O., Denis, N., Dewilde, S. *et al*. (2022) *A Reflection on Global Food Security Challenges Amid the War in Ukraine and the Early Impact of Climate Change*. McKinsey and Company.

Becker, G.S. (1962) Investment in human capital: a theoretical analysis. *Journal of Political Economy* 70, 9–49. DOI: 10.1086/258724.

Chigudu, D. (2015) Towards improvement of ethics in the public sector in Zimbabwe. *Journal of Governance and Regulation* 4(1), 103–111. DOI: 10.22495/jgr_v4_i1_c1_p2.

Chitiyo, K., Dengu, C., Mbae, D. and Vandome, C. (2019) Briefing note forging inclusive economic growth in Zimbabwe insights from the Zimbabwe futures 2030 roundtable series. Chatham House, The Royal Institute of International Affairs.

Collett, K. and Gale, C. (2009) Training for Rural Development: Agricultural and Enterprise Skills for Women Smallholders. City & Guilds Centre for Skills Development, London.

Davis, K., Gammelgaard, J., Preissing, J., Gilbert, R. and Ngwenya, H. (2021) *Investing in Farmers: Agriculture Human Capital Investment Strategies*. FAO and IFPRI, Rome.

Dunlop, J.T. (1958) *Industrial Relations Systems*. Harvard Business School, Boston.

FAO (2021) The Impact of Disasters and Crises on Agriculture and Food Security.

FAO (2015) Zimbabwe Country Programme Framework 2012–2015.

Farr, J.V. and Faber, I. (2019) The basic theory of interest. In: Farr, J.V. and Faber, I.J. (eds) *Engineering Economics of Life Cycle Cost Analysis*. pp. 93–112. DOI: 10.1201/9780429466304.

Fix, B. (2021) The rise of human capital theory. *Real-World Economics Review* 95, 29–41.

Fleischhauer, K.-J. (2011) A Review of Human Capital Theory: Microeconomics. University of St. Gallen, Department of Economics Discussion Paper No. 2007-01.

Global Network Against Food Crisis (2021) New and old challenges: conflict, climate change and COVID-19 impacts on rising acute food insecurity.

GoZ (2022) Ministry of Lands, Agriculture, Fisheries, Water and Rural Development. Available at: http://www.zim.gov.zw/index.php/en/my-government/government-ministries/lands,-agriculture,-water,-climate-and-rural-resettlement (accessed 8 July 2023).

Hassanzoy, N. (2019) What is agribusiness? DOI: org/10.13140/RG.2.2.23776.33285.

Holden, L. and Biddle, J. (2017) The introduction of human capital theory into education policy in the United States. *History of Political Economy* 49(4), 537–574. DOI: 10.1215/00182702-4296305.

Ju, B. (2019) The roles of the psychology, systems and economic theories in human resource development. *European Journal of Training and Development* 43(1/2), 132–152. DOI: 10.1108/EJTD-02-2018-0020.

Kabonga, I. (2020) Analysis of the Fast Track Land Reform Programme (FTLRP) contribution to access to natural, financial and physical capital in Norton, Zimbabwe. *Cogent Social Sciences* 6(1).

Katz, D. and Kahn, R. (1966) *The Social Psychology of Organizations*. Wiley, Hoboken.

LEDRIZ (2020) Review of the Transitional Stabilisation Programme.

Makumbe, J. (1994) Bureaucratic corruption in Zimbabwe: causes and magnitude of the problem. *Journal of African Development* 19, 45–60.

Manida, M. and Ganeshan, M. (2021) New agriculture technology in modern farming. *International Journal of Management Research and Social Science* 8, 3. DOI: 10.30726/ijmrss/v8.i3.2021.83016.

Masaka, D. (2012) Paradoxes in the "sanctions discourse" in Zimbabwe: a critical reflection. *African Study Monographs* 33, 49–71.

McCracken, M., McIvor, R., Treacy, R. and Wall, T. (2017) Human capital theory: assessing the evidence for the value and importance of people to organisational success. Available at: https://www.semanticscholar.org/paper/Human-capital-theory%3A-assessing-the-evidence-for-of/25dfff34d21f34ad77e341f5eaf977650c70784d (accessed 8 July 2023).

Merwe, A. Van der. (2010) Does human capital theory explain the value of higher education? A South African case study. *American Journal of Business Education* 3(1), 107–118. DOI: 10.19030/ajbe.v3i1.378.

Mincer, J. (1958) Investment in human capital and personal income distribution. *Journal of Political Economy* 66(4), 281–302. DOI: 10.1086/258055.

Mlambo, O.B. and Gwekwerere, T. (2019) Names, labels, the Zimbabwean Liberation War veteran and the third Chimurenga: the language and politics of entitlement in post-2000 Zimbabwe. *African Identities* 17(2), 130–146. DOI: 10.1080/14725843.2019.1660619.

Moore, D. (2012) Progress, power, and violent accumulation in Zimbabwe. *Journal of Contemporary African Studies* 30(1), 1–9. DOI: 10.1080/02589001.2012.646748.

Mutambara, J. (2016) USAID strategic economic research and analysis – Zimbabwe (SERA) program: maize production and marketing in Zimbabwe: policies for a high growth strategy. USAID Strategic Economic Research and Analysis – Zimbabwe (SERA) Program.

Mutongoreni, N.A. (2015) An examination of human resource management strategies and their contribution to effective local government systems in Zimbabwe. Thesis, University of Fort Hare.

Narman, A. (1991) *Education, Training and Agricultural Development in Zimbabwe*. UNESCO, Paris.

Nelson, R. and Phelps, E. (1966) Investment in humans, technological diffusion, and economic growth. *The American Economic Review* 56, 69–75.

Ngara, T. (2017) *Climate-Smart Agriculture Manual Agriculture Manual Agriculture Manual for Agriculture Education in Zimbabwe*. Climate Technology Centre and Network, Denmark.

Organization for Economic Co-operation and Development (OECD) (2022) In: *Key Issues Paper – 2022 Ministerial Council Meeting*, Paris, June 9–10, 2022.

Panizzon, M. and Barcellos, P.F.P. (2020) Critical success factors of the university of the future in a society 5.0: a maturity model. *World Futures Review* 12(4), 410–426. DOI: 10.1177/1946756720976711.

Raikes, P. (1988) *Modenising Hunger*. Catholic Institute for International Relations.

Rutherford, B. (2001) *Working on the Margins: Black Workers, White Farmers in Post-Colonial Zimbabwe*. Zed Books and Weaver Press.

Schultz, T.W. (1961) Investment in human capital. *American Economic Review* 51, 1–17.

Schwab, K. (2016) *The Fourth Industrial Revolution*. World Economic Forum, Geneva, Switzerland.

Sneller, G. and Schelling, N. (2014) *Agribusiness in Zimbabwe: Opportunities for Economic Cooperation*. Ministry of economic affairs, The Netherlands.

The Zimbabwean Chronicle (2021) Minister fires manager on the spot. Available at: https://www.chronicle.co.zw/minister-fires-manager-on-the-spot/

USAID (2017) *Combatting Corruption Among Civil Servants: Interdisciplinary Perspectives on What Works*.

Van Fleet, D. (2016) What is Agribusiness? A visual description. *Amity Journal of Agribusiness* 1, 1–6.

Van Rijckeghem, C. and Weder di Mauro, B. (1997) Corruption and the rate of temptation: do low wages in the civil service cause corruption? *IMF Working Paper No.97/73*. DOI: 10.5089/9781451849424.001.

Waeterloos, E. and Rutherford, B. (2004) Land reform in Zimbabwe: challenges and opportunities for poverty reduction among commercial farm workers. *World Development* 32(3), 537–553. DOI: 10.1016/j.worlddev.2003.06.017.

Wangenge-Ouma, G. and Kupe, T. (2020) Uncertain times. *Physics World* 34(1), 15. DOI: 10.1088/2058-7058/34/01/17.

World Bank (2019) Zimbabwe Public Expenditure Review with a Focus on Agriculture. World Bank, Washington DC.

World Bank (2022) The World Bank Annual Report 2022. World Bank, Washington DC.

Zikhali, W. (2017) Impact of agricultural staff training and development on agricultural productivity: a case of Bubi district, Zimbabwe. *International Journal of Academic Research and Reflection* 5, 11–20.

Zinyama, T. (2021) Systemic corruption in Zimbabwe: is the human factor the missing link? *African Journal of Public Affairs* 12, 132–152.

6 Smallholder Farmers' Enterprising Tendencies and Collective Entrepreneurship Towards Increased Incomes and Poverty Elimination: A Case of Zimbabwe

Benjamin Mudiwa[1]* and Adulation Khayisano Ndlovu[2]

[1]*Zimbabwe Ezekiel Guti University (ZEGU), Bindura, Zimbabwe; [2]Independent Researcher*

Abstract

The study investigated how collective entrepreneurship by smallholder farmers in Zimbabwe unlocks market opportunities. It collected primary data from 62 smallholder farmers using quantitative and qualitative data collection techniques. To understand and collect empirical evidence on the entrepreneurial behaviours of smallholder farmers, a comprehensive desk research covering over 50 journal articles was conducted. A General Enterprising Test (GET2) was used to assess smallholder farmers' entrepreneurial behaviour. Data were analysed using SPSS and NVivo software. Results showed that smallholder farmers from the study area possess the following enterprising tendencies: innovativeness, achievement motivation, decision-making ability, risk-taking, coordination ability, information-seeking and -sharing behaviour, self-confidence, planning ability, and cosmopoliteness. Overall, the majority of grouped farmers (74%) and farmers operating individually (77%) showed medium and low levels of the enterprising tendency, respectively. Collective entrepreneurship helped beef farmers in groups to access formal beef markets in urban areas where transparent grading and pricing systems were guaranteed. By eliminating exploitative middlemen from the supply chain, farmer groups received almost double the prices offered to individual farmers and paid lower transaction costs. Smallholder farmers who operated outside groups failed to penetrate formal markets and ended up selling to local informal markets. Further, group formation was established not to be a natural process but facilitated by an internal response to an external stimulus such as the government, private sector companies, and non-governmental organizations (NGOs). It is concluded that collective entrepreneurship helps reduce transaction costs, increase farmers' bargaining power and improve formal market access.

Keywords: Smallholder farmers, Collective entrepreneurship, Enterprising tendency, Market opportunities, Poverty elimination

---

*Corresponding author: b.mudiwa@gmail.com

© CAB International 2023. *Sustainable Agricultural Marketing and Agribusiness Development: An African Perspective* (eds B. Nyagadza and T. Rukasha)
DOI: 10.1079/9781800622548.0006

51

6.1 Introduction

Three-quarters of Africa's poor live in rural areas and the majority are smallholder farmers (Moyo, 2006). In Zimbabwe, there are more than 1.5 million farmers and less than five thousand large-scale farmers (Moyo, 2010; Kuhudzayi and Mattos, 2018). About 92% of farmers are smallholders engaged in different crops and livestock enterprises and are scattered across the five agroecological regions in Zimbabwe. They own 80% of the livestock and 50% of the land in Zimbabwe (Kuhudzayi and Mattos, 2018). However, despite their ownership, access to and control over these productive economic resources, smallholder farmers fail to sustainably service the formal markets. They have struggled to penetrate and meet the formal market demands such as quality and quantity requirements, and consistency in supply, and hence have received lower prices from informal traders (Nyikahadzoi et al., 2013). Consequently, they have remained poor and in need of food aid.

In the face of these challenges, some smallholder farmers in Zimbabwe responded by forming or joining groups, engaging in group input procurement and group marketing. Due to low production and productivity, smallholder farmers scattered across the country are quickly adopting the group approach to meet the volumes of produce required by the formal markets. According to Nguyet (2002), the engagement of smallholder farmers in groups is a sustainable vehicle to solve their problems and increase production and productivity through knowledge and information sharing. Smallholder farmers were also able to ride on collective action for bulk input procurement, which fosters the benefits of economies of scale, and aggregating agricultural produce to consistently meet the market demands. Barham and Chitemi (2009) viewed smallholder farmer groups as entrepreneurial structures and collective action activities that are motivated by financial incentives, and external stimulus designed to leverage product sales, increase agricultural productivity, enhance farmers' social capital, positively impact the farmer's household food security, and eliminate poverty. Although farmer's group are informal, the entrepreneurial tendencies of the groups and the right group dynamics enable the farmers to realize good returns on investment and penetrate high-value markets (Swaminathan and Balan, 2013).

Given these smallholder farmers' circumstances, this study was commissioned to better understand the farmers' market penetration challenges with a view to proffering solutions. The study was guided by the following research questions: What enterprising tendencies do smallholder farmers possess? Is collective entrepreneurship an option for smallholder farmers in Zimbabwe to unlock market opportunities? and What is the influence of collective action and group leadership on smallholder farmers' sales?

6.2 Methods

The study adopted an amalgam approach, where both quantitative and qualitative data were collected. To understand and collect empirical evidence on the entrepreneurial behaviours of smallholder farmers, a comprehensive desk research covering over fifty journal and scholarly articles was conducted. The study further collected primary data by administering household questionnaires to 62 smallholder beef farmers. In this regard, primary data were collected from 31 farmers from Chipinge district, Manicaland province, affiliated with the producer and marketing groups. Musaoingura, Dzidzai, Kumboedza, Pepeukai-Kondo and Matikwa were the five cattle marketing groups used in the study, where six or seven farmers were purposively sampled per group. The farmers interviewed were in different group leadership positions such as the chairperson, secretary, one or two committee members, and three non-committee group members.

Smallholder farmers in different group leadership positions were purposively selected because they were likely to be more informed and knowledgeable about the group and its economic activities, while the non-committee members were randomly selected. Furthermore, 31 non-group affiliated smallholder beef farmers were randomly sampled in the same geographical locations where the groups were picked up.

The Statistical Package of Social Sciences (IBM-SPSS) version 25 was used to capture the data and for analysis purposes. Descriptive statistics such as measures of central tendency (means) and measures of dispersion (standard deviation from the mean, maximum and minimum values) were employed. A General Enterprising Test (GET2) was used to assess smallholder farmers' entrepreneurial behaviour. The research was the first to assess people's general enterprising tendencies in Zimbabwe and one of the few studies in Africa. The nature of the research problem required the researcher to address it from a multidisciplinary standpoint as far as the theoretical framework is concerned. This research also serves as a guide for future research toward a theory of collective entrepreneurship.

6.3   Results and Discussions

6.3.1   Farmers' entrepreneurial behaviours

6.3.1.1   Smallholder farmers' enterprising tendencies

The research revealed that smallholder farmers in lower Chipinge possess the following enterprising tendencies: innovativeness, achievement motivation, decision making ability, risk-taking, coordination ability, information-seeking and -sharing behaviour, self-confidence, planning ability, and cosmopoliteness.

6.3.1.2   Innovativeness

The innovativeness of farmers is subject to farmer's socio-economic characteristics such as age, marital status, education level, annual income, land holding, and livestock ownership (Wanyonyi and Bwisa, 2015). Empirical evidence from Ahmed *et al.* (2011), Patel *et al.* (2014), Kulkarni and Jahagirdar (2015), Porchezhiyan and Sudharshan (2016) and Boruah *et al.* (2015) found that most farmers in Sri Lanka and India had a medium level of innovativeness. On the contrary, Nyello *et al.* (2015) in a study entitled 'The effect of entrepreneurship education on the entrepreneurial

behaviour of graduates in Tanzania' found that 67% of farmers had a high level of innovativeness compared to 33% with a low level of innovativeness. Diverging research findings could be attributed to variations in farmer typology and differences in geographical locations of Asia vs Africa.

6.3.1.3   Achievement motivation

The farmer's ability to strive to increase their capabilities in all entrepreneurship activities in which the excellence standards are applied is known as achievement motivation (Heckhausen, 1967). The research findings revealed variations in the levels of achievement motivation by entrepreneurial farmers. The findings of Ahmed *et al.* (2011), Kulkarni and Jahagirdar (2015), Chaurasiya *et al.* (2016) and Patel *et al.* (2014) revealed that the majority of the agri-preneurs (70%) had medium levels of achievement motivation while 17 and 13% had high and low levels of achievement motivation, respectively.

Interestingly, Porchezhiyan and Sudharshan (2016) found that approximately 60% of farmers possess a high level of achievement motivation followed by 22 and 17% of farmers with medium and low levels of achievement motivation, respectively, which is a diversion from the findings aforementioned. Despite the difference in the results, the majority of the scholars concur that the farmers had a medium level of achievement motivation. The medium level of achievement motivation could be attributed to annual income and economic motivation, as they have a positive correlation.

6.3.1.4   Decision making ability

The decision making ability is key in agri-prenuership and agribusinesses. Literature has demonstrated two main dimensions of farmer decision making ability. The first dimension shows that the majority of the farmers (68%) have moderate decision making ability, followed by poor decision making ability, and the least proportion, 14% have high decision making ability (Boruah *et al.*, 2015). The second dimension concurs that the majority of the farmers are at a moderate level of decision making ability, however, it diverges where it states that the moderate level is followed by the high level and then low level of decision making ability (Vijaykumar, 2001). The latter dimension is in tandem with the findings by Chaudhari *et al.* (2007).

Deductively, from the findings, it is evident that the majority of the farmers possess moderate decision making ability as an entrepreneurial tendency. This level of decision making ability is attributable to the following covariates: education level of the farmers, farm income, improved and better communication behaviours, and medium-size land holding, as these were found to be positively correlated to decision making ability (Ahmed *et al.*, 2011; Tekale *et al.*, 2013).

6.3.1.5 Calculated risk-taking

Farmers, like any other value chain actor, do not operate in a vacuum but in a risky environment. Therefore, the smallholder farmer's ability to carry risk is of paramount importance to enterprising tendencies. This is supported by Misra and Kumar (2009), who argued that in order for an entrepreneur to guarantee success, they have to take the right level of risk and at the right time and place. Furby and Beyth-Marom (1992) defined risk-taking as involvement and participation in behaviours that have the potential of an undesirable outcome. Albeit, the aforementioned researchers did not quantify or qualify the acceptable level of risk, suggesting their argument dives into the risk continuum. In support of the risk continuum argument, Caird (2013) indicated that high scores of calculated risks are associated with the following traits: (i) self-awareness, which is the aptitude to assess one's capabilities accurately; (ii) decisiveness, the ability to comprehend information asymmetry and bounded rationality problems and the capacity to act on this information; and (iii) effective information management.

The study conducted in India on entrepreneurial characteristics of agri-preneurs under the scheme of agri-clinics and agribusiness centres by Ahmed *et al.* (2011), found that the majority of entrepreneurs possessed a medium level of risk-taking ability, shadowed by the high-level risk-taking ability, and the least was the low level. The findings were in tandem with what is in the literature from Rathod *et al.* (2012) and Boruah *et al.* (2015). However, the findings by Ahmed *et al.* (2011) were queried by Bhagyalaxmi *et al.* (2003), Suresh (2004) and Bheemappa *et al.* (2014), who argued that most farmers possess a medium level of risk-taking ability tailed by farmers with low level, and the trivial portion of farmers possess the high level of risk-taking ability. Kumar *et al.* (2012) argued that despite the calculated risk-taking ability importance, it is not the only important enterprising tendency; other traits such as self-confidence, the hope of success, and knowledge affinity influence entrepreneurship.

6.3.1.6 Coordination ability

The interrelation of activities, resources, and structures at different organizational levels is considered coordination, hence a critical enterprising tendency for smallholder farmers. The studies conducted in Sri Lanka and India showed that the majority of vegetable and dairy smallholder farmers had a high level of coordination ability tailed by medium-level coordination ability. The studies show that a trivial portion of the smallholder farmers possessed a low level of coordination ability (Turker and Sonmez Selcuk, 2009; Tekale *et al.*, 2013). Tekale *et al.* (2013) further argued that the high level of coordination ability of dairy farmers is attributed to the perishability nature of milk, the youthful age of farmers, higher level of education, and high income levels. Boruah *et al.* (2015) and Patel *et al.* (2014) found contradicting findings, which revealed that the majority of the smallholder farmers had a moderate level of coordinating ability, while a trivial proportion of smallholder farmers had a high level of coordinating ability. The findings are similar to findings by Rathod *et al.* (2012), Gamit *et al.* (2015), Chaurasiya *et al.* (2016) and Porchezhiyan and Sudharshan (2016). Gamit *et al.* (2015) attempted to qualify the medium level of coordination ability and argued that it was the result of a medium level of experience in dairy farming, a secondary level of education, and low to medium herd size. Despite the variations in the results of smallholder farmers' coordination ability, it was concluded that the majority of farmers possess a medium level of coordinating ability as an enterprising tendency.

6.3.1.7 Information-seeking behaviour

Manivannanan and Tripathi (2007) in a study conducted in Tamil Nadu, India, found that the higher proportion of dairy farmers was at a medium level of information-seeking behaviour followed by farmers at a high level and the least were farmers at a low level of information seeking behaviour. The findings were in tandem with the results from Boruah *et al.* (2015). The variation in the level of information-seeking behaviour was found by Lawrence and Ganguli (2012) in the study entitled 'Entrepreneurial behaviour of dairy farmers in Tamil Nadu, Indian Res'. The scholars argued that approximately 56% of the farmers had medium-level information-seeking behaviour while 26 and 18% were at low level and high level of information-seeking behaviour, respectively. Information-seeking behaviour is of critical importance in the daily activities of the farmer as decision making ability and calculated risk-taking hinge on the ability to seek enough information to avert moral hazards and adverse selection behaviour in the market.

6.3.1.8 Self-confidence

Ahmed *et al.* (2011) found that about 69% of the smallholder farmers had a medium level of self-confidence in executing agri-preneural activities. The findings conform with the literature reported by Boroah *et al.* (2015), Wankhede *et al.* (2013) and Murali and Jamtani (2003). The findings reported by Rathod *et al.* (2012) and Porchezhiyan and Sudharshan (2016) vary from the findings discussed above, as the authors found that approximately 75% of respondents had high levels of self-confidence, followed by farmers with low levels of self-confidence and farmers with medium levels of self-confidence. Chaudhari *et al.* (2007) argued that self-confidence is a function of achievement motivation level, economic motivation level, and decision making ability, hence the medium level of these results in the farmer settling at a medium level of self-confidence. Furthermore, Ahmed *et al.* (2011) reasoned that the medium level of self-confidence might be influenced by the literacy levels of the farmer and exposure to extension services.

6.3.1.9 Planning ability

The anchor of entrepreneurship also involves planning ability, which is associated with the skill set of forward-looking and strategizing to achieve the organizational goal. Empirical evidence from Chauhan and Patel (2003), Chaudhari *et al.* (2007) and Turker and Sonmez Selcuk (2009) shows that farmers possess a medium level of planning ability. However, those integrated into groups have a higher planning ability amplified by high levels of information-seeking and -sharing behaviour. It is noted that the majority of scholars found that individual farmers had a medium level of planning ability, while the conflicting findings were in the proportion of farmers with high-level and low-level planning ability (Patel *et al.*, 2014; Boruah *et al.*, 2015; Porchezhiyan and Sudharshan, 2016). The variations in the findings on high and low levels of planning ability were attributed to the difference in farmer typology.

6.3.1.10 Cosmopoliteness

Cosmopoliteness is the notch to which the agri-preneur is oriented outside their community for seeking information. The studies conducted revealed that the majority of farmers possess a medium level of cosmopoliteness, followed by a low level, and lastly a high level of cosmopoliteness (Ahmed *et al.*, 2011; Kulkarni and Jahagirdar, 2015; Chaurasiya *et al.*, 2016). Notably, different findings from Porchezhiyan and Sudharshan (2016) showed that most of the farmers had a medium level of cosmopoliteness. However, from the empirical evidence, the researcher concluded that the majority of the farmers fall in the medium level of cosmopoliteness because of their medium level of economic motivation, planning ability, information-seeking ability, and innovativeness.

6.3.2 Overall entrepreneurial behaviour of smallholder beef farmers

The smallholder beef farmers belonging to various groups possessed a medium level of overall enterprising tendency (74%) as shown in Table 6.1. Of the group farmers, 13% attained high and low levels of overall entrepreneurial

Table 6.1. Overall entrepreneurial behaviour of respondents. From survey data, 2017.

	Grouped farmers		Individual farmer	
Categories	Frequency (n = 31)	Percentage	Frequency (n = 31)	Percentage
Low (0–26)	4	13	24	77
Medium (27–43)	23	74	7	23
High (44–54)	4	13	0	0
Mean	2		1.23	
S.D	0.516		0.425	

tendency. Compared to individual smallholder beef farmers, the majority (77%) possessed a low level of entrepreneurial behaviour and only 23% were at a medium level. Notably, there were no individual smallholder beef farmers with a high level of enterprising tendency. The findings of the study for farmers affiliated with groups conform with empirical evidence from several studies in India (Tekale *et al.*, 2013; Wankhade and Manakar, 2013; Gamit *et al.*, 2015; Porchezhiyan and Sudharshan, 2016). The scholars found out that most of the dairy cattle and vegetable farmers possessed a moderate level of entrepreneurial behaviour.

Although the findings support each other, there is a difference with individual smallholder beef farmers who were found to have low levels of enterprising tendency. Farmers operating as individuals are less entrepreneurial in comparison to farmers affiliated with cattle producer groups.

6.3.3 Smallholders farmers' collective entrepreneurship

The GET2 test results revealed that the smallholder farmers outside the groups had low enterprising tendencies, which resulted in accessing the local informal beef market. The informal buyers comprise middlemen, other farmers, local shops and butcheries. The smallholder individual farmers were selling cattle off the rangeland, without any value addition such as pen fattening, hence paying lower prices relative to farmers in formal groups who access formal markets. The farmers and the buyers used the eyeball pricing methodology, where the buyer uses visual assessment and estimates the cattle price; however, the price is subject to negotiation.

The study showed that the individual farmers receive a minimum of US$150 to a maximum of US$600, with average cattle buying price of US$365, which is relatively low compared to grouped farmers receiving a minimum of US$234 to a maximum of US$1078 and the average price of US$614 per animal.

The smallholder farmers affiliated with cattle producer groups engaged in improved marketing practices such as aggregation and group marketing, hence a medium level of enterprising tendency. The smallholder farmers were selling beef cattle to formal distant markets such as abattoirs (Koala Park, Sabie Meats, and Montana Caswell Meats) from Chiredzi and Masvingo towns of Zimbabwe. The group marketing strategy enabled the farmers to realize an average price that was 83% higher than the prices realized by an individual farmer. The findings were in tandem with the findings of Shiferaw *et al.* (2009), who found that formal buyers offer higher prices (20–25%) compared to informal beef cattle buyers. The price difference was enacted from the Cold Dry Mass (CDM) pricing method, where the cattle price is a function of cattle grade and weight.

The study established that the smallholder farmers operating in groups and aggregating at least 20–30 beef cattle for sale negotiated with buyers to meet transport costs to ferry cattle to abattoirs. The smallholder farmers indicated that in some instances the abattoirs would send trucks to cattle at low-to-no cost, hence reducing the farmers' cost. The study, therefore, deduced that collective entrepreneurship is an option for smallholder farmers in Zimbabwe to unlock market opportunities. The assertion is supported by Shiferaw *et al.* (2009) and Nyikahodzoi *et al.* (2013), who argued that collective action is a vehicle for smallholder farmers to circumvent barriers to entry to lucrative markets through riding on economies of scale and size, hence increasing farmers' bargaining power and improving farmers' acceptability and access to formal markets and profits.

6.3.4 Impact of collective action and group leadership on smallholder farmers' sales

Results also showed that group establishment by smallholder farmers is not a natural process, formation is facilitated by an internal response to an external stimulus such as the government, private sector companies and NGOs. Perret and Mercoiret (2003) reported similar findings, where farmers' group formation is influenced by external stimulus and motivation by the development programmes. The findings of the study indicated that the farmers engaged in collective action through groups driven by the incentive of high-value enterprise and the potential of beef entrepreneurial activity, hence profit motive is one of the drivers for group formation. It was evident that collective action had a positive influence on smallholder farmers' sales and has shown the farmers where there is money, hence promoting the sustainability of the innovation.

6.3.4.1 Group leadership

Empirical evidence has demonstrated that group dynamics and group success are glued on the group leadership quality, which include but are not limited to trustworthiness, inspiration, innovativeness, information-seeking and -sharing behaviour, honesty, and conflict resolution ability (Mgbada and Agumagu, 2007). The smallholder farmers cited that group leadership plays a significant role in group success and sustainability. The group members vouched for the leadership as the enabling forces to increase cattle sales to beef markets and improvement on the easiness of selling cattle, as they are critical in forging sustainable market linkages.

6.3.4.2 Innovativeness

The results demonstrated that the smallholder farmers that organized had medium levels of entrepreneurial behaviour, hence cascading to improved innovativeness and creative tendency. The farmers in the group alluded that group leadership is key in boosting cattle sales and ensuring the growth and sustainability of the group. The leaders are tasked with creativity and innovation in the ways of conducting business and group activities, which leads to group success, increased sales and farmers' revenue, and group sustainability.

6.4 Conclusion

The research revealed the following enterprising tendencies: innovativeness, achievement motivation, decision making ability, risk-taking, coordination ability, information-seeking behaviour, self-confidence, planning ability, and cosmopoliteness. The literature review and results from the study showed that smallholder farmers possess low, moderate and high levels of these enterprising tendencies. Although there were variations in findings among scholars on the levels of entrepreneurial behaviour, the majority of the research work in this field revealed that smallholder farmers fall under the medium-level category. Collective action coupled with components of entrepreneurial behaviour such as creative tendency and calculated risk (collective entrepreneurship) helped smallholder farmers operating in groups to access formal beef markets in urban areas where transparent grading and pricing systems were guaranteed. By eliminating exploitative middlemen from the supply chain, farmer groups received almost double the prices offered to individual farmers and paid lower transaction costs as unit costs decrease with increasing volumes. Smallholder farmers who operated outside groups, operated as individuals and could not penetrate formal markets; as a consequence, they ended up selling to local informal markets within their local periphery. Collective entrepreneurship was found to be more rewarding for grouped farmers than for smallholder farmers operating as individuals, regardless of the value chain(s) of their interest. Increased sales and incomes have been identified as a sustainable pathway out of poverty. Thus, collective entrepreneurship influences reducing transaction costs, increasing farmers' bargaining power and improving farmers' acceptability and access to formal markets. However, results also showed that group establishment by smallholder farmers is not a natural process, formation is facilitated by an internal response to an external stimulus such as the government, private sector companies and NGOs. Also, group leadership, innovativeness, and information-seeking and -sharing behaviour were found to influence group success and sustainability.

References

Ahmed, T., Hasan, S. and Haneef, R. (2011) Entrepreneurial characteristics of the Agripreneurs under the scheme of agriclinics and agri-business centres. *Journal of Community Mobilization and Sustainable Development* 6, 145–149.

Barham, J. and Chitemi, C. (2009) Collective action initiatives to improve marketing performance: lessons from farmer groups in Tanzania. *Food Policy* 34(1), 53–59. DOI: 10.1016/j.foodpol.2008.10.002.

Bhagyalaxmi, K., Gopalakrishna Rao, V. and Sudarshanreddy, M. (2003) Profile of rural women micro-entrepreneurs. *Journal of Research* 31, 51–54.

Bheemappa, R.A., Natikar, K.V., Birradar, N., Mundinamani, S.M. and Havaldar, Y.N. (2014) Entrepreneurial characteristics and decision-making behaviour of farm women in livestock production activities. *Karnataka Journal in Agriculture Sciences* 27, 173–176.

Boruah, R., Borua, S., Deka, C.R. and Borah, D. (2015) Entrepreneurial behavior of trival winter vegetables growers in Jorhat District of Assam. *Indian Research Journal of Education* 15, 65–69.

Caird, S. (2013) General measure of enterprising tendency test. The Open University. Available at: www .get2test.net (accessed 28 June 2022).

Chaudhari, R.R., Hirevenkanagoudar, V., Hanchinal, S.N., Mokashi, A.A., Katharki, P.A. *et al.* (2007) A scale for the measurement of entrepreneurial behaviour of dairy farmers. *Karnataka Journal of Farm Sciences* 20, 792–796.

Chauhan, N.B. and Patel, R.C. (2003) Entrepreneurial uniqueness of poultry entrepreneurs. *Rural India* 66, 236–239.

Chaurasiya, K.K., Babodiya, S.K., Somvanshi, S.P.S. and Gaur, C.L. (2016) Entrepreneurial behaviour of dairy farmers in Gwelior district of Madhya Pradesh. *Indian Journal of Dairy Science* 69, 112–115.

Furby, L. and Beyth-Marom, R. (1992) Risk taking in adolescence: a decision-making perspective. *Developmental Review* 12(1), 1–44.

Gamit, M.P., Rani, D.V., Bhabor, I.N., Tyagi, K.K. and Rathod, A.D. (2015) Entrepreneurial behaviour of dairy farmers in Surat District of South Gujarat. *International Journal of Advanced Multidisciplinary Research* 2, 50–56.

Heckhausen, H. (1967) *The Anatomy of Achievement Motivation*. Academic Press, New York.

Kuhudzayi, B. and Mattos, D. (2018) A model for farmer support in Zimbabwe- opportunity for change. *Cornhusker Economics*. Available at: https://agecon.unl.edu/cornhusker-economics/2018/farmer -support-model-zimbabwe.pdf (accessed 28 June 2022).

Kulkarni, N.P. and Jahagirdar, K.A. (2015) Entrepreneurial behaviour of rose growers. *International Journal of Multidisciplinary Research, Studies, and Developments* 1, 1–5.

Kumar, S.R., Ramakumar, D., Babu, D., Babu, V.D. and Jaishridhar, P. (2012) Socio-economic analysis, and it correlates with entrepreneurial behaviour among dairy farmers in Tamil Nadu. *Journal in Dairying, Foods and Home Sciences* 31, 108–111.

Lawrence, C. and Ganguli, D. (2012) Entrepreneurial behavior of dairy farmers in Tamil Nadu. *Indian Research Journal Extension of Education* 12, 66–70.

Manivannanan, C. and Tripathi, H. (2007) Management efficiency of dairy entrepreneurs. *Indian Research Journal of Extension Education* 7, 44–51.

Mgbada, J.U. and Agumagu, A.C. (2007) Role of local leaders in sustainable agricultural production in Imo State implication for youth in agriculture. *Journal of Economics Theory* 1, 1–5.

Moyo, S. (2006) The evolution of Zimbabwe's land acquisition. In: Rukuni, R., Tawonezvi, P., Eicher, C., Munyukwi-Hungwe, M. and Matondi, P. (eds) *Zimbabwe's Agricultural Revolution Revisited*, 2nd edn. University of Zimbabwe Publication, pp. 143–164.

Moyo, T. (2010) Derminants of participation of smallholder farmers in the marketing of small grains and strategies for improving their participation in the Limpopo River Basin of Zimbabwe. MSc (Agricultural Economics) Thesis, University of Pretoria, South Africa. Available at: https://repository.up.ac.za/bitstream/handle/2263/27365/dissertation.pdf?sequence=1 (accessed 27 June 2022).

Misra, S. and Kumar, E.S. (2009) Resourcefulness: a proximal conceptualisation of entrepreneurial behaviour. *The Journal of Entrepreneurship* 9(2), 15–154. DOI: 10.1177/097135570000900201.

Murali, K. and Jamtani, A. (2003) Entrepreneurial characteristics of floricultural farmers. *Indian Research Journal of Extension Education* 39, 19–25.

Nguyet, N.T.K. (2002) Establishment and maintenance of farmers' groups (FGs). *Agricultural Extension Network Updates* 5, 1.

Nyello, R., Kalufya, N., Rengwa, C., Nsolezi, M.J. and Ngirwa, C. (2015) Effect of entrepreneurshipeducation on the entrepreneurial behaviour: the case of graduates in the higherlearning institutions in Tanzania. *Asian Journal of Business Management* 7, 37–42. DOI: 10.19026/ajbm.7.5167.

Nyikahadzoi, K., Siziba, S., Mango, N., Zamasiya, B. and Adekunhle, A.A. (2013) The impact of integrated agricultural research for development on collective marketing among smallholder farmers of Southern Africa. *Asian Journal of Agriculture and Rural Development* 3, 321–336.

Patel, P., Patel, M.M., Badodia, S.K. and Sharma, P. (2014) Entrepreneurial behavior of dairy farmers. *Indian Research Journal of Extension Education* 14, 46–49.

Perret, S.R. and Mercoiret, M.R. (2003) Supporting small-scale farmers and rural organisations: learning from experiences in West Africa. *A handbook for development operators and local managers*. Available at: https://www.semanticscholar.org/paper/Supporting-small-scale-farmers-and-rural-learning-A-Perret-Mercoiret/36c0a7bd7b64e48361c3c59c512fa4cc6e9f4614 (accessed 18 September 2023).

Porchezhiyan, S., Sudharshan, A., and Umamageswari (2016) Entrepreneurial behavioural index of dairy farmers in the northern districts of Tamil Nadu. *Indian Journal of Economics and Development* 4, 1–5.

Rathod, P.K., Nakim, T.R., Landge, S. and Hatey, A. (2012) Entrepreneurial behaviour of dairy farmers in Western Maharashtra, India. *International Journal of Commerce and Business Management* 5, 115–121.

Shiferaw, B., Obare, G., Muricho, G. and Silim, S. (2009) Leveraging institutions for collective action to improve markets for smallholder producers in less-favoured areas. *Afjare* 3, 1–18.

Suresh, J. (2004) Entrepreneurial behaviour of milk producers in Chittoor district of Andhra Pradesh: A critical study. M.V.Sc. Thesis, Acharya N. G. Ranga Agricultural University, Hyderabad.

Swaminathan, B. and Balan, K.C.S. (2013) An inquiry into the role of group dynamics in enhancing farm remuneration. *American International Journal of Research in Humanities, Arts, and Social Sciences* 4, 41–44.

Tekale, V.S., Bhalekar, D.N. and Shaikh, J.I. (2013) Entrepreneurial behaviour of dairy farmers. *International Journal of Extension Education* 9, 32–36.

Turker, D. and Sonmez Selcuk, S. (2009) Which factors affect entrepreneurial intention of university students? *Journal of European Industrial Training* 33(2), 142–159. DOI: 10.1108/03090590910939049.

Vijaykumar, K. (2001) *Entrepreneurship behaviour of floriculture farmers in Ranga Reddy district of Andhra Pradesh*. MSc (Agri.) Thesis, Acharya N.G. Ranga Agricultural University, Hyderabad.

Wankhade, R.P., Sagane, M.A. and Manakar, D.M. (2013) Entrepreneurial behaviour of vegetable growers. *Agricultural Science Digest* 33, 85–91.

Wanyonyi, N.J. and Bwisa, H.M. (2015) Factors influencing entrepreneurial behaviour among farmers: a case of cabbage farmers in Kiminini ward. *International Journal of Technology Enhancements and Emerging Engineering Research* 23, 143–148.

7 Factors Influencing Rural Female Entrepreneurs in Enhancing Livelihoods: A Global Perspective

Rahabhi Mashapure[1]*, Purity Hamunakwadi[2] and Julius Tapera[3]

[1]*Marondera University of Agricultural Sciences and Technology (MUAST), Zimbabwe; [2]Nelson Mandela University (NMU), South Africa; [3]Lupane State University (LSU), Zimbabwe*

Abstract

Rural female entrepreneurs are driven by various positive and negative factors in a quest to achieve sustainable livelihoods for themselves. Different contexts (developed or developing countries) in which these rural female entrepreneurs are operating are based on societal, education and social skills available in the country. This study drew its arguments from secondary data, which included documentary and conceptual analysis of relevant sources to conceptualize and contextualize rural female entrepreneurs and livelihoods. Hence, data from secondary sources were collected and reviewed from published journal articles, books and relevant databases. It is shown that in developed countries rural entrepreneurs are influenced mostly by pull factors to obtain independence, individual freedom, self-fulfillment, individual security and self-confidence. In contrast, in developing countries they are largely influenced by push factors associated with social pressures, which force them to venture into entrepreneurship to obtain income and employment to support their families. It is recommended that since rural entrepreneurship inspires innovation and creativity in both female and male entrepreneurs it is very significant because it generates and encourages wealth distribution and increases wellbeing. The authors envision that rural female entrepreneurs can enhance sustainable livelihoods and that assists in informing and positioning international entrepreneurial agencies, regional bodies, national and local spheres to divert their resources in support of such initiatives.

Keywords: Rural female entrepreneurs, Rural livelihoods, Globalization

7.1 Introduction

Worldwide, entrepreneurship is regarded as the economic engine necessary for job creation, economic growth, poverty alleviation and reduction. This chapter interrogates on myriad factors that influence female entrepreneurs in rural areas that may support or hinder their abilities to enhance their livelihoods. We seek to parse out the understanding and significance of the role that women are playing in enhancing their livelihoods and the barriers that hinder such great interventions. Stifel (2010) asserts that the rural setting in which these females are located is characterized by various activities comprised of both agricultural activities

*Corresponding author: cmashapure29@gmail.com

(cropping, livestock husbandry, fishing and forestry), non-farm industry activities (including mining, wood products, energy, food and beverages, textiles and leather and construction materials) and services (commerce, handicrafts, hotels and restaurants, transport, public works and private health). All these various activities show that rural dwellers can venture into diverse livelihood strategies depending on where they are situated. Hence, this paper starts with an introduction, a literature review that discusses entrepreneurship, factors influencing rural entrepreneurship, and then theories contributing to sustainable livelihoods in rural areas, and ends with conclusions and recommendations that can be adopted by countries in addressing the factors that may hinder rural female livelihood initiatives. The section below provides the conceptualization of rural entrepreneurship based on different scholars.

7.2   Conceptualizing Rural Entrepreneurship

Most often, the terms entrepreneur, entrepreneurship and entrepreneurial tend to be used interchangeably, whereas each concept denotes something specific about the individual, group or society undertaking resource utilization to create satisfaction. The term entrepreneur as posited by Clark (2004) is applicable to individuals, organizations and projects. Kuratko (2006) describes the outcome of entrepreneurship as being strategic change management encompassing vision, mission and core values determination and implementation mechanisms. In effect, the entrepreneur pursues novel processes and methods to take advantage of opportunities through creation of value (Brown and Ulijn, 2004). The entrepreneurial process is considered a key aspect of the dynamism of economic crisis (renewal) during which less efficient firms fail and are replaced by the more efficient ones in the continuous process of creative destruction (Schumpeter, 1976). Entrepreneurship as a process gives people more jobs, creates new inventions and ideas, increases and stimulates national income, consequently having the potential of affecting economic development (Abosede and Onakoya, 2013).

Hoyle *et al.* (2022) argue that entrepreneurship includes the possession of managerial skills, and an entrepreneur must be able to coordinate and combine the factors of production.

Korsgaard *et al.* (2015) define rural entrepreneurship as the various forms of entrepreneurship that are undertaken in socio-spatial zones characterized by large open spaces and small population settlements relative to the national contexts. Fortunato *et al.* (2014) assert that rural entrepreneurship is a peculiar discipline and has its own unique developmental opportunities, but it is not without its own challenges as it is suggested from former research that rural entrepreneurship can be influenced by various factors and these factors affect rural entrepreneurs differently. Of note, different theories of entrepreneurship already give us a rich background to proceed from in empirical analysis but different theories and different methodologies should be chosen according to context of the study. A consistent universal theory does not exist in entrepreneurship, but rather it consists of several different approaches including psychology, sociology, anthropology, regional science and economics. No common theoretical framework, even if demanded for rigorously, exists to synthesize the different points of views. Some trials have been made to develop a multidimensional approach to entrepreneurship study but there are some problems, mainly emanating from the differing perspectives of the above-mentioned well-established disciplines (Johnson, 1990). Given the differing perspectives, this chapter adopts a sustainable livelihoods approach to problem solving regarding various factors influencing rural female entrepreneurs towards their wellbeing. In this regard, the following sections provide an understanding of the theory for sustainable rural livelihoods with the aim of positioning female entrepreneurs' capabilities in sustaining their livelihoods.

7.2.1   Theory for sustainable rural livelihoods

We pivot our paper based on the theoretical lens of the sustainable livelihoods framework (SLF). This theoretical lens underpins and shapes the current research with the intention to frame the argument that borders around the sustainable rural livelihoods of female entrepreneurs across the world. The SLF assumes that people live within a vulnerability environment made up of shock waves, developments and seasonality (Chitongo, 2019; Kabonga, 2020). The vulnerability environment is determined by converting structures and methods, which include levels of government regulations, rules and culture that govern the livelihood strategies (Tabares *et al.*, 2022) that people use to meet their desired livelihood outcomes such as food security. The concept assumes that individuals move in and out of deficiency and the perception releases the progressions of modification better than insufficiency line dimensions (Serrat, 2017).

Financial capital defines the set of monetary resources that are required by individuals to meet livelihood goals (income, employment and savings) (Kabonga, 2020). Physical capital comprises rudimentary infrastructures and possessions vital for encouraging and supporting sustainability of rural livelihoods (infrastructure, marketplace accessibility and transport). Human capital denotes the knowledge, skills training, education and the ability to work in good health that allows people to pursue livelihood strategies (Zhou *et al.*, 2016). Social capital characterizes the features of a social organization that function to manage actions driven by community, experience and inner drive. These are social resources people draw on in a quest for livelihood objectives consisting of social prestige, cooperation and decision making power (Fafchamps, 2001). Natural capital comprises tangible and intangible goods, such as air quality, soil quality and lodging for a healthy atmosphere (Erenstein *et al.*, 2010). The SLF framework describes conditions that govern an individual's access to possessions and livelihood opportunities which can be converted into sustainable rural livelihood outcomes thus driving individuals out of the deprivation trap of poverty (Zhou *et al.*, 2016). Knutsson (2006) posited that the sustainable livelihoods outcome consists of five indicators, which are more income, increased wellbeing, reduced vulnerability, improved food security and more sustainable use of natural resources. More income or sufficient income allows people to meet their needs.

To facilitate an easy analysis, we look at different contexts to derive lessons that can suit each setting from various international, regional and African experiences on factors influencing rural female entrepreneurship livelihood strategies. The SLF has been effectively used to appreciate and promote rural development (Baker *et al.*, 2018). In this regard, the SLF was selected to inform this chapter because rural female entrepreneurs are within the space that is vulnerable to trends and shocks which are shaped by policies, laws, culture and the institutions in which they are situated in. Hence, it is suggested that human, financial, social, natural and physical capital can assist in navigating and reducing negative implications on rural female entrepreneurs.

Thus, rural female entrepreneurs can sustain their livelihoods from intended outcomes of more income and increased wellbeing in this case. The following section provides deeper instances shown in the literature on diverse drivers of entrepreneurship in rural areas basing the arguments on pull and push factors that influence female entrepreneurs to venture into businesses worldwide.

7.3   Drivers of Entrepreneurship in Rural Areas

Rural entrepreneurs' motivation for engaging in livelihoods is not always driven by positive factors but is also due to negative circumstances. Thom (2015) defined entrepreneurial drive as the procedure that drives or activates entrepreneurs to put high effort levels to attain their entrepreneurial objectives. This drive has been given various names or categories by various authors. According to Fosić *et al.* (2017), entrepreneurs engage into entrepreneurship motivated by pull and push factors. Rey-Martí *et al.* (2015) view rural entrepreneurship drivers as internal and external factors, which may also be referred to as pull or push factors. The internal influences consist of demographic features such as age, marriage position, and total number and surviving young children in a household, whereas external factors are perceived as business growth and funding. For the internal factors, married women have lower rates for engaging into entrepreneurship because of the cultural myths whereby they are expected to spend much of their time with family. Pull factors can be defined as factors that influence an individual to start a new business venture for the need to change their status quo (Cabrera and Mauricio, 2017). Ismail *et al.* (2012) establish that push factors are connected with necessities, for instance joblessness, insufficient family income and dissatisfaction with current employment. Literature proposes that females in developed countries are mainly inspired by push influences such as need for individuality, providing and maintaining family, reducing poverty and difficulty in securing appropriate work (Bullough *et al.*, 2015). Rodríguez-Pose (2018) emphasizes that opportunity-driven entrepreneurs are vital to the economic development of a country. However, drivers of rural entrepreneurship vary across countries and regions.

7.4   Pull Factors of Rural Entrepreneurship

Rural entrepreneurship can be derived from pull factors (Ali and Mahamud, 2013). Pull factors inspire rural entrepreneurs to engage in entrepreneurship whereas push factors force them to start enterprises (Mordi *et al.*, 2010). Pull factors are more likely motivated by positive circumstances such as one's desire for independence, self-motivation or achievement, need for adequate finance, increased profit and wealth, self-actualization, personal development, power among others and social status (World Bank, 2010). Entrepreneurship can be driven by the quest for satisfaction and independence such as self-sufficiency, creativeness, self-fulfilment, financial gains and individual success. To add, rural entrepreneurs are motivated by the need to intensify their savings that they use to sustain the family and household thereby cultivating their household values of living and attaining self-sufficiency (Bullough *et al.*, 2015). In a study within Salim (2017) it was found that factors influencing entrepreneurship were assistance from third parties, initiatives, inspiration from family and friends, ability and previous experiences, and the desire to be independent. Additionally, Swinney and Runyan (2007) assert that rural entrepreneurs are motivated to be successful entrepreneurs by creating employment for themselves, generating income and the support they get from the society.

Grigore and Mitroi (2012) assert that many rural entrepreneurs choose an entrepreneurial career to demonstrate that they are more proficient at attaining additional/supplementary income than their male counterparts and the society expect them to produce.

The most common encouragements motivating rural entrepreneurs into entrepreneurship are societal and educational reasons (Gibb, 1993). Societal influences depend on experiences, stage of employment, age and environment in which one is exposed to (Copp and Ivy, 2001). Furthermore, among others, Lent *et al.* (2019) suggest that often times rural entrepreneurs may be more driven by social pressures. Pertinent to this issue, educational factors add value to females' decisions to engage in entrepreneurship and to their choice of business. These factors consist of: educational level; educational type, which is 'formal' or 'informal'; type of abilities established, that is professional or educational; and the college surroundings (Jiménez *et al.*, 2015).

International examples, from countries such as Singapore, the USA, Pakistan, Norway and Germany, show that entrepreneurs are influenced, encouraged and motivated by the quest for independence, individual freedom, self-fulfilment, and individual security (Klapper and Parker, 2011). On a different note, Benzing *et al.* (2009) establish that generally entrepreneurs in developing nations are greatly expected to be encouraged by need for income, while those in developed nations are encouraged by advanced order necessities such as self-fulfilment and self-confidence. Evidence from literature shows that rural entrepreneurs who are pulled into private enterprise are much more accountable and focused on growth compared to females that are pushed into private enterprise by extrinsic situations (Iacob and Nedelea, 2014). In this respect, women private enterprises in developing countries are usually regarded as an engine of poverty reduction. With the discussion so far, push factors also play a significant role in influencing rural entrepreneurship.

7.5   Push Factors of Rural Entrepreneurship

According to Vossenberg (2013) entrepreneurs engage into entrepreneurship for survival or by being pushed by the status quo. In transitional countries and emerging economies, rural entrepreneurship is driven by the idea that creating your own employment provides flexibility and allows rural entrepreneurs to have a decent equilibrium among employment and household care duties. Rural entrepreneurs suffer from multitasking whereby they are expected to spend most of their time with the family yet on the other hand they are supposed to be at work. Thus, rural entrepreneurs end up pushed into entrepreneurship since it allows them to set their own timetable. However, this sometimes affect sustainability of rural entrepreneurs' livelihoods as more time and effort can be drained by the family compared to business.

Conclusions of McGowan *et al.* (2012) show that rural entrepreneurs in Northern Ireland are motivated by influences which are alike to those of their men comprising a need for individuality and monetary increase, not like the many male entrepreneurs who decide on private enterprise to balance job accountabilities and earning potential with household obligations. On the other hand, female businesspersons in developing countries are encouraged by the necessity to provide sufficient food daily in their households, not for pleasure of work or free-time benefits. In most cases they engage in entrepreneurial activity as a survival strategy out of necessity since there is shortage of employment with no other options for generating income (Cullen and Archer-Brown, 2020), while in developed countries rural entrepreneurs become entrepreneurs because they have vast opportunities (Vossenberg, 2013). This comparison clearly shows that what pushes people into rural entrepreneurship differs with context, place, necessities and resources available in that country.

Bullough *et al.* (2015) posit that rural entrepreneurs become entrepreneurs to efficiently overcome poverty, generate value, and encourage social and financial development, approving the conclusions of Benzing *et al.* (2009) who found that revenue levels and employment opportunities are the driving factors for rural entrepreneurship. Bullough *et al.* (2015) and Benzing *et al.* (2009) perceive that increase in income has a positive impact on providing security for rural entrepreneurs and their household. Therefore, if rural entrepreneurs are not employed, they end up creating employment

for themselves to overcome poverty and meet family basic needs. This type of driver is regarded as a push factor as rural entrepreneurs engage into entrepreneurship due to the discomfort of the present status quo.

Researchers in Tanzania on small and medium enterprises show that most females in emerging nations engage in entrepreneurship because they have a lower level of education and they have found it difficult to get formal paid jobs (Magigaba and Jili, 2019). According to Buttner and Moore (1997) rural businesspersons are mostly driven by existence pressures, disapproving circumstances and loss of occupations. This implies that the majority venture into entrepreneurship as the only option available to them for survival purposes. Most rural entrepreneurs are forced by lower levels of training and lack of occupational education, lack of previous industry experience, discrimination, poor social networking and because they are not ready to take risks (Jagero and Kushoka, 2011).

The salary gap between males and females is also another influencing factor that accounts for rural entrepreneurs engaging in entrepreneurship. This was supported by Nicolás and Rubio (2016) and Kobeissi (2010) who state that gender disparity/unfairness in salary has an imperative influence on females' choice to engage in private enterprise. Study findings show that in established nations females receive about 77% of males' salaries and in emerging nations females receive only 73% (Kobeissi, 2010). Holmen *et al.* (2011) analysed motivations of eight Afghan rural entrepreneurs, and they found that revenue creation was the greatest significant push influence while need for independence and autonomy were considered as pull factors. Nevertheless, in dissimilarity to researches in other nations, the need for attainment existed but was not highlighted. An unforeseen conclusion was on the importance of the need to help extended household members. Accessing adequate finance to kickstart business was regarded as the major problem as well as absence of associates and safety throughout business procedures. Sexual characteristics, explicit difficulties, including partial marketplace, flexibility restrictions and undesirable arrogance and absence of societal recognition for woman businesspersons were problems faced by rural entrepreneurs in Afghanistan.

Conclusions of the study by Chan and Quah (2012) in Malaysia show that key entrepreneurs are motivated by the previous business experience from family business, obtainability of resources such as land and buildings, market potential, strategic location, the positive influence of friends and family, and profitability of the business venture. Conversely, the key factors pushing entrepreneurs to business emerging in the Chan and Quah (2012) study include individual attitude towards entrepreneurship, previous employment experience, individual interest and retirement/retrenchment. Results from three different groups (2013, 2014 and 2015) of rural entrepreneurs in South Africa shows that most rural entrepreneurs desired to be entrepreneurs for the need to be independent and have freedom (Meyer and Landsberg, 2015). However, Kirk and Belovics (2006) found that females are motivated into entrepreneurship by the need to balance their business and household responsibilities.

Research by Langowitz and Minniti (2007) establishes that females who are highly educated are highly likely to start businesses. Similarly, Hui-Chen *et al.* (2014) and De Wit and Van Winden (1989) report that people with higher levels of training, previous employment experience, and information of marketplace and work experience have a greater tendency to see an opportunity for engaging in entrepreneurial activities. Also, Edewor *et al.* (2014) state that entrepreneurship success was influenced by the dominant social financial influences such as training background, training received and skills. Likewise, Rose *et al.* (2006) establish that ability, training background and life involvements and adequate monetary sustenance are part of the main influencing factors driving the enthusiasm for entrepreneurs to venture into business. To add, Nziku (2012) indicates that high levels of training, possession of the business and presence of role models are some of the main influences for the development of income in business. Furthermore, Bhola *et al.* (2006) illustrate that individuals with advanced education are greatly expected to pursue opportunity-based businesses, unlike those with a lower level of education, who are more involved in necessity (problem solving) ventures; therefore, it shows that level of training has a critical part in making a decision to start-up and sustainability of business ventures.

In a way to conclude on the driving influences for rural private enterprise, it is conceivable to assert that there are diverse sets of motives and encouraging factors motivating rural entrepreneurs to venture into entrepreneurship and these depend on nations, beliefs and stereotypes (Hayrapetyan, 2016). These factors can be categorized as pull or push factors (Fosić *et al.*, 2017) whereas Suchart (2017) preferred to use the terms opportunity-driven and necessity-driven. However, these categories emerge due to positive factors (pull or presence of opportunities) or negative influences (discomfort in living) that force rural entrepreneurs into entrepreneurship.

7.6 Effects of Rural Entrepreneurship on the Enhancement of Rural Livelihoods

Globally, entrepreneurship is widely known as an important component for economic development and growth in all nations (Meyer and Synodinos, 2019). Entrepreneurship is measured as an important tool that drives job formation and fuels economic growth and development (Ribeiro-Soriano, 2017). Entrepreneurship can add to the formation of employment and wealth, may nurture innovation, and offers self-sufficiency and a sense of individual achievement (Nieva, 2015). Accordingly, entrepreneurs both male and female are recognized as economic actors who create business ventures, resulting in employment opportunities that lead to the economic development and success of a country (Sadaf *et al.*, 2018). Internationally, females are displaying a significant interest in entrepreneurship, resulting in many rural entrepreneurs creating new business ventures (Meyer, 2018). Female businesses consist of one of the global fastest increasing business populaces (Brush *et al.*, 2006), accounting for up to a third of all companies functioning in the official international economy (Di Fabio and Blustein, 2016), and playing a significant role in economic development at both the public and nation level (Muñoz-Fernández *et al.*, 2019). Accordingly, female entrepreneurs are acknowledged as key contributors to economic development (Kalinic *et al.*, 2014).

Wage and salary occupation is very limited in most emerging nations. Cho and Kim (2017) allude that it is popular to note that in emerging nations workers in low-paying work usually own businesses to thwart the insufficiency from their salaries. Therefore, nurturing rural private enterprise is generally observed to be very important for increasing employment thereby producing chances for decreasing insufficiency through sustainable rural livelihood engagements. Female businesspersons in lower-middle-revenue markets find it as a fast suitable significant sector; it is a business and income-creating division that upsurges the gross domestic product of a nation (Hammawa and Hashim, 2016). Female businesspersons play an important role in cultivating a diverse understanding in cultures and private businesses and have over the years developed a cumulative thoughtfulness that capacitates them to seize entrepreneurial opportunities. This is the reason why they now play an important role in numerous markets (Monitor, 2012).

Female entrepreneurs are the strength of rural financial prudence in emerging nations especially in Africa, and play an important part in making sure of their relatives' wellbeing (Farah, 2014). This is realized in terms of providing shelter, sustenance, education and wellbeing for the household. Rural entrepreneurs play crucial and vibrant parts in the financial development, they easily adjust to change and are highly innovative. As mediators of change and growth, female entrepreneurs offer significant roles in all societies, both in formal and informal sectors. Rural entrepreneurs have gained substantial significance in the socio-economic growth of not only the developed economies but also the developing economies, because they constitute a substantial share of micro-, small and medium-sized enterprises (SMEs) (Kjeldsen and Nielson, 2000). Female entrepreneurship has countless influences on the socio-economic domain which assist females to earn some extra income for the household, generating job opportunities for them and for others, relishing financial independence, and gearing up the process of financial development. In this research the impacts of rural entrepreneurship on sustainable rural livelihoods were recognized as dependent and independent variables based on an underlying relationship.

Even though female entrepreneurs mainly engage in small unnoticed enterprises they have delivered occupations to hundreds of uneducated, unskilled, unemployable and less advantaged females living in insufficiency circumstances in their indigenous community (Kalinic *et al.*, 2014). Female entrepreneurs make significant inputs to gross national product, employment formation, invention, and universal communal wellbeing (Brush and Cooper, 2012), which legislators have recognized (Bullough *et al.*, 2015). For instance, Carter and Scarbrough (2001) specified that female entrepreneurs constitute 26% of the 3.2 million workers who managed to create their own employment (n=824,659) for the UK. In 2002, the International Test Pilot School established that females were generating 28% of new companies and creating employment of an 'average of 0.6 permanent workers in Sweden compared to 1.7 for men'. Kay *et al.* (2003) revealed that in Germany there is a total of 1.03 million rural entrepreneurs who own successful companies. Companies managed by females have profits of at least €16,620 (n=522,000) signifying 18% of the total in that group and created employment for a million workers, both males and females. These businesses make a total income of €232 billion, or about 6% of the total and 11% of that produced by companies managed by female entrepreneurs. In 2016, the 6th Economic Research Report by the Indian Government indicated that the total number of people employed in female-owned businesses were 13.45 million. Additionally, 8.2 million people were working in female-owned businesses situated in rural areas and only 5.18 million in businesses situated in urban areas, confirming the above observations. Thus, if rural entrepreneurship is taken seriously in Zimbabwe and other African countries it may reduce the unemployment rate, which will have a positive impact to the livelihoods of families, communities and nations at large. Rural entrepreneurship does not only create job opportunities but also makes their employees understand the relevance of savings, hygiene, physical wellness and education. This socio-economic integration helps the people living at the grassroots level to join entrepreneurship activities, boost their self-image and confidence and provide an intellect of financial independence and security. The growth of entrepreneurial activities generates demand for public utilities such as road transport, health, education and even rural electrification among others. This promotes regional development and reduces the urge in the residents to migrate to the city.

Muhammad Yunus, founder of Grameen Bank, states that income proves to be the best medicine for people living at the grassroots level. Rural entrepreneurs generate income, which maintains their families and increases their household wellbeing (Solesvik, 2013). Rural entrepreneurs provide for themselves and their children through the returns they generate from their businesses (Kantar, 1999). Steinem (1992) distinguishes that female business owners tend to take a general approach to life as they try to balance work, domestic, economic and society morals. Rural entrepreneurs have a right to control income and working conditions, since they will be managing their own enterprises. This empowers them and encourages their participation and contribution in economic, societal and political events. This results in reduced gender disparity and discernment, particularly in the labour market. Being entrepreneurs helps females to gain self-confidence, self-respect and management experience, which gives them greater control over their lives in society, economic and political circles (Kantar, 1999).

According to Allen et al. (2007) it was statistically proven that more than 30% of the rural entrepreneurs play an important part within the context of economic growth. Rural entrepreneurs are crucial for overall wealth creation and creation of employment in all economies (De Bruin et al., 2006). Many studies indicate the prominent role rural entrepreneurship plays in the reduction of poverty (Yunus, 2007) and emphasize the profound impact it has on local communities and surroundings (Morgan, 2012). Rural entrepreneurs increase significantly in the economies across the globe. The unseen business potential of females has slowly grown, with the increasing understanding of their important influence to the economy and world at large (Agarwal et al., 2017). Hanushek and Woessmann (2008) highlight that some of the factors influencing female entrepreneurship include skill, knowledge and adaptability.

Female businesspersons have a significant influence on countrywide frugalities through their rural livelihoods. Their livelihoods enable people living at the grassroots level to avail basic necessities of life, which consequently result in improvement of their standard of living.

In 2005, Global Entrepreneurship Monitor established that females across the world took part in wide-ranging entrepreneurial initiatives and their actions resulted in establishment of numerous business enterprises, for wealth creation and generation of jobs. The role of entrepreneurs as agents for resource utilization, wealth creation, job creation, alleviation of poverty and development of human capital enormously increased the number of rural entrepreneur-owned ventures across the world (Sarfaraz et al., 2014). The effect of female enterprise has robust effects for household structures, consumption patterns and future decisions of the household associates, such as the choice for education or health concerns, sometimes they could even be the motives to become a businessperson. Rural private enterprise inspires innovation and creativity. Female entrepreneurship improves novel goods or services for the marketplace to achieve human requirements. It also motivates asset attentiveness in the novel businesses being formed. Rural private enterprise through its procedure of novelty generates novel ventures of novel schemes. The more schemes being generated, the more original employment will be created, thereby reducing the joblessness rate. That will generate and encourage wealth distribution and increase wellbeing.

Rural entrepreneurship assists in sustainability of livelihoods in indirect ways (Perl-Kot, 2011). Once females are financially mobilized through self-governing enterprise ownership they are inclined to transform the consumption patterns of the family in a good way, compared to their male counterparts, which is favourable and encouraging to family and child welfare. Females spend their business profits on clothing, sustenance and inevitabilities for the family whereas males might spend business profits on alcohol or entertainment (Kantor, 2001). Rural entrepreneurship supports sustainability of livelihoods by creating wealth for many people looking for entrepreneurial opportunities. Even if this is not the major motivation for people to engage in entrepreneurship, entrepreneurship plays a crucial part in the economy. Both a new industry and the business titleholder can get wealth, which will benefit the economy by providing innovative products as well as the expenditure influence generated for the businessperson. Deprived of businesspersons, our frugality would not profit from the improvement they provide from additional commercial products and concepts.

7.7 Empirical Evidence From the Zimbabwean Context

Previous research on Zimbabwe found that socio-cultural factors impact on women's participation in entrepreneurship (Chigudu, 2018). The findings are also in line with Cejka and Eagly (1999) who found that stereotypical characteristics ascribed to women and men by society impact on the grouping of various occupations as feminine or masculine. The findings also agree with Powell and Graves (2003) who revealed that gender-related characteristics are associated with the gender-role stereotypes identified as either feminine or masculine. Women also agreed that other business activities were the main sources of financial capital to start their business. The study concludes that family duties and needs, and women's marital status and education are the individual factors that influence women participation in SME activities. Also, socio-cultural values contribute to the effect of women participation in SME activities in which the differences between women and men's activities in entrepreneurship are related to gender characterization. Dzapasi (2020) concurred that family background in business as well as confidence and self-esteem were cited as key to successful women entrepreneurship in Zimbabwe. Mashapure *et al.* (2022) suggested that the government must avail special grants or awards to motivate Zimbabwean women to start and stay in the entrepreneurship field. Derera *et al.* (2020) found that women in Zimbabwe are mostly driven into entrepreneurship because of harsh economic, political and social circumstances. Drawing from the literature on women entrepreneurship in Zimbabwe, it can be argued that the majority of women are pushed into entrepreneurship because of the suffering from the adverse effects of the economic and political crises that the country is experiencing (Mutsagondo and Chaterera, 2016). To illustrate this, Osirim's study (2003) reveals that women with no academic qualifications were forced to establish micro-enterprises because of limited employment opportunities. In another study,

Chamlee-Wright (2002) observed that more than 25% of women were forced to develop income-generating projects as a means for survival because of some financially devastating events (death or illness of a spouse, divorce and other unforeseen personal circumstances). Today, many Zimbabwean women bear a disproportionate share of the burdens of economic and social deprivation both as breadwinners and as caretakers (Mutsagondo and Chaterera, 2016). The income generated by women from entrepreneurial activities not only contributes towards household resources but also raises their self-esteem (Chamlee-Wright, 2002).

7.8 Conclusion

In conclusion, being entrepreneurs helps rural entrepreneurs to be their own bosses thereby having more control over their hours at work and working environments than they would have if they were employed by someone else. Entrepreneurship helps rural entrepreneurs not to stress since if they cannot find the formal employment they want, they can go into industry to create their business. If rural entrepreneurs consider that others would be concerned in it, they can go into industry for themselves. They may create an income, which is the currency left over after paying their bills, from being inventive and achieving what they enjoy. This will help to increase their wellbeing thereby reducing poverty. This will enable rural entrepreneurs to get sustainable income, formal employment by increasing their skills, and thus can turn out to be a major contributor to the monetary growth of the country. It is essential to replace conventional professional courses by developing technological ones for encouraging as well as sustaining rural livelihoods through rural entrepreneurship. Thus, from the discussion so far it can be deduced that both push and pull factors influence rural female entrepreneurs globally. However, the severity and settings differ between countries when comparing developed countries, emerging countries and developing nations as explained in the literature. In this regard, in the African context rural female entrepreneurship is very significant in promoting sustainable livelihoods for the poor and the less privileged in the society.

References

Abosede, A.J. and Onakoya, A.B. (2013) Intellectual entrepreneurship: theories, purpose and challenges. *International Journal of Business Administration* 4(5), 30–37. DOI: 10.5430/ijba.v4n5p30.

Agarwal, N., Gneiting, U. and Mhlanga, R. (2017) *Raising the Bar: Rethinking the Role of Business in the Sustainable Development Goals*. Oxfam.

Ali, A.Y.S. and Mahamud, H.A. (2013) Motivational factors and performance of women entrepreneurs in Somalia. *Journal of Education and Practice* 4, 47–53.

Allen, I.E., Langowitz, N. and Minniti, M. (2007) Global entrepreneurship monitor. *2006 Report on Women and Entrepreneurship* 3, 54–88.

Baker, D.M., Murray, G. and Agyare, A.K. (2018) Governance and the making and breaking of social-ecological traps. *Ecology and Society* 23, 38. DOI: 10.5751/ES-09992-230138.

Benzing, C., Chu, H.M. and Kara, O. (2009) Entrepreneurs in Turkey: a factor analysis of motivations, success factors, and problems. *Journal of Small Business Management* 47(1), 58–91. DOI: 10.1111/j.1540-627X.2008.00262.x.

Bhola, R., Verheul, I., Thurik, R. and Grilo, I. (2006) *Explaining Engagement Levels of Opportunity and Necessity Entrepreneurs*. EIM Business and Policy Research, Zoeterneer.

Brown, T.E. and Ulijn, J.M. (eds) (2004) *Innovation, Entrepreneurship and Culture: the Interaction Between Technology, Progress and Economic Growth*. Edward Elgar Publishing.

Brush, C.G. and Cooper, S.Y. (2012) Female entrepreneurship and economic development: an international perspective. *Entrepreneurship & Regional Development* 24(1–2), 1–6. DOI: 10.1080/08985626.2012.637340.

Brush, C.G., Carter, N.M., Gatewood, E.J., Greene, P.G. and Hart, M.M. (2006) The use of bootstrap-ping by women entrepreneurs in positioning for growth. *Venture Capital* 8(1), 15–31. DOI: 10.1080/13691060500433975.

Bullough, A., de Luque, M.S., Abdelzaher, D. and Heim, W. (2015) Developing women leaders through entrepreneurship education and training. *Academy of Management Perspectives* 29(2), 250–270. DOI: 10.5465/amp.2012.0169.

Buttner, E.H. and Moore, D.P. (1997) Women's organizational exodus to entrepreneurship: self-reported motivations and correlates with success. *Journal of Small Business Management* 35, 34–46.

Cabrera, E.M. and Mauricio, D. (2017) Factors affecting the success of women's entrepreneurship: a review of literature. *International Journal of Gender and Entrepreneurship* 9(1), 31–65. DOI: 10.1108/IJGE-01-2016-0001.

Carter, C. and Scarbrough, H. (2001) Towards a second generation of KM? The people management challenge. *Education + Training* 43(4/5), 215–224. DOI: 10.1108/EUM0000000005483.

Cejka, M.A. and Eagly, A.H. (1999) Gender-stereotypic images of occupations correspond to the sex segregation of employment. *Personality and Social Psychology Bulletin* 25(4), 413–423. DOI: 10.1177/0146167299025004002.

Chamlee-Wright (2002) *The Cultural Foundations of Economic Development: Urban Female Entrepreneurship in Ghana*. Routledge, London. DOI: 10.4324/9780203448335.

Chan, J.K.L. and Quah, W.B. (2012) Start-up factors for small and medium-sized accommodation busi-nesses in Sabah, Malaysia: push and pull factors. *Asia Pacific Journal of Tourism Research* 17(1), 49–62. DOI: 10.1080/10941665.2011.610150.

Chigudu, D. (2018) Strength in diversity: an opportunity for Africa's development. *Cogent Social Sciences* 4(1), 1558715. DOI: 10.1080/23311886.2018.1558715.

Chitongo, L. (2019) Rural livelihood resilience strategies in the face of harsh climatic conditions. The case of ward 11 Gwanda, South, Zimbabwe. *Cogent Social Sciences* 5(1), 1617090. DOI: 10.1080/23311886.2019.1617090.

Cho, S. and Kim, A. (2017) Relationships between entrepreneurship, community networking, and economic and social performance in social enterprises: evidence from South Korea, human service organizations. *Management, Leadership & Governance* 41(4), 376–388. DOI: 10.1080/23303131.2017.1279094.

Clark, D.M. (2004) Developing new treatments: on the interplay between theories, experimental science and clinical innovation. *Behaviour Research and Therapy* 42(9), 1089–1104. DOI: 10.1016/j.brat.2004.05.002.

Copp, C.B. and Ivy, R.L. (2001) Networking trends of small tourism businesses in post-socialist Slovaki. *Journal of Small Business Management* 39(4), 345–353. DOI: 10.1111/0447-2778.00031.

Cullen, U. and Archer-Brown, C. (2020) Country-specific sociocultural institutional factors as determinants of female entrepreneurs' successful sustainable business strategies within the context of Turkey and the UK. In: Ratten, V. (ed.) *Entrepreneurial Opportunities*. Emerald Publishing Limited. DOI: 10.1108/9781839092855.

De Bruin, A., Brush, C.G. and Welter, F. (2006) Introduction to the special issue: towards building cumulative knowledge on women's entrepreneurship. *Entrepreneurship Theory and Practice* 30(5), 585–593. DOI: 10.1111/j.1540-6520.2006.00137.x.

Derera, E., Croce, F., Phiri, M. and O'Neill, C. (2020) Entrepreneurship and women's economic empowerment in Zimbabwe: research themes and future research perspectives. *The Journal for Transdisciplinary Research in Southern Africa* 16(1), 13. DOI: 10.4102/td.v16i1.787.

De Wit, G. and Van Winden, F.A. (1989) An empirical analysis of self-employment in the Netherlands. *Small Business Economics* 1(4), 263–272. DOI: 10.1007/BF00393805.

Di Fabio, A. and Blustein, D.L. (2016) From meaning of working to meaningful lives: the challenges of expanding decent work. *Frontiers in Psychology* 7, 1119. DOI: 10.3389/fpsyg.2016.01119.

Dzapasi, F.D. (2020) The impact of liquidity management on bank financial performance in a subdued economic environment: a case of the Zimbabwean banking industry. *PM World Journal* 9, 1–20.

Edewor, P., Imhonopi, D. and Amusan, T.A. (2014) Socio-cultural and demographic dynamics in sustainable entrepreneurial development in Nigeria. *Developing Country Studies* 4, 58–64.

Erenstein, O., Hellin, J. and Chandna, P. (2010) Poverty mapping based on livelihood assets: a meso-level application in the Indo-Gangetic Plains, India. *Applied Geography* 30(1), 112–125. DOI: 10.1016/j.apgeog.2009.05.001.

Fafchamps, M. (2001) Networks, communities and markets in Sub-Saharan Africa: implications for firm growth and investment. *Journal of African Economics* 10, 109–142. DOI: 10.1093/jae/10.Suppl2.109.

Farah, A.I. (2014) Factors influencing women participation in entrepreneurial activities in Mandera township, Mandera central division, Kenya. University of Nairobi, Nairobi.

Fortunato, V.C.R., Giraldi, J.D.M.E. and De Oliveira, J.H.C. (2014) A review of studies on neuromarketing: practical results, techniques, contributions and limitations. *Journal of Management Research* 6(2), 201. DOI: 10.5296/jmr.v6i2.5446.

Fosić, I., Kristić, J. and Trusić, A. (2017) Motivational factors: drivers behind women entrepreneurs' decision to start an entrepreneurial venture in Croatia. *Scientific Annals of Economics and Business* 64(3), 339–357. DOI: 10.1515/saeb-2017-0022.

Gibb, A.A. (1993) Enterprise culture and education: understanding enterprise education and its links with small business, entrepreneurship and wider educational goals. *International Small Business Journal* 11, 11–34. DOI: 10.1177/026624269301100301.

Grigore, A.M. and Mitroi, A. (2012) Romanian culture and its attitude towards entrepreneurship. *Revista de Management Comparat International/Review of Comparative Management* 13, 149–157.

Hammawa, Y.M. and Hashim, N.B. (2016) Women-micro entrepreneurs and sustainable economic development in Nigeria. *Journal of Business and Management* 18, 27–36.

Hanushek, E.A. and Woessmann, L. (2008) The role of cognitive skills in economic development. *Journal of Economic Literature* 46(3), 607–668. DOI: 10.1257/jel.46.3.607.

Hayrapetyan, M. (2016) *Factors that Drive Female Entrepreneurship in Armenia*. Instituto Politecnico de Braganca (Portugal), ProQuest Dissertations Publishing.

Holmen, M., Min, T.T. and Saarelainen, E. (2011) Female entrepreneurship in Afghanistan. *Journal of Developmental Entrepreneurship* 16(3), 307–331. DOI: 10.1142/S1084946711001860.

Hoyle, V., Flasco, M.T., Choi, J., Cieniewicz, E.J., McLane, H, *et al.* (2022) Transmission of grapevine red blotch virus by Spissistilus festinus [Say, 1830] (Hemiptera: Membracidae) between free-living vines and Vitis vinifera 'Cabernet Franc' *Viruses* 14(6), 1156. DOI: 10.3390/v14061156.

Hui-Chen, C., Kuen-Hung, T. and Chen-Yi, P. (2014) The entrepreneurial process: an integrated model. *International Entrepreneurship and Management Journal* 10(4), 727–745. DOI: 10.1007/s11365-014-0305-8.

Iacob, V.S. and Nedelea, A. (2014) Entrepreneurship, support of the economic changes in China. *The USV Annals of Economics and Public Administration* 14, 15–28.

Ismail, H.C., Shamsudin, F.M. and Chowdhury, M.S. (2012) An exploratory study of motivational factors on women entrepreneurship venturing in Malaysia. *Business and Economic Research* 2(1), 1–13. DOI: 10.5296/ber.v2i1.1434.

Jagero, N. and Kushoka, I. (2011) Challenges facing women micro entrepreneurs in Dar es Salaam, Tanzania. *International Journal of Human Resource Studies* 1(2), 1–9. DOI: 10.5296/ijhrs.v1i2.1023.

Jiménez, A., Palmero-Cámara, C., González-Santos, M.J., González-Bernal, J. and Jiménez-Eguizábal, J.A. (2015) The impact of educational levels on formal and informal entrepreneurship. *Business Research Quarterly* 18(3), 204–212. DOI: 10.1016/j.brq.2015.02.002.

Johnson, M.H. (1990) Cortical maturation and the development of visual attention in early infancy. *Journal of Cognitive Neuroscience* 2(2), 81–95. DOI: 10.1162/jocn.1990.2.2.81.

Kabonga, I. (2020) Reflections on the 'Zimbabwean crisis 2000–2008' and the survival strategies: the sustainable livelihoods framework (SLF) analysis. *Africa Review* 12(2), 192–212. DOI: 10.1080/09744053.2020.1755093.

Kalinic, I., Sarasvathy, S.D. and Forza, C. (2014) 'Expect the unexpected': implications of effectual logic on the internationalization process. *International Business Review* 23(3), 635–647. DOI: 10.1016/j.ibusrev.2013.11.004.

Kantar, M. (1999) Entrepreneurship and rural women. *Journal of Agricultural Economics* 29–42.

Kantor, S. (2001) *River Ecology and Management: Lessons from the Pacific Coastal Ecoregion*. Springer Science & Business Media.

Kay, R., Günterberg, B., Holz, M. and Wolter, H.J. (2003) *Female Entrepreneurs in Germany*. Institut für Mittelstandsforschung, Bonn, Germany.

Kirk, J. and Belovics, R. (2006) Counseling would-be entrepreneurs. *Journal of Employment Counseling* 43(2), 50–61. DOI: 10.1002/j.2161-1920.2006.tb00006.x.

Kjeldsen, J. and Nielson, K. (2000) *The Circumstances of Women Entrepreneurs*. Danish Agency for Trade and Industry.

Klapper, L.F. and Parker, S.C. (2011) Gender and the business environment for new firm creation. *The World Bank Research Observer* 26(2), 237–257. DOI: 10.1093/wbro/lkp032.

Knutsson, P. (2006) The sustainable livelihoods approach: a framework for knowledge integration assessment. *Human Ecology Review* 13, 90–99.

Kobeissi, N. (2010) Gender factors and female entrepreneurship: international evidence and policy implications. *Journal of International Entrepreneurship* 8(1), 1–35. DOI: 10.1007/s10843-010-0045-y.

Korsgaard, S., Müller, S. and Tanvig, H.W. (2015) Rural entrepreneurship or entrepreneurship in the rural – between place and space. *International Journal of Entrepreneurial Behavior & Research* 21(1), 5–26. DOI: 10.1108/IJEBR-11-2013-0205.

Kuratko, D.F. (2006) A tribute to 50 years of excellence in entrepreneurship and small business. *Journal of Small Business Management* 44(3), 483–492. DOI: 10.1111/j.1540-627X.2006.00185.x.

Langowitz, N. and Minniti, M. (2007) The entrepreneurial propensity of women. *Entrepreneurship Theory and Practice* 31(3), 341–364. DOI: 10.1111/j.1540-6520.2007.00177.x.

Lent, M., Anderson, A., Yunis, M.S. and Hashim, H. (2019) Understanding how legitimacy is acquired among informal home-based Pakistani small businesses. *International Entrepreneurship and Management Journal* 15(2), 341–361. DOI: 10.1007/s11365-019-00568-7.

Magigaba, M.F. and Jili, N.N. (2019) Using entrepreneurial networks in enhancing the growth of women-owned businesses in KwaZulu-Natal, South Africa. *Ubuntu: Journal of Conflict and Social Transformation* 8(2), 93–111. DOI: 10.31920/2050-4950/2019/8n2a5.

Mashapure, R., Nyagadza, B., Chikazhe, L., Msipa, N., Ngorora, G.K.P. *et al.* (2022) Challenges hindering women entrepreneurship sustainability in rural livelihoods: case of Manicaland province. *Cogent Social Sciences* 8(1), 2132675. DOI: 10.1080/23311886.2022.2132675.

McGowan, P., Redeker, C.L., Cooper, S.Y. and Greenan, K. (2012) Female entrepreneurship and the management of business and domestic roles: motivations, expectations and realities. *Entrepreneurship & Regional Development* 24(1–2), 53–72. DOI: 10.1080/08985626.2012.637351.

Meyer, N. (2018) Research on female entrepreneurship: are we doing enough? *Polish Journal of Management Studies* 17(2), 158–169. DOI: 10.17512/pjms.2018.17.2.14.

Meyer, N. and Landsberg, J. (2015) Motivational factors influencing women's entrepreneurship: a case study of female entrepreneurship in South Africa. *International Journal of Economics and Management Engineering* 9, 3864–3869.

Meyer, N. and Synodinos, C. (2019) Entrepreneurial skills, characteristics and intentions amongst unemployed individuals in the Vaal-Triangle Region of South Africa. *Journal of Contemporary Management* 16(2), 1–22. DOI: 10.35683/jcm198.0024.

Monitor, G.E. (2012) *Empreendedorismo no Brasil*. Relatório Executivo.

Mordi, C., Simpson, R., Singh, S. and Okafor, C. (2010) The role of cultural values in understanding the challenges faced by female entrepreneurs in Nigeria. *Gender in Management: An International Journal* 25(1), 5–21. DOI: 10.1108/17542411011019904.

Morgan, N.A. (2012) Marketing and business performance. *Journal of the Academy of Marketing Science* 40(1), 102–119. DOI: 10.1007/s11747-011-0279-9.

Muñoz-Fernández, N., Ortega-Rivera, J., Nocentini, A., Menesini, E. and Sánchez-Jiménez, V. (2019) The efficacy of the "dat-e adolescence" prevention program in the reduction of dating violence and bullying. *International Journal of Environmental Research and Public Health* 16(3), 408. DOI: 10.3390/ijerph16030408.

Mutsagondo, S. and Chaterera, F. (2016) Mirroring the national archives of Zimbabwe act in the context of electronic records: lessons for ESARBICA member states. *Information Development* 32(3), 254–259. DOI: 10.1177/0266666914538272.

Nicolás, C. and Rubio, A. (2016) Social enterprise: gender gap and economic development. *European Journal of Management and Business Economics* 25(2), 56–62. DOI: 10.1016/j.redeen.2015.11.001.

Nieva, F.O. (2015) Social women entrepreneurship in the Kingdom of Saudi Arabia. *Journal of Global Entrepreneurship Research* 5(1), 1–33. DOI: 10.1186/s40497-015-0028-5.

Nziku, D.M. (2012) Tanzanian education and entrepreneurial influence among females. *Journal of Women's Entrepreneurship and Education* 52–73.

Osirim, M.J. (2003) Crisis in the state and the family: violence against women in Zimbabwe. *African Studies Quarterly* 7, 154–169.

Perl-Kot, M.R. (2011) Entrepreneurship, women and development: case studies of women entrepreneurs in Tanzania. *Penn Journal of Philosophy, Politics and Economics* 6, 3.

Powell, G. and Graves, L. (2003) *Women and Men in Management*, 3rd edn. SAGE, Thousand Oaks.

Rey-Martí, A., Porcar, A.T. and Mas-Tur, A. (2015) Linking female entrepreneurs' motivation to business survival. *Journal of Business Research* 68(4), 810–814. DOI: 10.1016/j.jbusres.2014.11.033.

Ribeiro-Soriano, D. (2017) Small business and entrepreneurship: their role in economic and social development. *Entrepreneurship & Regional Development* 29(1–2), 1–3. DOI: 10.1080/08985626.2016.1255438.

Rodríguez-Pose, A. (2018) The revenge of the places that don't matter (and what to do about it). *Cambridge Journal of Regions, Economy and Society* 11(1), 189–209. DOI: 10.1093/cjres/rsx024.

Rose, R.C., Kumar, N. and Yen, L.L. (2006) Entrepreneurs success factors and escalation of small and medium-sized enterprises in Malaysia. *Journal of Social Sciences* 2(3), 74–80. DOI: 10.3844/jssp.2006.74.80.

Sadaf, R., Oláh, J., Popp, J. and Máté, D. (2018) An investigation of the influence of the worldwide governance and competitiveness on accounting fraud cases: a cross-country perspective. *Sustainability* 10(3), 588. DOI: 10.3390/su10030588.

Salim, S. (2017) Oxidative stress and the central nervous system. *The Journal of Pharmacology and Experimental Therapeutics* 360(1), 201–205. DOI: 10.1124/jpet.116.237503.

Sarfaraz, L., Faghih, N. and Majd, A.A. (2014) The relationship between women entrepreneurship and gender equality. *Journal of Global Entrepreneurship Research* 4(1), 1–11. DOI: 10.1186/2251-7316-2-6.

Schumpeter, J.A. (1976) The process of creative destruction. In: *Capitalism, Socialism, and Democracy, 1942*. Manhattan Institute for Policy Research, Routledge, New York.

Serrat, O. (2017) *Knowledge Solutions: Tools, Methods, and Approaches to Drive Organizational Performance*. Springer Nature. DOI: 10.1007/978-981-10-0983-9.

Solesvik, M.Z. (2013) Entrepreneurial motivations and intentions: investigating the role of education major. *Education + Training* 55(3), 253–271. DOI: 10.1108/00400911311309314.

Steinem, G. (1992) Creating jobs we can't be fired from. *Morgan Stanley Investment Management Independent Liquid Fund* 5, 21.

Stifel, D. (2010) The rural non-farm economy, livelihood strategies and household welfare. *African Journal of Agricultural and Resource Economics* 4, 82–109.

Suchart, T. (2017) Factors Influencing opportunity driven nascent entrepreneurs in Europe and Asia. *European Research Studies Journal* 20, 774–782. DOI: 10.35808/ersj/745.

Swinney, J. and Runyan, R. (2007) Native American entrepreneurs and strategic choice. *Journal of Developmental Entrepreneurship* 12(3), 257–273. DOI: 10.1142/S1084946707000678.

Tabares, A., Muñoz-Delgado, G., Franco, J.F., Arroyo, J.M. and Contreras, J. (2022) Multistage reliability-based expansion planning of AC distribution networks using a mixed-integer linear programming model. *International Journal of Electrical Power & Energy Systems* 138, 107916. DOI: 10.1016/j.ijepes.2021.107916.

Thom, M. (2015) The entrepreneurial value of arts incubators: why fine artists should make use of professional arts incubators. *Artivate* 4(2), 51–75. DOI: 10.1353/artv.2015.0007.

Vossenberg, S. (2013) *Women Entrepreneurship Promotion in Developing Countries: What Explains the Gender Gap in Entrepreneurship and How to Close It*. Maastricht School of Management, Maastricht.

World Bank (2010) *The Changing Wealth of Nations: Measuring Sustainable Development in the New Millennium*. The World Bank. DOI: 10.1596/978-0-8213-8488-6.

Yunus, M. (2007) *Banker to the Poor: Micro-lending and the Battle Against World Poverty*. Public Affairs.

Zhou, B., Lu, Y., Hajifathalian, K., Bentham, J. and Di Cesare, M. (2016) Worldwide trends in diabetes since 1980: a pooled analysis of 751 population-based studies with 4· 4 million participants. *The Lancet* 38, 1513–1530. DOI: 10.1016/S0140-6736(16)00618-8.

8 Sustainable Agricultural Supply Chains on Food Security: A Systematic Literature Review

Benson Ruzive[1]*, Reason Masengu[2] and Gibson Muridzi[3]

[1]Modern College of Business and Science, Muscat, Sultanate of Oman; [2]Middle East College, Sultanate of Oman; [3]Manicaland State University of Applied Science, Zimbabwe

Abstract

Agriculture is an important sector in any economy as it offers food security. Food security can best be attained by understanding available resources and how to distribute them, procure them in times of shortages, and preserve the total reserves. This study aimed to establish supply chains that are synonymous with the demand and supply of agricultural products. Research questions for the study are: i) what is the impact of sustainable agricultural supply chains on food security in Zimbabwe; ii) what are the challenges that agricultural supply chains face in Zimbabwe; iii) what risks and safety measures are faced in agricultural supply chains; and iv) what solutions may be proposed to challenges faced in agricultural supply chains in Zimbabwe. Literature was reviewed to establish models that are used in food supply chains. Reviewed sources from journals, books and articles written in English to avoid translation were used as sources of information. The study sought to use the findings to inform stakeholders on weaknesses and areas where opportunities can be exploited to increase food security in Zimbabwe. To the best knowledge of the research team, the study has not been carried out anywhere in Zimbabwe and is expected to contribute to the body of knowledge. The study is limited to desk research of food supply chains, which cannot be generalized. This study is also limited to the study of supply chains of finished products without investigating the supply chains of raw materials (inputs) of the same, which might influence the finished products.

Keywords: Sustainable supply chain, Agricultural supply chains, Food supply chain, Sustainability

8.1 Introduction

Integration, readiness and coordination of partners in food supply chains is critical for the survival of people in any nation. There are several risks that food supply chains are vulnerable to, starting from the supply of raw materials side, through to the production processes and the distribution stages before the product reaches the final consumer. This chapter gives a detailed background to the study of food supply chains. Literature has been reviewed from different credible sources such as journal articles, books and other scholarly sources written in English language to avoid translation of other languages. An overview of food supply chains, risks and safety measures, and risk mitigation strategies are some of the topics that have been subjected to vigorous review of literature. Ali *et al.* (2017)

---

*Corresponding author: benson.ruzive@mcbs.edu.om

© CAB International 2023. *Sustainable Agricultural Marketing and Agribusiness Development: An African Perspective* (eds B. Nyagadza and T. Rukasha)
DOI: 10.1079/9781800622548.0008

posit that the integrity of food is very broad and multifaceted because it encapsules the whole supply chain.

8.1.1 Background to the study

Agriculture is one of the significant fields in the world, affecting all human existence (Ganeshkumar *et al.*, 2017). The agricultural sector is responsible for the production of food to ensure food security and enough nutrients are within the reach of any country's populace. Agro-based supply chains are responsible for the production and distribution of food and stockfeeds (Gupta, 2022). The major partners in agricultural supply chains are composed of suppliers, farmers, transporters, warehouses, retailers and consumers (Ganeshkumar *et al.*, 2017) in addition to government and pressure groups. Coordinated relationships, effective communication and information sharing are fundamental traits of agricultural supply chains. Education increased disposable income and global exposure have shifted the focus of the ordinary food basket to a more sophisticated one that includes high-value products such as fruits and vegetables, as well as animal protein.

Gupta (2022) posits that governments have started initiatives aimed at boosting investment in the agricultural sector. Contract farming is one of the models that has been used globally, especially in developing countries. Multi-faceted efforts by governments to promote investment have witnessed a tremendous growth in the agricultural market. Visibility in food supply chains is a must as it fosters integration of supply chain management systems and infrastructural development. Infrastructure may be in the form of farming activities, ripening rooms, storage facilities with sufficient and ideal atmosphere chambers, transportation modes and information communication technology (ICT). Digitalized technologies can capture current information about weather, soil types, logistics, price fluctuations and many more activities that facilitate transportation and production of food items.

Supply chains in agro-based businesses are affected by several challenges that include mishandling of produce and involvement of intermediaries such as transporters who may mislead farmers or even be selective in the produce they want to carry (Bhatia and Bhat, 2020). These challenges culminate in lower prices for agricultural produce because of delayed deliveries and lower quality. There is lack of skilled manpower in agricultural supply chains because of the seasonal nature of the business and because labour is highly migratory due to the seasonal nature. Nier *et al.* (2019) posit that many decisions that are strategic, tactical or operational in nature are taken at different levels right from the procurement of seed to satisfy customers through supply chains that are robust and provide above level quality service. Attributes of sound service levels include speed of response, friendliness, convenience, courtesy and product knowledge (Wicaksono and Illés, 2022).

8.2 Literature Review

8.2.1 Overview of agricultural food supply chains

The Indian economy thrives on agriculture as a major source of employment (Ganeshkumar *et al.*, 2017). Although India is ranked second globally in the production of agro-based products, its share in agricultural food products trading is merely 0.2%, well below the proportions of other developing countries such as Vietnam and Brazil. Rice, maize, wheat and potato are the four major agricultural food crops that are grown globally, and India is regarded as one of the top producers (Ganeshkumar *et al.*, 2017). Several food items have complicated multi-step vertical and horizontal branching supply networks, relying on the one up/down strategy making their supply chains unsustainable. Admittedly, it is hard to get some precision regarding the exact upstream supplier or downstream partners of some agricultural products such as milk, which is often mixed in a dairy. Milk comes from several farms and is mixed in a dairy, and therefore one up/one down traceability can be easily lost. With this level of intricacy, verifying the provenance and quality requirements of individual items becomes impossible. The food sector is largely dominated and relies much on paper records, and the present food traceability systems are not integrated nor linked. This silo mentality produces a negative information imbalance among all supply chain stakeholders, a lack of traceability across numerous supply chain processes, and the potential for fraud is promoted or concealed. Although several technologies for product traceability have been employed, such as bar codes, radio frequency identification (RFID) tags and Electronic Data Interchange (EDI), Bosona and Gebresenbet (2013) posit that there is need for more technical applications in the food supply chain that enable traceability.

8.2.2 Risks and safety measures faced in agricultural supply chains

Food safety refers to food that is hazard free, be it biological, chemical or physical, thereby rendering it free from any effects that are related to health. Contamination of food can either be physical or chemical rendering the food unfit for consumption. Risks faced in food supply chains range from natural, such as the recent Covid-19 pandemic, to human actions (Wicaksono and Illés, 2022). Risk of food contamination is increased by the number of interacting elements in the food supply chain whose presence expose food to both physical and chemical contamination, and the risk is proportionate to the number of elements in a supply chain (Sloane and Oreilly, 2013). The World Health Organization (WHO) reports that annually 600 million people fall sick after consuming contaminated food and 420,000 die because of food contamination which further causes more than 200 diseases (WHO, 2019). The safety of food supply chains has come under the spotlight because of global scandals (Auler *et al.*, 2017). Luo *et al.* (2022) posit that consumer behaviour and consumption pattern was recently changed by the Covid-19 pandemic, which exposed how fragile global supply chains are to risks such as this natural one.

8.2.3 Food traceability

CAC/GL 60–2006 lays forth the basics of food traceability. To be acceptable to the targets of the food testing and enforcement scheme, traceability must be able to distinguish the upstream and downstream partners at any given stage of the food supply chain. The focus areas of food traceability has been suggested as: the process of evaluating implementation procedures; drivers for adaptation of the system; and the competitive advantage that accrues from traceability systems (Machado Nardi *et al.*, 2020). It is essential that the design of the model of traceability should spell out strategies, challenges and solutions identified in the assessment process (Saak, 2016).

8.2.3.1 Reasons for tracking and tracing

The European Union and African Union put in place legislative frameworks that require that all food should be traceable to its origins of production. The rationale for mandatory traceability is to enable timely identification of the source of contamination or defects in supply chains where inputs from different suppliers are commingled during the processing stage (Roth *et al.*, 2008). Traceability is used in agriculture systems as a means of gaining competitive advantage, for organizational sustainability and to convey information on the product characteristics along the chain (Gichure *et al.*, undated). Traceability can be used to maintain trust within a supply chain and build a reputation for producing high quality products when firms' behaviour is not perfectly observed by consumers (Marucheck *et al.*, 2011). Certification of agricultural products by quality management systems had better traceability and quality verification systems. Traceability systems in agro-food industries collect information about the origin of the inputs and the processing history and vary in the number of product attributes (breadth) and production stages (depth) covered by the system as well as the accuracy of information about product movement and origin (Galliano and Orozco, 2011). Traceability allows matching the levels of upstream and downstream efforts put into products. This makes it possible to vertically coordinate downstream and upstream shirking with a subset of suppliers (Saak, 2016). Saak

(2016) further indicated that traceability can increase the effectiveness of audits and reduce the costs of providing inter-firm incentives to provide high quality. A traceable environment can also be seen as a tool to maintain trust within a supply chain and build a reputation for producing high-quality products when firms' behaviour is not perfectly observed by consumers (Marucheck et al., 2011).

8.2.3.2 Modern traceability technologies

The supply chains of consumer goods are complex, spanning dozens of countries and actors. As a result, most firms cannot trace the products they produce and source beyond one or two upstream and downstream tiers (Blaettchen et al., 2022). This challenge therefore calls for modern traceability technologies. This limited traceability has negative consequences (ISO, 2005). Modern traceability technologies promise to improve supply chain management by simplifying recalls, increasing visibility, or verifying sustainable supplier practices (Blaettchen et al., 2022). Lack of traceability and visibility is a major barrier to building disruption-resilient supply chains and can lead to difficulties in supply chain coordination. The solution to these difficulties is to apply modern traceability technologies which can assist in removing such disruptions.

The success of a traceability technology for a product depends on its adoption by most or even all firms involved in the product's supply chain. As supply chains become more complex and globalized, there is a need for initiatives to consider economies of scale from highly interlinked supply chains forming a 'supply chain network' through technology and subsequently make traceability technology dissemination profitable to all stakeholders. Traceability technology enhances information sharing and disclosure thereby increasing trust among stakeholders along value chains (Gichure et al., 2014). Traceability technologies enable systems in food industries to collect information about the origin of the inputs and the processing history and vary in the number of product attributes (breadth) and production stages (depth) covered by the system as well as the accuracy of information about product movement and origin (Charlier and Valceschini, 2008).

8.2.4 Supply chain visibility

Literature reveals that blockchain technology has been conceptualized as a tool for better visibility in supply chains (Madhawal and Panfilov, 2017; Tian, 2018; Saberi et al., 2019; Rogerson and Parry, 2020). The application of blockchain technology in supply chains requires the end consumer to be prepared to pay the premium price that comes with the technology being used (Rogerson and Parry, 2020). Technology such as RFID has been widely used and accepted as it can quickly identify affected products promptly mitigating risks of further contaminating food (Pouliot and Sumner, 2012). This visibility can identify products from their sources in case of labour abuses (Jones et al., 2019).

Decision making in supply chains demands generation of large volumes of data which is used by managers (Williams et al., 2013). Previously, supply chains faced information asymmetry because of centralization (Michalski et al., 2018). Francis (2008) defines supply chain visibility as 'the identity, location and status of entities transiting the supply chain, captured in timely messages about events, along with the planned and actual dates/times for these events'. Supply chain visibility is regarded as an enabler of inter-company collaboration (Francis, 2008) that permits integration among supply chain partners including the end consumer (Schoenherr and Swink, 2012), building trust (Johnson et al., 2013) as well as improving the efficiency of the operations (Bartlett et al., 2007). The importance of visibility in supply chains cannot be underestimated as it facilitates action (Delen et al., 2009) and leads to risk reduction (Christopher and Lee, 2004). Kwon and Kim (2018) posit that successful supply chain management elements depend largely on visibility. Several concepts have evolved over the years to assist supply chain visibility such as bill of materials (Molla and Bhalla, 2006), materials resource planning (Parry et al., 2003), enterprise resource planning (Parry and Graves, 2008), enterprise resource management (Chuang and Shaw, 2008), and customer relationship management (Lambert and Schwieterman, 2012). The use of barcodes has made them widespread because of their low cost and simplicity (Apiyo and Kiarie, 2018). RFID tagging has become popular in supply chains to offer real-time information of products, even though the technology application is a bit expensive (Wang *et al.*, 2017). Internet of Things (IoT) has further reduced human error and given greater visibility to the entire supply chain (Majeeb and Rupasinghe, 2017). Visibility offers what traceability cannot offer (Rogerson and Parry, 2020).

8.2.5   Supply chain risk mitigation measures

The importance of risk management in any business setup cannot be overlooked especially in environments that are rapidly changing. Afifa and Santoso (2022) posit that managers of supply chains are mandated towards the improvement, determination and continuity of resilient food supply chains. A risk management culture is required to provide value for the companies whose strength is anchored on four elements, namely supply chain re-engineering, collaboration, agility and risk awareness. Risk assessment paves the way for risk prioritizing and mitigation, and assures full employment of limited resources. Generally, risk mitigation strategies do not eliminate risks, but they reduce the effect in the event of the expected mishaps taking place (Rahmatin *et al.*, 2018).

Risks in agribusiness supply chains are categorized into three groups namely procurement risk, production risk and shipping risk (Sreedevi and Haritha, 2017). The supply and demand sides of the supply chain are affected by seasonality and consumer taste, which may culminate in the bullwhip effect. These sudden changes in food agrobusiness cannot adjust quickly because it takes much time to grow and harvest crops thus affecting the delivery lead times (Behzadi *et al.*, 2018). Risks in agrobusiness include natural disasters and weather changes, biological and environmental risks, logistical and infrastructure risks, management and operational risks, public policy and institutional risks, and political and security risks (Yazdani *et al.*, 2019) while others state economic risks, social risks related to public policies, and law and regulatory changes (Jianying *et al.*, 2021).

The decision to mitigate risks in food supply chains is made difficult because it involves external parties whose goals are different. If they decide to cooperate, they might not embrace everything 100%. Mensah *et al.* (2017) posit that such challenges become the pushing factor for coming up with a model at strategic level to mitigate these risks. Risk mitigation strategies are either proactive or reactive (Ghadge *et al.*, 2012) wherein the former entails coming up with mitigation strategies before a risk occurs and the latter is more reacting to the aftereffects of a risk after its occurrence (Wieland and Wallenburg, 2013).

8.2.6   Supply chain resilience in food agribusiness

Parast *et al.* (2018) define supply chain resilience as the ability of a robust supply chain operation to maintain that system by going back to its desirable and original position or embracing a new and more desirable position after encountering some disturbances and to be able to avoid failure of the operation. This is supported by Belhadi *et al.* (2021) who point out that supply chain resilience is concerned with the ability of supply chains to subdue the impact of unavoidable risks in supply chains and be able to go back to their envisioned initial plans or better move on to more advanced positions after experiencing these disruptions.

Coopmans *et al.* (2021) posit that mixed methods are used to describe supply chain resilience by making use of both the qualitative and quantitative assessments of a firm's ability to recover quickly, whereas Soni *et al.* (2019) point out that deterministic modelling approaches are used to measure resilience. Increasing supply chain flexibility is another way of improving supply chain resilience, however this requires reduction in the complexity of the supply chain as a whole (Rajesh, 2021). For example, a contract can be more flexible regarding the quantities that are delivered by a supplier or buyer giving room for either increased or reduced demand (Li *et al.*, 2021). Partners in the supply chain can build more resilient supply chains by collaborating on several areas that include information dissemination, decision making, effective communication, and aligning goals within those chains (Wicaksono and Illés, 2022).

8.3 Methodology

The study used a systematic literature review from journal-indexed articles to identify discussions on sustainable agricultural supply chain management. Systematic literature review was used due to its ability to summarize and synthesize studies from a global perspective. It provides a clear picture of the study, and it combines findings from studies which used a wide range of methodologies in the data collection process (Tranfield et al., 2003). Systematic literature review uses key words in the identification of relevant articles (Kumah et al., 2022). The following key phrases were used in this study: Sustainable Supply Chain, Agricultural Supply Chains, Food Supply Chain, Sustainability, Supply Chain Management. Articles were also excluded from the study based on the relevance of the abstract and subject area (Crossan and Apaydin, 2010). The study used a total of 58 articles to understand the impact of agricultural supply chain management on food security. Relevance, context and scope were the key parameters used in the final inclusion process of the articles.

8.4 Discussion

The current world population of approximately 7.3 billion people is projected to reach 9.7 billion by the year 2050 and a further projected figure of 11.2 billion by the year 2100 (United Nations, 2017). That growth in population is proportionate to the growth in demand for food and therefore proactivity in food supply chains is a must to reduce shortages and impacts caused by supply chain risks. Technology adoption by farmers is critical for the improvement of food production, storage and distribution leading to food sustainability. Internet of things (IoT), big data, digital technologies, blockchain, and many more such as cyber security are technological developments that have recently changed the face of the world and can be used in food supply chains as well, such as in food tracing, monitoring, information gathering and providing transparency within the food supply chain. IoT in food supply chains can capture the complex activities that are required for product traceability (Abideen et al., 2021). The use of blockchain in any business environment makes it impossible for any partner to alter or delete any transaction because all transactions are kept and validated by other members of that supply chain. Agri-food supply chains affect human life (Ganeshkumar et al., 2017), so it is no wonder the attention that this topic is gaining. Food supply chains are responsible for the production, storage and distribution of food (Gupta, 2022). These combined processes are vulnerable to a number of challenges, among them mishandling of food and involvement of intermediaries (Bhatia and Bhat, 2020). Nier et al. (2019) posit that decision making in food supply chains, like any other business models, are strategic, tactical and operational. For example, at strategic level top management decides which products to deal with, at tactical level mangers decide how many trucks to hire or dispatch and at operational level route planning and order planning are executed. Wicaksono and Illés (2022) posit that good service levels are shown by the speed of response to deliveries, friendliness, courtesy and knowledge of the products. Food contamination cannot be avoided but proactive actions improve the traceability and help reduce the cases. Food contamination causes illnesses and death (WHO, 2019) and these are unwelcome developments in any society. The ability of a supply chain to trace food improves the competitiveness of its members (Machado Nardi et al., 2020). It is not always easy to trace all sources of food within the supply chain, for example in dairy farming commingling is common (Gichure et al., undated) and the originator of such products as milk cannot be easily identified once the product is mixed in a storage tank. Nevertheless, traceability technologies improve food supply chain management and simplify processes such as product recalls (Blaettchen et al., 2022). Food items within supply chains that can be easily traced enhance trust among the partners and the end consumers (Marucheck et al., 2011).

References

Abideen, A.Z., Sundram, V.P.K., Pyeman, J., Othman, A.K. and Sorooshian, S. (2021) Food supply chain transformation through technology and future research directions: a systematic review. *Logistics* 5(4), 83. DOI: 10.3390/logistics5040083.

Afifa, Y.N. and Santoso, I. (2022) Proactive risk mitigation strategies and building strategic resilience in the food supply chain: a review. *Food Research* 6(2), 9–17. DOI: 10.26656/fr.2017.6(2).257.

Ali, M.H., Tan, K.H. and Ismail, M.D. (2017) A supply chain integrity framework for halal food. *British Food Journal* 119(1), 20–38. DOI: 10.1108/BFJ-07-2016-0345.

Apiyo, R.O. and Kiarie, D. (2018) Role of ICT tools in supply chain performance. *International Journal of Supply Chain Management* 3, 17–26.

Auler, D.P., Teixeira, R. and Nardi, V. (2017) Food safety as a field in supply chain management studies: a systematic literature review. *International Food and Agribusiness Management Review* 20(1), 99–112. DOI: 10.22434/IFAMR2016.0003.

Bartlett, P.A., Julien, D.M. and Baines, T.S. (2007) Improving supply chain performance through improved visibility. *The International Journal of Logistics Management* 18(2), 294–313. DOI: 10.1108/09574090710816986.

Behzadi, G., O'Sullivan, M., Olsen, T. and Zhang, A. (2018) Agribusiness supply chain risk management: a review of quantitative decision models. *Omega* 79, 21–42. DOI: 10.1016/j.omega.2017.07.005.

Belhadi, A., Kamble, S., Jabbour, C.J.C., Gunasekaran, A., Ndubisi, N.O., *et al.* (2021) Manufacturing and service supply chain resilience to the COVID-19 outbreak: lessons learned from the automobile and airline industries. *Technological Forecasting and Social Change* 163, 120447. DOI: 10.1016/j.techfore.2020.120447.

Bhatia, M. and Bhat, G.M.J. (2020) Agriculture supply chain management-an operational perspective. *Brazilian Journal of Operations & Production Management* 17(4), 1–18. DOI: 10.14488/BJOPM.2020.043.

Blaettchen, P., Calmon, A. and Hall, G. (2022) *Traceability Technology Adoption in Supply Chain Networks.* DOI: 10.2139/ssrn.3805040.

Bosona, T. and Gebresenbet, G. (2013) Food traceability as an integral part of logistics management in food and agricultural supply chain. *Food Control* 33(1), 32–48. DOI: 10.1016/j.foodcont.2013.02.004.

Charlier, C. and Valceschini, E. (2008) Coordination for traceability in the food chain. A critical appraisal of European regulation. *European Journal of Law and Economics* 25, 1–15. DOI: 10.1007/s10657-007-9038-2.

Christopher, M. and Lee, H. (2004) Mitigating supply chain risk through improved confidence. *Journal of Physical Distribution and Logistics Management* 34(5), 388–396. DOI: 10.1108/09600030410545436.

Chuang, M.L. and Shaw, W.H. (2008) An empirical study of enterprise resource management systems implementation: from ERP to RFID. *Business Process Management Journal* 14(5), 675–693. DOI: 10.1108/14637150810903057.

Coopmans, I., Bijttebier, J., Marchand, F., Mathijs, E., Messely, L. *et al.* (2021) COVID-19 impacts on flemish food supply chains and lessons for agri-food system resilience. *Agricultural Systems* 190, 103136. DOI: 10.1016/j.agsy.2021.103136.

Crossan, M.M. and Apaydin, M. (2010) A multi-dimensional framework of organizational innovation: a systematic review of the literature. *Journal of Management Studies* 47(6), 1154–1191. DOI: 10.1111/j.1467-6486.2009.00880.x.

Delen, D., Hardgrave, B.C. and Sharda, R. (2009) RFID for better supply chain management through enhanced information visibility. *Production and Operations Management* 16(5), 613–624. DOI: 10.1111/j.1937-5956.2007.tb00284.x.

Francis, V. (2008) Supply chain visibility: lost in translation? *Supply Chain Management : An International Journal* 13(3), 180–184. DOI: 10.1108/13598540810871226.

Galliano, D. and Orozco, L. (2011) The determinants of electronic traceability adoption: a firm-level analysis of French agribusiness. *Agribusiness* 27(3), 379–397. DOI: 10.1002/agr.20272.

Ganeshkumar, C., Pachayappan, M. and Madanmohan, G. (2017) Agri-food supply chain management: literature review. *Intelligent Information Management* 9(2), 68–96. DOI: 10.4236/iim.2017.92004.

Ghadge, A., Dani, S. and Kalawsky, R. (2012) Supply chain risk management: present and future scope. *The International Journal of Logistics Management* 23(3), 313–339. DOI: 10.1108/09574091211289200.

Gichure, J., Wahome, R., Njage, P.M.K., Karuri, E.G., Nzuma, J.M. *et al.* (2014) Factors influencing extent of traceability along organic fresh produce value chains: case of Kale in Nairobi, Kenya. *Journal of Organic Agricfulture* 7, 293–302.

Gichure, O., Wahome, R., Kuriri, E. and Karantininis, K. (undated) Traceability among smallholders in organic fresh produce value chain: case of Nairobi "Building Organic Bridges." In: Building Organic Bridges. Proceedings of the 4th ISOFAR Scientific Conference, Instanbul, Turkey.

Gupta, S. (2022) Management of agriculture supply chain: digital advancements offers farmers, transporters, warehouses, retailers and suppliers ways to optimise the Agriculture supply chains. *Food Logistics*. Available at: https://www.foodlogistics.com/sustainability/agriculture/article/22159660/allied-market-research-management-of-agriculture-supply-chain

ISO (2005) Traceability in the food chain. General principles and basic requirements for system design and implementation. ISO 22005:2005.

Jianying, F., Bianyu, Y., Xin, L., Dong, T. and Weisong, M. (2021) Evaluation on risks of sustainable supply chain based on optimized BP neural networks in fresh grape industry. *Computers and Electronics in Agriculture* 183. DOI: 10.1016/j.compag.2021.105988.

Johnson, N., Elliott, D. and Drake, P. (2013) Exploring the role of social capital in facilitating supply chain resilience. *Supply Chain Management* 18(3), 324–336. DOI: 10.1108/SCM-06-2012-0203.

Jones, K., Visser, D. and Simic, A. (2019) Fishing for export: calo, recruiters, informality, and debt in international supply chains. *Journal of the British Academy* 7, 107–130. DOI: 10.5871/jba/007s1.107.

Kumah, E.A., McSherry, R., Bettany-Saltikov, J., van Schaik, P., Hamilton, S. *et al.* (2022) Evidence-informed vs evidence-based practice educational interventions for improving knowledge, attitudes, understanding and behaviour towards the application of evidence into practice: a comprehensive systematic review of undergraduate students. *Campbell Systematic Reviews* 18(2). DOI: 10.1002/cl2.1233.

Kwon, I.W. and Kim, S.H. (2018) Framework for successful supply chain implementation in healthcare area from provider's prospective. *Asia Pacific Journal of Innovation and Entrepreneurship* 12(2), 135–145. DOI: 10.1108/APJIE-04-2018-0024.

Lambert, D.M. and Schwieterman, M.A. (2012) Supplier relationship management as a macro business process. *Supply Chain Management: An International Journal* 17(3), 337–352. DOI: 10.1108/13598541211227153.

Li, J., Luo, X., Zhang, Q. and Zhou, W. (2021) Coordination through capacity reservation contract and quantity flexibility contract. *Omega* 99, C. DOI: 10.1016/j.omega.2020.102195.

Luo, J., Leng, S. and Bai, Y. (2022) Food supply chain safety research trends from 1997 to 2020: a bibliometric analysis. *Frontiers in Public Health* 9. DOI: 10.3389/fpubh.2021.742980.

Machado Nardi, V.A., Auler, D.P. and Teixeira, R. (2020) Food safety in global supply chains: a literature review. *Journal of Food Science* 85(4), 883–891. DOI: 10.1111/1750-3841.14999.

Madhawal, Y. and Panfilov, P.B. (2017) Blockchain and supply chain management: aircraft parts business case 28th DAAAM. In: *International Symposium on Intelligent Manufacturing and Automation*. DAAAM International, Vienna, pp. 1051–1056. DOI: 10.2507/28th.daaam.proceedings.

Majeeb, M.A.A. and Rupasinghe, T.D. (2017) Internet of things (IoT) embedded future supply chains for industry 4.0: an assessment from an ERP-based fashion apparel and footwear industry. *International Journal of Supply Chain Management* 6, 25–40.

Maruicheck, A., Greis, N., Mena, C. and Cai, L. (2011) Product safety and security in the global supply chain: issues, challenges and research opportunities. *Journal of Operations Management* 29, 707–720. DOI: 10.1016/j.jom.2011.06.007.

Mensah, P., Merkuryev, Y., Klavins, E. and Manak, S. (2017) Supply chain risks analysis of a logging company: conceptual model. *Journal of Procedia Computer Science* 104, 313–320. DOI: 10.1016/j.procs.2017.01.140.

Michalski, M., Montes-Botella, J.L. and Narasimhan, R. (2018) The impact of asymmetry on performance in different collaboration and integration environments in supply chain management. *Supply Chain Management* 23(1), 33–49. DOI: 10.1108/SCM-09-2017-0283.

Molla, A. and Bhalla, A. (2006) Business transformation through ERP: a case study of an asian company. *Journal of Information Technology Case and Application Research* 8(1), 34–54. DOI: 10.1080/15228053.2006.10856081.

Nier, S., Klein, O. and Tamásy, C. (2019) Global crop value chains: shifts and challenges in South-North relations. *Social Sciences* 8(3), 85. DOI: 10.3390/socsci8030085.

Parast, M.M., Sabani, S. and Kawalahmadi, M. (2018) The relationship between firm resilience to supply chain disruptions and firm innovation. *Revisiting Supply Chain* 279–298. DOI: 10.1007/978-3-030-03813-7.

Parry, G. and Graves, A. (2008) The importance of knowledge management for ERP systems. *International Journal of Logistics: Research and Applications* 12, 427–441. DOI: 10.1080/13675560802340992.

Parry, G., Price, P. and James-Moore, M. (2003) *ERP Implementation and Maintenance in a Lean Enterprise.* University of the West of England Bristol. Research Repository.

Pouliot, S. and Sumner, D.A. (2012) Traceability, recalls, industry reputation and product safety. *European Review of Agricultural Economics* 40(1), 121–142. DOI: 10.1093/erae/jbs006.

Rahmatin, N., Santoso, I., Suti, R. and Shint, W. (2018) Integration of fuzzy failure mode and effect analysis (Fuzzy FMEA) and the analytical network process (ANP) in marketing risk analysis and mitigation. *International Journal of Technology* 9(4), 809–818. DOI: 10.14716/ijtech.v9i4.2197.

Rajesh, R. (2021) Flexible business strategies to enhance resilience in manufacturing supply chains: An empirical study. *Journal of Manufacturing Systems* 60, 903–919. DOI: 10.1016/j.jmsy.2020.10.010.

Rogerson, M. and Parry, G.C. (2020) Blockchain: case studies in food supply chain visibility. *Supply Chain Management* 25(5), 601–614. DOI: 10.1108/SCM-08-2019-0300.

Roth, A., Tsay, A., Pullman, M. and Gray, J. (2008) Unravelling the food supply chain: strategic insights from chain and the 2007 recalls. *Supply Chain Management* 44(1), 22–39. DOI: 10.1111/j.1745-493X.2008.00043.x.

Saak, A.E. (2016) Traceability and reputation in supply chains. *International Journal of Production Economics* 177, 149–162. DOI: 10.1016/j.ijpe.2016.04.008.

Saberi, S., Kouhizadeh, M., Sarkis, J. and Shen, L. (2019) Blockchain technology and its relationships to sustainable supply chain management. *International Journal of Production Research* 57(7), 2117–2135. DOI: 10.1080/00207543.2018.1533261.

Schoenherr, T. and Swink, M. (2012) Revisiting the arcs of integration: cross-validations and extensions. *Journal of Operations Management* 30(1–2), 99–115. DOI: 10.1016/j.jom.2011.09.001.

Sloane, A. and Oreilly, S. (2013) The emergence of supply network ecosystems: a social network analysis perspective. *Production Planning & Control* 24(7), 621–639. DOI: 10.1080/09537287.2012.659874.

Soni, U., Jain, V. and Kumar, S. (2019) Measuring supply chain resilience using a deterministic modeling approach. *Computers & Industrial Engineering* 74, 11–25. DOI: 10.1016/j.cie.2014.04.019.

Sreedevi, R. and Haritha, S. (2017) Uncertainty and supply chain risk: the moderating role of supply chain flexibility in risk mitigation. *International Journal of Production Economics* 193, 332–342. DOI: 10.1016/j.ijpe.2017.07.024.

Tian, F. (2018) An information system for food safety monitoring in supply chains based on HACCP, block chain internet of things. Doctoral Thesis submission to Wirtschafts University, Wien, Vienna University of Economics and Business.

Tranfield, D., Denyer, D. and Smart, P. (2003) Towards a methodology for developing evidence-informed management knowledge by means of systematic review. *British Journal of Management* 14(3), 207–222. DOI: 10.1111/1467-8551.00375.

United Nations (2017) World Population Projected to reach 9.8 Billion in 2050 and 11.2 Billion in 2100. World Population Prospects: The 2017 Revision. Available at: https://www.un.org/en/desa/world-population-projected-reach-98-billion-2050-and-112-billion-2100 (accessed 14 July 2023).

Wang, Z., Hu, H. and Zhou, W. (2017) RFID enabled knowledge-based precast construction supply chain. *Computer-Aided Civil and Infrastructure Engineering* 32(6), 499–514. DOI: 10.1111/mice.12254.

WHO (2019) Food safety: key facts.

Wicaksono, T. and Illés, C.B. (2022) From resilience to satisfaction: defining supply chain solutions for agri-food SMEs through quality approach. *PloS ONE* 17, 1–20. DOI: 10.1371/journal.pone.0263393.

Wieland, A. and Wallenburg, C.M. (2013) The influence of relational competencies on supply chain resilience: a relational view. *International Journal of Physical Distribution & Logistics Management* 43(4), 300–320. DOI: 10.1108/IJPDLM-08-2012-0243.

Williams, B.D., Roh, J., Tokar, T. and Swink, M. (2013) Leveraging supply chain visibility for responsiveness: the moderating role of internal integration. *Journal of Operations Management* 31, 543–554. DOI: 10.1016/j.jom.2013.09.003.

Yazdani, M., Gonzalez, E. and Chatterjee, P. (2019) A multi-criteria decision-making framework for agriculture supply chain risk management under a circular economy context. *Management Decision* 59(8), 1801–1826. DOI: 10.1108/MD-10-2018-1088.

9 Jack of All Trades: Explicating the Eminence of Music in Agricultural Marketing for Sustainable Development in Zimbabwe

Praise Zinhuku*

Great Zimbabwe University, Masvingo, Zimbabwe

Abstract

Music performs a fundamental role towards promoting and fortifying agricultural marketing and sustainable development. This chapter seeks to demonstrate that there is a strong connection between music and marketing, which could be explored to enable effective agricultural marketing. Therefore, this research is strategic as it seeks to provide clear guidelines and clarifications on how music could also be roped in as an active marketing strategy for agribusinesses. The chapter utilizes an exploratory study as a research design. Data was congregated through interviews of selected music experts and analyses of selected songs by music artists. Analysis of data was achieved through thematic analysis and 'thick descriptions'. The study established that music marketing in agribusiness could be used as a motivator in the purchase decision of consumers. Music is a stimulus in the retail environment as well as in radio and television advertising. Marketing through music has the potential to launch the products, amenities and philosophies of an agricultural business. Music marketing also enables structuring, targeting, authority establishment, branding and brand identity of an agribusiness. Moreover, music could also provide entertainment in advertisements and gain the consideration of buyers. Thus, on the basis of this study's findings, the study will propose an agro-based music marketing model to address how music marketing could be roped in to become fit for purpose in advancing sustainable agricultural development.

Keywords: Music marketing, Songs, Sustainable agricultural development, Agribusiness

9.1 Introduction

Music permeates and dominates all sectors of human lives (Opondo, 2000; Agawu, 2007; Mbaegbu, 2015). Music performs a crucial role in the psychological, physical, social, economic, cultural and religious aspects of life, as well as in the agricultural sector, which is the focus of this study. This study intends to examine conceptualizations of sustainability of agricultural marketing through music within Zimbabwe's agribusiness, which is the basis of many livelihoods. Current debates in the academic field focus on sustainability issues as scholars debate on how to successfully attain this concept in all areas and, in particular, the agribusiness sector. UNESCO (2021) posits that the global attention being given to the concept of 'sustainability' is traced to the publication of the Brundtland Report (1987, p. 8), which defined

*pzinhuku@gzu.ac.zw

© CAB International 2023. *Sustainable Agricultural Marketing and Agribusiness Development: An African Perspective* (eds B. Nyagadza and T. Rukasha)
DOI: 10.1079/9781800622548.0009

'sustainable development' as 'meeting the needs of the present without compromising the ability of future generations to meet their own needs'. This definition emphasizes the essence of upholding, conserving and continuity in providing needs of both the current generation as well as the future generation. Sustainability means a capacity to maintain some entity, outcome or process over time for the betterment of a community (Basiago, 1999). In agreement, DESA-UN (2018) posits that the ultimate goal of the concept of sustainability is to ensure appropriate alignment and equilibrium in society, in terms of the regenerative capacity of the planet's life-supporting ecosystems.

For Zimbabwe, achieving sustainable agricultural development is crucial in advancing the nation and achieving Vision 2030 (Murwira, 2019). Sustainable agriculture development is defined as the process of satisfying society's current food and textile needs without compromising the aptitude of future generations to meet their own necessities (UC Sustainable Agriculture Research and Education Program, 2021). In this case, sustainable agricultural development demands all stakeholders in the agriculture sector to ensure a sustainable agricultural system, including: growers, food processors, distributors, retailers, consumers and waste managers. Therefore, this study argues that in order for the nation to achieve sustainable agricultural development, there is a need to build a strong agricultural marketing strategy.

Agriculture and marketing are the two terms that coin the term agricultural marketing. Agriculture refers to activities that exploit the use of natural resources for human benefit. On the other hand, the word 'marketing' refers to the processes involved in transferring products from the point of manufacture to the point of intake (Agribusiness Education and Research International, 2022). Thus, agricultural marketing encompasses a series of activities involved in moving goods from the point of production to the point of consumption (Srivastava, Undated). In addition, agricultural marketing is also defined as a process that involves all the aspects of market structure or system and pre- and postharvest operations, assembling, grading, storage, transportation and distribution (Report of the National Commission on Agriculture, 1976). Thus, this study argues that

incorporating music into agricultural marketing would ensure sustainability of agricultural development.

Several studies on music and agriculture have been conducted. However, a gap still exists as no direct studies have been conducted on the relationship between music marketing and agricultural marketing. Therefore, this study seeks to contribute to existing literature on music and the agricultural processes, by focusing on the issue of music and sustainable agricultural marketing. In support of this dimension, Wolcott (2016) asserts that music has a critical role to play in the transition towards a culture of sustainability; in this case, towards agribusiness. Music is essential to human survival and human development. It is not an 'add on'; instead, it is the lifeblood. Its ability to induce symbols, emotions and social-knowledge structures make it a refreshing element for current efforts towards sustainability.

9.2 Unpacking the Concept of Music Marketing: Theories and Concepts

The world of marketing is both competitive and gigantic. However, music plays an imperative role in marketing. When effectively employed and applied, music could contribute greatly to the marketing effort of a company or product (Neese, 2015). Meler and Škor (2013) assert that it is necessary to emphasize that music is really the reproduction of art, vocalization and/or instrumental. The foregoing perspective rests on the idea that a concrete music product presents a piece of music or composition in various forms; consequently, with differing tones, rhythms and dynamics, melodies and harmonies, which are the fundamental elements of music.

Thus, music marketing is manifested through its influence on the purchasing practices of consumers in advertising communications. The observed interdependence between marketing and music provides the term commonly understood as music marketing. Music marketing should fundamentally concern itself with establishing products, services and ideas which offer or search out a specific market. Neese (2015) argues that an increasing number of marketing strategies involve the use of music and this is for a good reason as it is highly effective. The influence of music on marketing is both obvious and discreet. Thus, the behavioural effects of music in humans are predictable, and various studies have shown a strong correlation between emotions and music. Additionally, an increasing number of organizations have discovered that sound is important for understanding arguments, opinions and feelings (Neese, 2015).

9.3   Music and the Agriculture Sector: A Review

Music infiltrates all sectors of human lives and its influence in agricultural activities is not surprising. Chowdhury and Gupta (2015) carried out a very insightful research study that explores how plants respond to music, which is considered as a stimuli. The study specifically examined the response of *Tagetes sp.* (marigold) to light Indian music and meditation music as well as to noise. The results revealed that music promoted growth and development of plants whereas noise hindered the growth of the plant. The study shows the amazing capabilities of music in the production of agricultural products for future marketing and consumption. Alavijeh *et al.* (2016) examined the effect of music on the growth and yield of cowpeas in the summer of 2013 in Tehran. The results of the study showed that playing different types of music had an impact on plant growth. Classical music treatment improved cowpea yield as a marked growth was witnessed. However, in an apparent departure from the above views, this study seeks to marry the concept of agribusiness and music marketing.

Dhungana *et al.* (2018) revealed that music has analgesic and anxiolytic properties; hence, animals are affected by the music. The objective of this study was to review the influence of music in animal behaviour and discuss its usefulness for stress relief. The result of the study exposed that music has a number of benefits in animals, through auditory enrichment, which modifies the behaviour of animals. Interestingly, milking behaviour and milk yield of farm animals such as cattle and buffalo were also affected by music. Bhadiru and Akande (2018) evaluated the success of 'Cocoa na Chocolate', a musical intervention for agricultural mobilization in Oyo State, Nigeria. Responses from selected students

on awareness, exposure, retention, perception, and constraints to utilizing its message using structured questionnaires were analysed. Results of the study revealed that respondents confirmed that the message for mobilizing the community was spread through the music, as many heard the song several times and recollected its contents.

Renaud (2017), in an article entitled 'For Me, Music and Agriculture Go Hand In Hand', discovered that in Canada music played a critical role in farm activities conducted at *Médé Langlois'* farm and the store called *L'Économusée de la conserverie*. Interestingly, the Médé, whose family members have been proud vegetable and milk producers for eleven generations, are also punk musicians. Their ancestors, who used to live around here in the old days, played folk music and were surrounded by folk music. There has always been folk music here and it is an integral part of agricultural activities. Hence, cows at the farm are now used to punk music, as music rehearsals are conducted in the barn or the farming life.

Ciborowska *et al.* (2021) presented literature findings regarding the influence of music on cattle, poultry and pigs. According to the author, intensified livestock production requires methods that reduce stress, as it has an adverse impact on the health and welfare of their animals. Therefore, the study considered the possibility of using various musical genres to alleviate stress in chickens, cattle or pigs. Results indicate that choosing a musical item is important, as it can positively affect the health and production performance of animals by increasing the feeling of relaxation (Ciborowska *et al.*, 2021).

Boyce (2018), in a research titled 'Roots of Country Music are Planted Firmly in Agriculture', revealed that the roots of country-western music and American agricultural are inseparable. For instance, in 1985, country-western and folk rock artists joined forces for the inaugural Farm Aid Concert series. The flight from farms by younger people was sung about by Waylon Jennings in 1990, in the song, 'Where Corn Don't Grow', as an explanation of younger rural Americans' desire to go to the city, only to find that life was not quite what they thought it would be like. Chris LeDoux's song, 'Cadillac Ranch', describes the modern ag-tourism as a struggling family farm converts itself into a music hall and bar, and Montgomery

Gentry celebrated stubborn farmers who refused to surrender their heritage with 'Daddy Won't Sell the Farm'.

An article by the Food Tank (2014) indicated that John McDowell, a musician, film composer, and biodynamic farmer, established Music for Farms, an international initiative concerned with organic agriculture and farming communities. The purpose of establishing music for farms was to renew and sustain local organic agriculture through the arts. Music for Farms tours farms and venues throughout North America and Europe, performing music that reflects farming. Like the Music for Farms initiative, the Harvest Sun Music Festival in Canada brings awareness and support to family farms and agriculture. The Harvest Sun Music Festival deals directly with Manitoba Farm and Rural Support Services (MFRSS), which provides telephone and online support to farm and pastoral families throughout Manitoba. The yearly exhibition features band, chorus, and talent performances.

Thompson (2022) equally researched on how American music was born on a farm. The cultures of American agriculture provided the myths, economic realities, and artists that shaped much of the nation's popular music. In particular, country music and the blues owe their creation to the ways people worked in and desired an escape from agrarian labour. Songwriters have long found inspiration for their music in farming. For example, Zimbabwe has its own artist, Oliver Mutukudzi, whose song 'Murimi tora kapadza urime' encourages people to get into agricultural endeavours. Sometimes the songs reflected direct experience and their imagined conceptions of agricultural work and rural life. Hence, the blues and country music offered an interdisciplinary approach to the history of American agriculture.

Environment and Science News Agency (2020) examined how playing music to plants quickens their growth or helps them develop. Findings indicate that there are some farmers who believe that classical music has a positive effect on plants, through playing those pieces of the well-known composers, such as Wolfgang Amadeus Mozart. According to the research, playing classical music to watermelons yielded the sweetest fruits in the entire region for the particular farmer. Musician Mort Garson, upped the ante when he wrote music exclusively for plants. The album, 'Mother Earth's Plantasia', released in 1976, with the title, 'Warm Earth Music for Plants ... and People Who Love Them, contained songs with names such as 'Symphony for the Spider Plant' and 'Concerto for Philodendron and Photos'. In contrast, this study seeks to examine how music can be incorporated in agricultural marketing to achieve sustainable agricultural development.

Bricout (2017) wrote a fascinating article titled 'What if the debate about pesticides could be silenced by music?' This was an unusual claim of the French company Genodics in 2008. The article revealed that this solution of replacing pesticides with music was successful. This method has so far been adopted by 130 farmers in France, and has recently aroused the curiosity of INRA (the French National Institute for Agricultural Research), which plans to put tests in place. Therefore, music already has a firm role to play in the production, retailing and consumption of agricultural products. Accordingly, this study seeks to determine how this relationship may be explored further by exploring how music could be linked to agricultural marketing in the agribusiness sector.

9.4 Methodology

The study adopted and applied an exploratory approach that analysed how music marketing could be incorporated into agricultural marketing to achieve sustainable agricultural development. An exploratory design study results in the provision of alternative options for a solution of how music could be incorporated in agricultural marketing for sustainable agricultural development (Sandhusen, 2000). Interviews were conducted in the Masvingo District to discuss agricultural marketing through music and the sustainability of the agribusinesses.

In this qualitative study, the population was derived from music artists, music experts and agriculture business experts. The sample of this study included three music artists, three business experts and three music experts. According to Holloway and Wheeler (2002), sample size does not influence the quality, significance, reliability or validity of the study and there are no guidelines in determining sample size in qualitative research. Thus, snowball sampling was utilized to select the sample that had traits rare to find for this study. Kirchherr and Charles (2018) define snowball sampling as a non-random sampling method that makes use of a few subjects who will provide required data. Thus, the sample for this study was selected because of its vast experience and commitment in the teaching, performance and management of musical arts in Zimbabwe. The sample also included business practitioners to provide insight on the agribusiness sector. The participants provided primary data on how music marketing could be roped in the agricultural industry in Zimbabwe.

Participants were interviewed to get an insider's view designed for exploring sustainable agriculture, music marketing and agribusiness. A tape recorder was utilized to record interviews with all selected participants for future analysis through thick descriptions. The use of thick descriptions is what Geertz (1973) considers essential in an analysis of qualitative data of this nature.

This study engaged observation as a technique for assembling data. Observation allows immersion in the daily lives of the subjects under study, which resulted in deepening the understanding of the research problem (Stone, 2008, p. 13). The approach of 'being in the field', which is participating in the daily lives, activities and practices of the intended community of study, is further emphasized by Titon (2009, p. 25). Therefore, the researcher observed the agricultural music performances for future analysis. Observations were made on how music impacts on agricultural marketing efforts. The observations provided researchers with insights into the role of music marketing in sustainable agricultural development.

According to Roshaidai and Kulliyyah (2018), the protection of human subjects, through the execution of appropriate ethical codes, is imperative in a research inquiry. Ethics refers to 'the branch of philosophy which deals with human conduct in respect to the rightness or wrongness of certain actions and the badness or goodness of the motives and ends of such actions' (Shumbayawonda, 2011, p. 28). Thus, consent to participate in this research was given freely and voluntarily as participants were made to understand what was being required from them.

Responses to interviews would be presented in the indigenous language of the participants and translated to English verbatim, to accommodate a wider readership. The analysis of research results was done through thick descriptions to assemble the final manuscript. The employment of thick descriptions facilitated the generation of a rich and detailed ethnographic research.

9.5 Linking Music Marketing and Agribusiness for Sustainable Development

9.5.1 Music marketing and the launching of agribusinesses

Champions Music and Entertainment (2021) reveals that launching a new product can be stressful. Participant one had this to say: *'Tinganyeperane hama, kuti munhu ashambadzire zvitengeswa zvake anoda mumimhanzi kuti ibetsere kuzivisa vanhu zvinekunakidza mukati hazvina mubvunzo.'* ['We cannot lie to one another, in order for a person to successfully launch his/her products one needs music to advertise in an entertaining way no question about it.']

Launching an agricultural product is an exciting event as it can generate attention and interest for the product or service (Fusion Music Marketing, 2018). A clearly defined product launch will add revenue streams. Thus, launching a product utilizing music is an effective way to put an agribusiness on top as music can inspire the fans, create desire to buy, and then convert those potential fans into loyal lifetime customers (Fusion Music Marketing, 2018).

9.5.2 Music marketing increases customer base

Business moguls know that for any successful business venture to materialize, a firm customer base is essential. Music marketing has the potential to generate a firm customer base for the agribusiness sector. Participant eight observed that: *'Vanhu vanofarira music saka ukairidza mushop mako uchitengesa vanhu vanouya kuzonzwa vopedzisira vatotenga nezvavanga vasingadi kutenga.'* ['If you play music in your shop people will come and listen to it, in the end they buy some goods even if they were not prepared to buy that stuff.']

In agreement, Audiosocket (2022) argues that scientific research has shown that music has a powerful effect on our psychology, altering human moods and producing corresponding changes in human behaviour. Therefore, choosing the right music can boost sales. Milliman (1982) found that when background music was faster, customers bought less as they walked more quickly, picked up only what they came for, and spent little to no time browsing. However, when the tempo slowed down, customers' movements slowed down, they browsed more and spent more. In agreement, Kotler (1973, p. 1) asserts that 'In some cases, the place, the atmosphere of the place, is more influential than the product itself in the purchase decision. In some cases, the atmosphere is the primary product. And when the right music fills the space, it creates an atmosphere that encourages sales'.

9.5.3 Music marketing and agricultural product advertising

Music marketing performs an effective role in agricultural product advertising. According to participant five:

'*Kana uchida kuti zvinhu zvaunotengesa zvitengwe unogona kushandisa mimhanzi inoshambadzira zvinhu zvacho nenzira inotora vanhu moyo. Zvitengeswa zvako zvinotengwa chose.*' ['If you want your products to sell effectively you can use music to advertise yore products in a way that is attractive to customers. Your products will sell like hot cakes.']

From the foregoing perspective, choosing a compelling song will enable your advertisement to stand out. For example, popular songs are overplayed on the radio and in retail stores resulting in an increase in sales.

9.5.4 Music marketing and branding in agribusiness

Neese (2015) posits that marketers employ different techniques in order to create the ideal brand image for products and services. However, music is becoming increasingly important in marketing decisions. The types of music playing in a store, on a commercial, on a website, or at a worksite, could affect the image of a product and, ultimately, whether or not a product sells. Therefore, branding is a crucial aspect of a successful agribusiness corporation. Branding is the process of creating a strong and positive perception of a company and its products or services in the customer's mind, by combining such elements as logo, design, mission statement and a consistent theme throughout all marketing communications. Effective branding helps companies differentiate themselves from their competitors and build a loyal customer base (Oberlo, undated).

Branding is a multifaceted process that requires careful planning and a premeditated approach. A strong brand is easy to relate to and draws on values that resonate well with the target audience. According to Olenski (2014), music brings value to a brand in three ways: identity, engagement and currency. Engaging music to establish an emotional connection with a brand increases brand respect and creates enthusiasm and thrill beyond the brand's core products or services. Music creates the value that brands need to win the war for attention and develop a genuine connection with their consumers. When used correctly, music not only creates loyalty, but true advocacy (Olenski, 2014). In the same vein, Mulcay (2019) posits that music takes the centre stage as an effective means of promoting a brand. As such, it makes sense that brands wishing to create a positive impression of their brand would try to collaborate with attention-grabbing music to benefit from the positive associations. Neese (2015) asserts that designing specific lyrical music to develop an appropriate brand image is necessary and music selection can provide customers with an increased understanding of a product.

9.6 Music and Agricultural Sustainable Development: Suggesting an Agri-based Music Marketing Model

On the basis of this study's findings and studies by Neese (2015) and Olenski (2014), the study

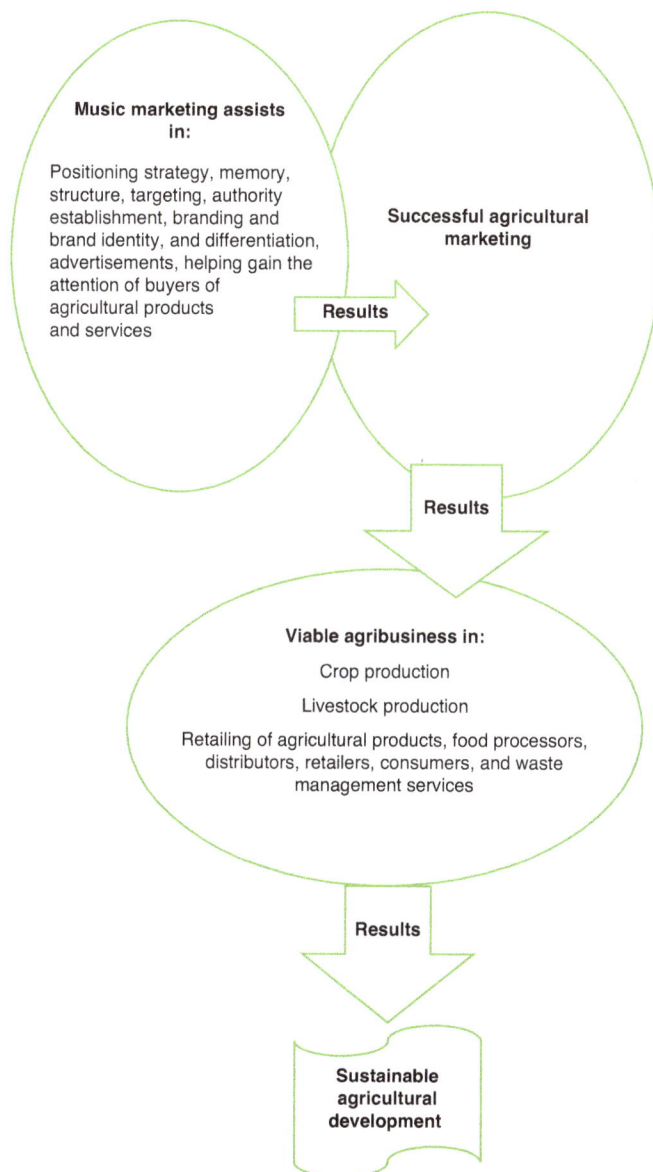

Music marketing assists in:

Positioning strategy, memory, structure, targeting, authority establishment, branding and brand identity, and differentiation, advertisements, helping gain the attention of buyers of agricultural products and services

Results

Successful agricultural marketing

Results

Viable agribusiness in:

Crop production

Livestock production

Retailing of agricultural products, food processors, distributors, retailers, consumers, and waste management services

Results

Sustainable agricultural development

Fig. 9.1. Conceptual model.

proposes an agro-based music marketing model to address how music marketing could be roped into the agricultural sector, to become fit for purpose in advancing sustainable agricultural development. The proposed model is shown in Fig. 9.1.

Based on the proposed model, this study argues that incorporating music into the agricultural marketing process will result in sustainable agricultural development. Music can both consciously and unconsciously play a large part in positioning strategy, memory, structure, targeting, authority establishment, branding and brand identity, and differentiation. It can also provide entertainment in advertisements, help gain the attention of buyers, and can even affect the mood of the buyer and the buyer's willingness to make a purchase. Thus, the incorporation of music marketing will result in the creation of successful retailing of agricultural products, food processors, distributors, retailers, consumers, and waste management services. This is the ultimate goal of sustainable marketing for the survival of the agribusiness sector. Given that the Zimbabwean economy has an inclination towards agriculture for food security and promotion of the health and wellbeing of the nation, the proposed model could be a prop to the agribusiness sector. Accordingly, there is need for synergies between agribusiness players and the music and arts industry for mutual and symbiotic relationships.

9.7 Conclusion

Music is an essential component of agricultural marketing as its incorporation in the agribusiness process promotes sustainable agricultural development. Music is a valuable tool for agricultural sustainability to materialize through its ability to create an enabling environment for promoting, for branding advertising and for launching agro-based industries and activities. The study, thus, presented a music-based agricultural marketing model that could be implemented in the agricultural sector for promoting agro-based business initiatives for the development of the nation. The study recommends that stakeholders in the agricultural business should value the contribution of music marketing in their businesses and explore its full potential in achieving sustainable agriculture development.

References

Agawu, K. (2007) *Representing African Music: Postcolonial Notes, Queries, Position*. New Routledge, New York and London.

Agribusiness Education and Research International (2022) What is the scope and importance of agricultural marketing. Available at: https://agribusinessedu.com/what-is-the-scope-and-importance-of-agricultural-marketing/ (accessed 22 May 2022).

Alavijeh, R.Z., Omid, S., Hasan, R. and Sayed, V. (2016) The effect of sound and music on some physiological and biochemical traits, leaf nutrient concentration and grain yield of Cowpea. *A Journal of Multidisciplinary Science and Technology* 7, 447–458.

Audiosocket (2022) Music and customer behavior: how retailers increases sales with music. Audiosocket. Com. Available at: https://blog.audiosocket.com/small-business (accessed 3 November 2022).

Basiago, A.D. (1999) Economic, social, and environmental sustainability in development theory and urban planning practice. Kluwer Academic Publishers, Boston, *The Environmentalist* 19, 145–161. DOI: 10.1023/A:1006697118620.

Bhadiru, I. and Akande, O. (2018) Assessing the success of "Cocoa na Chocolate" musical intervention in mobilizing the youth for agriculture in Oyo State, Southwestern Nigeria. *Journal of Agricultural & Food Information* 1–11. DOI: 10.1080/10496505.2017.1319769.

Boyce, B. (2018) Roots of country music are planted. Available at: https://supperclub.paris/en/weeklynews/agriculture (accessed 23 October 2022).

Bricout, J. (2017) Agriculture: replacing pesticides with music. Available at: https://supperclub.paris/en/weeklynews/agriculture-replacing-pesticides-with-music/ (accessed 28 October 2022).

Brundtland Report (1987) Sustainable development. Available at: https://www.are.admin.ch/are/en/home/medid/publication (accessed 22 May 2022).

Champions Music and Entertainment (2021) Product launching guide. Available at: https://musicandbands.co.uk/product-launches (accessed 24 June 2022).

Chowdhury, A.R. and Gupta, A. (2015) Effect of music on plants: an overview. *International Journal of Integrative Sciences, Innovation and Technology* 4, 30–34. Available at: https://www.researchgate.net/publication (accessed 2 November 2022).

Ciborowska, P., Michalczuk, M. and Bień, D. (2021) The effect of music on livestock: cattle, poultry and pigs. *Animals (Basel)* 11(12), 3572. DOI: 10.3390/ani11123572.

DESA-UN (2018) The Sustainable Development Goals Report 2017. Available at: https://undesa.maps. arcgis.com/apps/MapSeries/index.html (accessed 24 October 2022).

Dhungana, S., Khanal, D.R., Sharma, M., Bhattarai, N., Tamang, D.T. *et al*. (2018) Effect of music on animal behavior: a review. *Nepalese Veterinary Journal* 35, 142–149. DOI: 10.3126/nvj.v35i0.25251.

Environment and Science News Agency (2020) Music for plants is it a sustainable way to boost crop production. Available at: https://jewishjournal.com/commentary/blogs/322483/ (accessed 10 November 2022).

Food Tank (2014) Music for farms. Available at: https://foodtank.com/news/2014/02/music-for-farms/ (accessed 12 November 2022).

Fusion Music Marketing (2018) Fusion music product launch for musicians: the CD launch. Available at: https://www.fusionmusicmarketing.com/product-launches/ (accessed 13 November 2022).

Geertz, C. (1973) *Description: Toward an Interpretive Theory of Culture. The Interpretation Culture*. Basic Books, USA.

Holloway, I. and Wheeler, S. (2002) *Qualitative Research in Nursing*. Blackwell Science, Oxford, London.

Kirchherr, J. and Charles, K. (2018) Enhancing the sample diversity of snowball samples: recommendations from a research project on anti-dam movements in Southeast Asia. *PloS One* 13, 8. DOI: 10.1371/journal.pone.0201710.

Kotler, P. (1973) Atmospherics as a marketing tool. *Journal of Retailing* 49, 48–64.

Mbaegbu, C.C. (2015) The effective power of music in Africa. *Open Journal of Philosophy* 05(3), 176–183. DOI: 10.4236/ojpp.2015.53021.

Meler, M. and Škor, M. (2013) (R)evolution of music marketing. In: *Proceedings of 23rd CROMAR Congress: Marketing in a Dynamic in a Dynamic Environment – Academic and Practical Insights*, Lovran. DOI: 10.13140/2.1.1186.5600.

Milliman, R.E. (1982) Using background music to affect the behavior of supermarket shoppers. *Journal of Marketing* 46(3), 86–91. Available at: https://doi.org/10.2307/1251706

Mulcay, E. (2019) When music and marketing come together in harmony. The drum news blog. Available at: https://www.thedrum.com/news/2019/06/24/when-music-and-marketing-come-together-harmony (accessed 29 November 2022).

Murwira, A. (2019) Towards revitalizing the roles of universities in development [Zimbabwe]. Available at: https://www.ruforum.org/AGM2019/sites/default/files/HON_MINISTER_PROF_MURWIRA_SPEECH.pdf (accessed 18 November 2022).

Neese, K.V. (2015) The role of music in the enhancement of marketing. Honours Thesis, Liberty University Spring.

Oberlo (undated) What is branding in marketing. Oberlo.com blog. Available at: https://www.oberlo.com/ecommerce-wiki/branding (accessed 18 September 2022).

Olenski, S. (2014) Why music plays a big role when it comes to branding. Available at: https://www.forbes.com/sites/steveolenski/2014/02/06/why-music-plays-a-big-role-when-it-comes-to-branding/ (accessed 29 October 2022).

Opondo, P.A. (2000) Cultural policies in Kenya. *Arts Education Policy Review* 101(5), 18–24. DOI: 10.1080/10632910009600269.

Renaud, P. (2017) For me, music and agriculture go hand in hand. Socan magazine. Available at: https://www.socanmagazine.ca/features/for-me-music-and-agriculture-go-hand-in-hand/ (accessed 14 November 2022).

Report of the National Commission on Agriculture (1976) National Commission on agriculture. Available at: https://agris.fao.org/agris (accessed 21 October 2022).

Roshaidai, S. and Kulliyyah, M.A. (2018) Ethical considerations in qualitative study. *International Journal of Care Scholars* 1(2), 30–33. DOI: 10.31436/ijcs.v1i2.82.

Sandhusen, R.L. (2000) *Marketing*. Barrons, New York.

Shumbayawonda, W.T. (2011) *Research methods postgraduate diploma in education (PGDE)*. Zimbabwe Open University, Harare, Zimbabwe.

Srivastava, S.K. (Undated) Notes prepared for Course AgEcon530 (Agricultural Marketing). Course instructor: Scientist (Agril. Economics). Available at: http://jnkvv.org/PDF/10042020083748concept%20of%20ag%20markeing_EgEcon530.pdf (accessed 24 October 2022).

Stone, R. (2008) *A Theory for Ethnomusicology*. Prentice Hall, USA.

Thompson, J.M. (2022) The Blues, Country Music, and American Agriculture. Available at: https://doi.org/10.1002/9781119632214.ch30 (accessed 2 October 2022).

Titon, J. (2009) Music and sustainability: an ecological viewpoint. *The World of Music* 51, 119–137.

UC Sustainable Agriculture Research and Education Program (2021) What is sustainable agriculture? UC Agriculture and Natural Resources. Sustainable Agriculture Research and Education Program. Available at: https://sarep.ucdavis.edu/sustainable-ag (accessed 17 November 2022).

UNESCO (2021) Sustainable development. Available at: https://en.unesco.org/ (accessed 12 December 2022).

Wolcott, S.J. (2016) Empowering sustainability. *International Journal* 3, 1–19.

10 Financial Management for Agricultural Marketing and Agribusiness Development

Dumisani Rumbidzai Muzira[1]* and Basil Shumbanhete[2]
[1]*Africa University (AU), Mutare, Zimbabwe;* [2]*Marondera University of Agricultural Sciences and Technology (MUAST), Zimbabwe*

Abstract

The purpose of this chapter is to explore financial management for agricultural marketing and agribusiness development. The chapter is based on a literature review as well as practical analysis of key areas of financial management. This chapter became a necessity realizing that finances are the lubricant for any business to run smoothly. More literature on financial management for agriculture is also essential since agriculture is the backbone of most economies in the world. Financial management is needed at every stage of the business from inception through all the functional departments such as human resources management, sales and marketing, accounting and finance, and information communication technologies up to the end product. The chapter covers the definition of financial management, sources of finance, budgeting, cashflow management, inventory management, agricultural marketing, agribusiness tax planning, records management, and business financial reviews. Good financial management would help farmers to take farming as a viable business. This would then help farmers to manage their farm activities for growth and high profitability. Surplus coming from the farm venture could then be invested to get more returns for business development.

Keywords: Financial management, Budgeting, Cash-flow management, Working capital management, Inventory management

10.1 Defining Financial Management

Management is planning, organizing, leading and controlling the use of resources to accomplish performance goals (Schermerhorn Jr, *et al.*, 2020). Financial management should therefore have these four elements of management applied on financial matters. This then leads to the operational definition for financial management as the planning, organizing, leading and controlling of the use of financial resources to accomplish organizational goals.

Planning would include profit planning, sales forecasts, capital investment and financing, product costing, cashflow, working capital and inventory budgets. Planning would then guide the organization to work as a coordinated system by *organizing* these activities. Product costing would feed into the sales forecasts that would feed into the profit

---

*Corresponding author: rdmuzira@gmail.com

© CAB International 2023. *Sustainable Agricultural Marketing and Agribusiness Development: An African Perspective* (eds B. Nyagadza and T. Rukasha)
DOI: 10.1079/9781800622548.0010

planning. The inventory budgets would affect the cashflow budgets that would in turn affect the working capital and the cashflows of the business. Without proper financial planning, all these activities would not run smoothly as there would be disruptions caused by shortages in some areas affecting the other activities, for example if goods are not costed well, the sales figures would be distorted since the sales value is based on the mark-up on cost. The distorted sales figures will also affect the profit figure since sales constitute revenue, which is then used to calculate profits. This therefore means that financial planning and organizing are critical in every business, agribusinesses included. After the planning and organizing, business plans need to be coordinated by the people in management who will *lead* or influence their subordinates to work with the view of attaining the organizational goals (Griffin, 2021). *Controlling* would then be the comparison of the actual performance with the budgets or standards. Controlling will reveal areas that need improvement and adjustment.

10.2   The Importance of Financial Management in Agriculture

For people to implement something, they need to understand its importance to them. Financial management in agriculture helps in the following ways:

1. To maximize profits.
2. For effective cost control.
3. Maintenance of cashflows.
4. Proper distribution of financial resources.
5. In decision making on pricing, expansion and restructuring.

10.3   Sources of Agricultural Finance

The size and nature of a business determines the source of financing of a business (Thilmany *et al.*, 2021). There are many sources of agricultural finance but this chapter will cover short-term and long-term sources that are commonly used for agricultural businesses.

10.3.1   Short-term sources

• Retained profits
• Bank overdrafts
• Bank loans
• Debt factoring
• Friends and relatives

10.3.2   Long-term sources

• Mortgages.
• Venture capital.
• Government.
• Non-government organizations (NGOs).

10.4   Budgeting

Budgeting is a plan that has been expressed quantitatively (Wood and Sangster, 2005). It serves as a road map to be followed to reach a predetermined destination. A budget can be operational or capital. An operational budget is an estimate of the revenues and expenses of the organization while a capital budget has to do with long-term investment budgets. Competition, the business climate and past experiences should be taken into consideration when formulating budgets.

10.4.1   Why should farmers budget?

An unknown author says 'the person who doesn't know where his next dollar is coming from usually doesn't know where his last dollar went', so budgeting is very crucial. It provides a coordinated allocation of the scarce resources, and helps farmers to think ahead and to identify possible constraints, thereby mapping ways of dealing with such constraints beforehand. Budgeting also provides a basis for comparison of the actual against the budget for variance analysis.

10.4.2   Types of budgets

There are many types of budgets, but this chapter will cover zero base budget, fixed budget and flexible budget. Each type has its advantages and disadvantages.

10.4.2.1 Zero base budget (ZBB)

Farmers usually struggle to come up with a budget when they begin to take farming as a business. There would be no records to base their budget on. It is generally easier to just adjust last year's budget with the inflation rate when coming up with the current year's budget. ZBB is a budget that is prepared from scratch. Each budget line is justified for it to be funded. This then means that every activity starts at zero and an amount will be allocated if the expense has been justified after doing a cost–benefit analysis. Wasteful expenditures will be eliminated; hence, the popularity of this budgeting system for organizations that seek to control wasteful spending (Ibrahim, 2019).

10.4.2.2 Fixed budget

Fixed budgets are based on some assumed condition such as production output or sales volume. The budget is not adjusted to the actual volume of production. If the budgeted cost of materials is $1000 for production of 10 units, even if actual production has increased to 20 units, material cost budget will remain at $1000. This might give misleading information since an increase in units produced would logically require more on cost of materials. This challenge is solved by flexible budgets.

10.4.2.3 Flexible budget

Flexible budgets are budgets that are adjusted to the actual volume. This is a budget that has been adjusted to match with the actual activity level. For example, if a farmer had budgeted to use 20 bags of fertilizer of his 10-hectare farm, when he then plants five hectares, his budget would then be 10 bags of fertilizer because the activity level has fallen by half.

10.5   Cashflow Management

Cash for a business can be likened to the blood that has to reach every organ of the body for life to continue. Cash can be said to be the lubricant that makes the business run smoothly; hence, the need for it to be managed well. Cashflow management involves monitoring and controlling how cash moves into the business (cash inflows) and how cash moves out of the business (cash outflows). Cashflow management is then defined as 'the corporate process of collecting and managing cash as well as using it for (short term) investing' (Pandey, 2019, p. 1). Therefore, there is need to balance the cashflow levels such that the business is not starved of cash nor has too much cash that it would not need. Cashflow management helps those in agriculture to determine the money required for planned livestock and crop production. The business would rather invest the excess cash where it can get interest or invest in buying assets for business expansion than keep it. The best way to deal with business cashflows is to manage the streams that bring and those that take out cash from the business. Working capital management can be employed to manage business cashflows as presented below.

10.6   Working Capital Management

Working capital is the difference between the current assets and current liabilities (Wood and Sangster, 2005). Working capital management is essential as it has a direct impact on liquidity and profitability of the business (Singh and Pandey, 2008). Liquidity is the ability of the company to meet its short-term obligations when they become due. The approach to use when managing working capital is to manage the different accounts that fall under current assets and current liabilities. A change in one or more of these accounts will ultimately affect the working capital of the organization. The working capital requirements are industry specific (Nazir and Afza, 2009) hence the focus on those common to agribusiness.

10.6.1   Bank overdrafts

Agribusinesses often negotiate for bank overdrafts especially during the farming seasons so as to buy agricultural inputs. If the overdraft facility is not abused, then its use would be worthwhile. Special care should be made to just take an amount that will cover the existing financial gap so as to limit the charges associated with these overdrafts.

10.6.2 Bank loans

Obtaining a bank loan is one way agribusinesses can meet their financial needs. Bank loans come with interest; hence, the need to shop for banks that charge low interest on loans. If the interest rates are more than the return that the use of that loan will bring to the business, then that loan is not worthwhile.

10.6.3 Account payables

In order to maintain a positive working capital, companies should keep their creditors low. However, where possible, the companies can delay paying their creditors provided that would not tarnish their relationship with the creditors or ruin the credit rating. Delays in paying would create some leeway for companies to invest that money or to possibly restock their inventory for sale.

10.6.4 Account receivables

Receivables are best managed at the time of granting the credit terms than at the collection point. There should be an assessment of the creditworthiness of the new customers before granting them the credit facility. Bank references, credit rating agencies, trade references and the customer's sales records are some of the methods that could be used to assess the credit status of the client. At times agribusinesses are forced to sell their goods on credit to avoid loss when their perishable goods go bad but it is advisable to discount the products rather than offering credit to customers who will eventually become bad debts. If credit is granted, the credit terms should be made clear to the customer. These credit terms may include the credit limit, number of days and whether there would be any discount for early payments or interest charged on overdue accounts. Managers can create value for their

firms if the number of accounts receivable days and inventories are reduced (García-Teruel and Martínez-Solano, 2007). Regular follow ups for payments should be made with the debtors reminding them of their outstanding balances and due dates. These can be done by sending statements, reminder letters and even phone calls or visits. Those debtors with overdue accounts could be sent legal action threat and then eventually legal action to recover the money. However, this should be done as the last resort because it has potential of ruining relationships with customers.

10.7 Inventory Management

Management of company inventory/stock is very important for every business. There should be inventory control systems in place such as: segregation of duties, physical controls, and inventory storage. Besides these physical controls, inventory management also involves knowing the order point, safety stock and also the economic order quantity.

10.7.1 Segregation of duties

The company should avoid having one person responsible for more than one process. This will help with accountability and also controlling loopholes in the system where one can avoid detection of mistakes by covering them up. As an example, if the person responsible for receipting is the same one responsible for banking and also for bank reconciliations, they might not bank the receipted money in time and cover that up when they do bank reconciliation statements. However, conniving can be there, but job rotations may be used to minimize it.

10.7.2 Physical controls

Physical controls involves having locks and cameras where inventory is kept as well as keeping the inventory under the conducive environment such as refrigerating perishable inventory as well as controlling temperatures and humidity levels where inventory is kept. The cameras will deter those who might want to steal the inventory for fear of detection.

10.7.3   Order point

Order point is the point at which an order should be raised to replenish the inventory. This point should be decided after considering the demand for the product and also the lag time. Lag time is the time between the pressing of an order and time that the goods are delivered. The order point will avoid stock shortages. Stock shortages affect the business' reputation due to customer disappointment when their orders are not met in time.

10.7.4   Safety stock

Safety stock is the minimum level of inventory that a business should keep as buffer stock to avoid stock out.

10.7.5   Economic order quantity (EOQ)

EOQ is the optimal quantity that the business should order or purchase to minimize the cost of both carrying the inventory and process-ing the purchase order (Kumar, 2016). This optimal quantity reduces chances of stock outs and limits stock carrying costs. This ensures that the quantities ordered are not too high such that carrying costs such as warehouse rental space is reduced. The formula for calcu-lating EOQ is:

$$\sqrt{\frac{2CN}{K}}$$

where C is the cost of placing an order, N is the number of units required annually and K is the annual carrying cost per unit of inventory.

their obligations. However, one should strike a balance so that the business does not keep much more cash than its requirements or lower cash levels than it needs. Theories such as the Miller–Orr model and Baumol model are some of the models used to measure the cash levels where companies have set upper limits, lower limits, return point and the spread that is within the organizational range. These models are not within the scope of this chapter. Current asset ratio and acid test ratio can be used to measure the liquidity of a company (Higgins, 2012).

Current ratio = Current assets/Current liabilities

Current ratio measures the capacity of the company to meet its short-term obligation using its current assets. The general rule of thumb is 2:1 meaning $2 of assets are available to meet each $1 of obligations.

Acid test ratio/quick ratio = (Current assets – Inventory)/Current liabilities

This is a more refined ratio than the current ratio in the sense that it takes away inventory from the current assets figure. The exclusion of the inventory is done since inventory is not easily convertible to cash; hence, illiquid. The rule of thumb for the quick ratio is 1:1 meaning that the business will be able to meet the existing liabilities when they become due.

10.9   Product Costing

The choice of which costing system one adopts depends on the nature of product or service that needs to be costed. If the product is unique and tailormade according to customer specification then job costing will be ideal. Where a product can be produced in batches, then batch costing will be ideal. There are situations where process costing, marginal costing or activity-based costing (ABC) could be more ideal.

10.8   Cash/Bank

Liquidity of a company is increased when companies have enough cash available to meet

10.10   Agricultural Marketing

Marketing, especially inbound marketing, can be used since this marketing strategy earns the customer's attention so that they go online in search of your product through the company's website or through channels such as blogs and social media platforms. When using inbound marketing, the product offers are shared online for their alignment with potential customer needs and formation of long-term relationships since the consumers will be the ones to initiate engagements with the suppliers (Patrutiu-Baltes, 2016). If marketing is not done well, the products will get spoiled while in the shelves when the customer who needs the product is not aware that the farmer has the solution to their need. A good marketing strategy should therefore be in place if agricultural products are to be valuable to the clients. The market should be aware of what you offer, offer it in time to meet the customer need, delight the customer by offering them more than their expectations, be open for feedback, learn and improve.

10.11   Agribusiness Tax Planning

Basic knowledge about how agribusinesses are taxed will assist in managing the business finances. There is also a need to know areas where the business can reduce their tax burden. Some of the ways that agricultural-based businesses can manage their tax expenses are by either postponing the payment of taxes or using the Income Tax Act and other Acts such as the Finance Act, the Capital Gains Acts and other government pronouncements to reduce their taxes. It is important to note that this is tax avoidance and it is legal. Tax evasion is the one that is not legal because of the use of illegal action such as falsification of records, failure to disclose income received, or providing false answers wilfully with the intention of reducing the tax or even not paying it when it is due. These tax evasions attract heavy penalties of 100% penalty plus interest, some might even lead to prosecution and serving a sentence (Tapera and Majachani, 2017). The way farmers are taxed does not differ much from the way other businesses are taxed, however, there are some ways that agricultural businesses can manage their taxes that are unique to farmers. Some of these methods are listed below.

10.11.1   Valuation of stock (Section 8(1)(h) of the Income Tax Act)

Farm trading stock is composed of livestock, crops and even tree plantations. Livestock is divided into two categories: stud livestock and ordinary livestock. The value of stock has an influence on the tax charged; therefore, a farmer should try to keep the value low. Stud livestock is mainly kept for breeding purposes. The three valuation methods in respect to livestock are the purchase price value (PPV), the fixed standard value (FSV), and the cost and maintenance value (CMV). The PPV is the cost at which the livestock was acquired by the farmer. This valuation method is only applicable to stud livestock. The FSV is the value that the farmer fixes for each category of the livestock with the commissioner's approval. Once the value has been fixed, it would need the consensus of both the farmer and the commissioner for it to be adjusted. This valuation method can be used for both stud and ordinary livestock. The CMV is the summation of the cost of acquiring the livestock and the cost of maintaining the same livestock up to the end of the year of assessment. Cost of labour, feed and dips are some of the costs that make up the maintenance cost of livestock. A farmer has to elect which valuation method to use for stud livestock and for ordinary livestock. However, only two methods, FSV and CMV, are available for ordinary livestock. For crops and other unutilized consumables, the value determined by the commissioner as a fair and reasonable value is used. The closing stock for farmers is treated as income for tax purposes. The stock disposed of through a donation or consumed by the farmer is treated as gross income in the hands of the farmer. As a way of reducing the farmer's tax burden, it is prudent to choose the stock valuation method that will minimize the livestock value.

10.11.2   Relief from enforced sales (Para 5A, 7th Schedule of the Income Tax Act)

This relief is only availed to farmers who have been forced to sell their livestock due to epidemic disease, drought or compulsory acquisition of the land by the state for resettlement. The proceeds from these three kinds of sale can be spread over 3 years if the farmer elects for this.

This election would postpone part of the farmer's tax burden to the future years. Considering the time value of money, a farmer who utilizes this relief would have saved a lot.

10.11.3   Allowable deductions (Section 15 of the Income Tax Act)

Allowable deductions are expenses that are allowed by the Act to reduce the taxable income. Farmers enjoy special deductions like the restocking allowance. The value of opening stock at the beginning of the year is allowed as a deduction for farmers. This has an effect of reducing their taxable income at year end.

Farmers also enjoy the restocking allowance when they restock livestock depleted by epidemic diseases or drought. Half of the purchase price is allowable for tax purposes. This is an additional deduction availed only to farmers as the cost of livestock purchased would have been in the livestock trading account. However, the restocking allowance has a limit, which is the assessed carrying capacity of the land (ACCL). As an example, if the ACCL is 500 beasts, if the farmer buys 600 cattle to restock, the cost of only 500 will qualify for the allowance. Using Para 6, 7th Schedule of the Income Tax Act (Sarbam, 2020), the formula for calculating restocking allowance is $A/2 \times B/C$ where A is the cost of the livestock purchased, B is the livestock that should have been purchased, i.e. ACCL less livestock on hand before restocking, and C is the number of livestock actually purchased for restocking purposes.

The allowable deductions and reliefs available for farmers can only be of benefit to the farmer if they elect for them and claim them. Tax planning is needed in order to minimize the farmer's tax burden.

10.12   Records Management

There are basic records that every business should keep and these are: the receipt and payment account; the income statement (trading and profit and loss account); the statement of financial position (balance sheet); and the asset register. Record keeping will assist in the determination of profits/losses for a particular period. The data in the records will also help farmers to prepare budgets.

10.12.1   The receipt and payment account

Businesses that need not maintain a full set of accounts, probably due to their low sales volumes, can maintain a receipt and payment account. Like its name, all receipts are recorded on the left side/debit side and all payments are recorded on the right side/credit side. At the end of the period, these two sides are balanced off.

10.12.2   The income statement

This is a combination of the trading and profit and loss account, the upper part being the trading account and the lower part the profit and loss account. The trading account shows the sales, the cost of sales and the gross profit while the profit and loss account shows the revenue and expenses of the business as well as the profit or loss for the period.

10.12.3   The statement of financial position

This is a snapshot of the business' financial position at a particular date. It shows the assets, liabilities and capital of the business at a particular date.

10.12.4   The asset register

Good accountability of assets starts with the knowledge and record of how many assets the organization has, how old they are and where they are generally kept. This and more information would be reflected in the asset register. An asset register is a list of the non-current assets of a business. These assets include but are not limited to computers, printers, photocopiers, desks, office furniture, equipment and motor vehicles. The asset register can be kept and updated periodically or when changes such as asset disposal, write offs or additions take place. The register can be manual or electronic though electronic copies are preferable to be in line with technology.

10.12.5   Livestock reconciliation

Livestock farmers keep a livestock reconciliation in order to compute the number of closing stock of livestock. This closing stock would then be used for tax purposes as it shows the closing value of the livestock.

10.12.6   Livestock production and feeding records

Agribusinesses should keep proper records for their livestock and feeding. This includes the date the livestock was bought or born, the quantity of food that they eat, the vaccinations bought and the dates to be administered among other things. These records will be used for calculating the livestock costs for tax calculations.

10.12.7   Labour records

Labour is an element of cost that is critical for agribusinesses, and can be split into direct and indirect labour. Direct labour is the labour that is used in the production of goods and services while indirect labour is labour that cannot be traced to the production process. Farmers usually do not include their labour input in the costing of goods and services they produce. This is an undercast on labour and would lead to inaccurate labour costs.

10.12.8   Agribusiness development

It is the vision for each business to make profits and to continue growing. Businesses will not develop if their performances are not reviewed for possible improvements. Reviews are a mirror that reflect what was done well, what was performed badly as well as where there is need for improvement. Reviews help businesses to continuously learn and improve their business

processes by adapting to the changes in the business environment. The areas that need reviewing could be financial or non-financial.

10.12.9   Business financial reviews

When doing business reviews, farmers should take into account controllable and uncontrollable costs. Controllable costs are those costs that can be influenced by management, while, uncontrollable costs are those that management has no influence over. Managers' performance can be measured using controllable costs since they are under their control. Most businesses use financial reviews to measure performance. These reviews mostly use profitability, investment and liquidity ratios to measure performance. These can then be compared to the same ratios in other companies within the same industry (cross-sectional analysis) or different periods within the same organization (serial analysis). However, there are some factors that cannot be captured by financial information and need non-financial factors for a more comprehensive evaluation.

10.12.10   Non-financial reviews

There are reviews that are non-financial which are also useful when evaluating performance. These include, control of quality, control of cycle time and control of productivity. Quality is defined as 'meeting customer requirements' (Oakland, 2004, p. 5). There are four categories of the cost of quality which are: prevention, appraisal, internal failure and external failure (Oakland, 2014). Prevention of poor quality requires the business to inject some resources such that a quality product is produced. Prevention costs include cost of design, implementation and maintenance of the quality management system. Appraisal is the evaluation done after the product has been made to check if the product conforms to quality requirements. Internal failures are the failures that are discovered before the product reaches the customers. These failures are usually identified by the quality control department. Faults or defects identified at this stage are easy to deal with as they have less chance of ruining the company's reputation than those identified by the external customers. If the failure is identified after the product has left the company then it is an external failure. The customer can put their complaint on public platforms such as Facebook, Twitter, WhatsApp among others. Once the complaint has gone public chances of it going viral would be high and damage control would be very difficult. It is therefore important for agricultural businesses to have good quality control systems so as prevent the external failures that could affect the organization's reputation.

Control of cycle time is there to reduce or minimize the time that is taken to complete a production cycle. Effectiveness and efficiency are the measurements used for the control of cycle. The time taken to deliver the service from the time an order is raised can be used as the cycle time for the service industry. The control of productivity measures the inputs against outputs.

The productivity measures include sales per employee, labour costs as a percentage of sales, and labour cost per hour.

10.13   Summary

This chapter covered the definition of financial management, sources of finance, budgeting, cashflow management, inventory management, agricultural marketing, agribusiness tax planning, records management, and business financial reviews. Agricultural businesses that employ good financial management are bound to be viable even under difficult conditions. Activities that promote business growth and minimization of losses are to be promoted for business development.

References

García-Teruel, P.J. and Martínez-Solano, P. (2007) Effects of working capital management on SME profitability. *International Journal of Managerial Finance* 3(2), 164–177. DOI: 10.1108/17439130710738718.

Griffin, R.W. (2021) *Management*, 13th edn. Cengage, Beijing, China.

Higgins, R. (2012) *Analysis for Financial Management*, 10th edn. McGraw-Hill, New York, NY.

Ibrahim, M.M. (2019) Designing zero-based budgeting for public organizations. *Problems and Perspectives in Management* 17, 323–333. DOI: 10.21511/ppm.17(2).2019.25.

Kumar, R. (2016) Economic order quantity (EOQ) model. *Global Journal of Finance and Economic Management* 5, 1–5.

Nazir, M.S. and Afza, T. (2009) Working capital requirements and the determining factors in Pakistan. *The Icfai Journal of Applied Finance* 15, 28–38.

Oakland, J.S. (2004) *Oakland on Quality Management*. Routledge, London, UK.

Oakland, S.R. (2014) *Total Quality Management. Text With Cases*, 3rd edn. Butterworth-Heinemann, Burlington, MA.

Pandey, M. (2019) Study on cash flow management: with reference to bharat heavy electrical limited, Haridwar. In: *Proceedings of 10th International Conference on Digital Strategies for Organizational Success*, Prestige Institute of Management, Madhya Pradesh, India. DOI: 10.2139/ssrn.3323743.

Patrutiu-Baltes, L. (2016) Inbound marketing: the most important Digital marketing strategy. *Bulletin of the Transilvania University of Brasov. Series V: Economic Sciences* 9, 61–68.

Sarbam (2020) Income Tax Act Chapter 23:06. Available at: https://www.law.co.zw/download/income-tax-act-chapter-2306/ (accessed 14 February 2023).

Schermerhorn Jr, J.R., Bachrach, D.G. and Wright, B. (2020) *Management*. John Wiley & Sons.

Singh, J.P. and Pandey, S. (2008) Impact of working capital management in the profitability of Hindalco Industries limited. *The Icfai University Journal of Financial Economics* 6, 62–72.

Tapera, M. and Majachani, A.F. (2017) *Unpacking Tax Law & Practice in Zimbabwe*. Matrix Tax School, Harare, Zimbabwe.

Thilmany, D., Bauman, A., Hadrich, J., Jablonski, B.B.R. and Sullins, M. (2021) Unique financing strategies among beginning farmers and ranchers: differences among multigenerational and beginning operations. *Agricultural Finance Review* 82(2), 285–309. DOI: 10.1108/AFR-05-2021-0070.

Wood, F. and Sangster, A. (2005) *Business Accounting*, 10th edn. Prentice Hall, New Delhi.

11 Investment Analysis in Agribusiness, Mitigating Capital Risk

Basil Shumbanhete[1]* and Dumisani Rumbidzai Muzira[2]

[1]Marondera University of Agricultural Sciences and Technology (MUAST), Zimbabwe; [2]Africa University (AU), Mutare, Zimbabwe

Abstract

The purpose of this chapter is to address issues on agribusiness investment analysis. The chapter will address recent developments in investment analysis and their relevance in agribusiness investment. Agribusiness investment may be investment in a biofuel plant, an animal or crop research facility, a new process or production method, or investigating new uses for existing products hence a new market or novel product development. The rationale for this chapter arises from the need to modernize farm activities through investment leading to increased output to address growing population and scarcity. To grow the wealth to meet current and future needs thus requires informed investment decisions, which lead to efficacy in both the project appraisal techniques to be applied and proposal creation leading to the achievement of the main objective for investors in an economy, especially an agro-based economy like Zimbabwe. We extend our discussion to consider the poor appraisal techniques set within the framework of an investment process, which do not ask the right questions and inevitably provide erroneous conclusions that will inexorably lead to adverse investment choices that destroy wealth. With most capital investment decisions, it is difficult to back out once we commit. The chapter covers the definition of investment analysis, recent developments in investment analysis and their relevance to agriculture, the payback method, the internal rate of return, the net present value (NPV), the modified NPV and the real options analysis, among others.

Keywords: Agriculture investment analysis, Interest rates, Weighted average cost of capital, Inflation, Net present value, Internal rate of return

11.1 Introduction

Due to climate change, conflicts, and the knowledge gaps compounded by global price shocks, farmers are realizing that meeting the demand for food locally, nationally, regionally and globally is becoming more and more difficult. Yields are going down as the available seed varieties cannot withstand the changing climate. Pests are mutating, becoming resistant to the known available control measures. Monoculture, which is driven by comparative advantages and the pursuit of economies of scale, has rendered land unfertile, and soils exhausted. The best and possibly the only way to counter the effects of the challenges cited is expanded global investment in agricultural research and development especially through collaborative efforts such as public–private partnerships (PPPs) (Gaffney *et al.*, 2019). Increasing agricultural produce with the current land

---
*Corresponding author: bshumbanhete@gmail.com

DOI: 10.1079/9781800622548.0011

available needs new methods, new processes and new technologies, all which require an informed investment drive. Investment decisions, therefore, take centre stage in agricultural initiatives and innovations that are meant to improve farming, enabling farmers to cope with increasing demands from the expanding population and industry. Investment decision making is a challenging activity for investors especially in the dynamic environment with multidimensional alternatives (Farooq and Sajid, 2015). Investment, simply stated, is putting money into a venture with the hope of a return. Investment analysis on the other hand refers to the evaluation of investments, for example investing in a crop or animal research facility, the planting of an orchard or construction of an ethanol plant, to determine the profitability and risks attached. Investment analysis is complex due to the uncertainty attached to the future. Capital projects present numerous variables as well as possible outcomes. Cash flows must be estimated, for example working capital requirements, project risk, tax considerations, expected inflation rates and disposal value. There is a need to understand the existing agricultural commodity markets to estimate project revenues, assess the competitive impacts of the undertaking, and determine the lifecycle of the project. According to Stelling *et al.* (2018) cash flow in finance plays an important role as an analysis of the financial feasibility of the investment plan as measured by the current cash flows. Investment analysis has a profound effect on the value of the agricultural concern, profitability, the value of the market share under its control, and the main objective of every undertaking is shareholder value maximization (Al-Ani, 2015). Investment analysis is the process of evaluating proposed or available investments for income, risk and resale value. Investment analysis methods calculate expected returns on proposed projects based on cashflow forecasts of many often inter-related project variables. Risk, cash flows and the resale value are the three aspects that most investment analysis methods consider. Risk is important in investment as it can take prospective returns from a lucrative amount to zero if not meticulously considered in investment; decision making emanates from the uncertainty encompassing the earlier alluded-to project variables (Savvides, 1994). When investors are risk averse they face more gains, when investors take more risks they face losses (Farooq

and Sajid, 2015). Which venture to pursue given the scarce capital resources is the major question for all farmers subsistent to commercial. The investments will differ in terms of cost, duration, returns and risks attached. Investors therefore need to understand what they are investing in otherwise it is mere speculation, gambling something akin to Russian roulette.

11.2 Methods Used to Analyse Investment

11.2.1 Accounting rate of return (ARR)

In accepting a project the farmer may consider the rate of return on the money invested. The ARR may be defined as follows:

$$ARR = \frac{\text{Average incremental net income}}{\text{Average investment or initial investment}}$$

The average incremental net income is found by dividing the total net income of an investment by its economic useful life. The net income is income after deducting depreciation and other non-cashflow expenses.

The following scenario relates to project A. The project cost is $2.2 million and is expected to have no residual value after 5 years. The cash flows for each year are as in Table 11.1.

Average net income = $(11,360 + 32,600 + 32,600 + 32,600 + 32,600) / 5 = 28,352$

Average investment = $(220,000 - 0)/2 = 110,000$.

$$ARR = \frac{28,352}{110,000}$$

$$= 0.2577$$

$$= 25.28\%$$

Accounting rate of return method makes use of information garnered from financial statements especially from the income statement and the statement of financial position to assess the feasibility of a proposed capital investment. The ARR divides the average income after taxes by average capital, that is average book value of invested capital after allowing for depreciation (Nyarombe and Gwaro, 2015). It is imperative to recognize that for analysis purposes, an investment that yields a return lower than the

Table 11.1. Project A scenario (values are in $).

Year	0	1	2	3	4	5
Cost	220,000					
Cash flows		60,000	90,000	90,000	90,000	90,000
Depreciation		44,000	44,000	44,000	44,000	44,000
Net income		16,000	46,000	46,000	46,000	46,000
Taxation (at 29%)		4640	13,340	13,340	13,340	13,340
Net income after tax	220,000	11,360	32,660	32,660	32,660	32,600
Net book value	220,000	176,000	132,000	88,000	44,000	0

prevailing interest rates being offered by the banks is not worthwhile as it would be rather prudent to deposit the money and earn interest than to invest in a risky venture. When using ARR the profit used should be profit after tax (Nyarombe and Gwaro, 2015). ARR is likely to be popular with the smallholder farmers who cannot afford financial advisers as it is simple to understand and easy to apply. The data that is used to compute the ARR is easily accessed as it is the farmer's own accounting data. It uses the entire return from a given capital; it gives a fairly accurate picture of the going concern status or viability of the undertaking and does not require the use of intricate computers to do the calculations.

However, it has the following disadvantages:

- It ignores the time value of money.
- It is not universally accepted.
- There are many ways of calculating ARR, it gives an alternative (there are two ways of calculating ARR one using average investment and the other using initial outlay only).
- It is highly subjective.
- It uses profits unlike all the other methods, which use cashflows.
- It ignores the fact that intermediary profits can be reinvested to generate the company extra return.

11.2.2 Profitability index (PI)

Sometimes the farmer has a choice to make among projects and needs an objective and efficient way to select the most profitable alternative. The profitability index provides a means for ranking competing projects (Thomas, 2017). The profit is the net present value of cashflows occurring in time periods after the investment divided by the initial net cashflows (initial investment).

The formula is:

$$\text{Profitability index} = \frac{\text{Present value of future cash flows}}{\text{Present value of cost of investment}}$$

For example, a farmer may be considering an investment with an initial cost of $10,000 with cash inflows of $8500 and $7500 after the second year. Using a 10% discount rate, the profitability index is calculated by summing up the discounted future cashflows, that is the $8500 and $7500, and dividing by the initial cash flows:

$$\text{PI} = \frac{(\$8\,500*0.909) + (\$7\,500*0.826)}{\$10\,000}$$

PI = 1.39

If the profitability index is greater than one then generally the investment is considered economical (Gurau, 2012). Higher values are preferred as they represent better investments. This method, like the payback discussed in the next section, is useful when comparing projects under conditions for capital rationing due to limited finances for project investments. The profitability index is best used as a complementary method to the net present value (NPV).

Gurau (2012) details the merits and demerits of the profitability index for an enterprise as follows:

- The profitability index tells about an investment increasing or decreasing the firm's value.

- The profitability index takes the time value of money into consideration.
- The profitability index is also helpful in ranking and picking projects while rationing capital.
- The profitability index also considers the risk involved in future cashflows with the help of the cost of capital.
- The profitability index takes into consideration all cashflows of the project.

In addition to the aforesaid advantages, there are also certain disadvantages featured by the profitability index:

- An estimate about the cost of capital is required so as to calculate the profitability index of an enterprise.
- The profitability index of an enterprise might not, sometimes, provide the correct decision while being used to compare mutually exclusive projects under consideration.

Thomas (2017) adds that the drawbacks of this method are it does not reveal project size and does not provide sound decision making in the case of capital rationing over multiple time periods.

11.2.3 Payback

Research suggests that companies in the late 19th century did not do comprehensive investment appraisals, although some used the payback technique – along with gut feeling – to decide which projects to pursue (Steven, 2019). The payback method of investment analysis is one of the most simple methods a farmer can employ in capital budgeting. It is easy to calculate as well as understand. Despite almost unanimous agreement by the theorists that payback has little value in measuring profitability (Rappaport, 1965), however, it remains the most popular method of capital budgeting.

Payback period is the period or period of time required to be able to recoup the expenditures on investments made through the profits derived from a project that has been run or operated, according to Botchkarev (cited in Stelling et al., 2018). The payback method measures how long it will take for an investor to recoup an outlay on a project. The farmer would want to know how long before he can enjoy benefits from his macadamia or avocado plantation or alternatively how long it will take for the abattoir to recoup the money spent on constructing and fitting it. The theory is that after the original cost of investment is recuperated the remaining cashflow goes towards creating shareholder value (Alkaraan and Northcott, 2006). In agriculture, payback makes a difference in whether the farming activities are sustainable or not as the variables are in a state of flux and the level of uncertainty is high where weather is concerned. However, when the firm opts to use this rule, they must decide an appropriate cut-off date, as the rule will ignore the cashflow after that date (Dai et al., 2022). This can be a disadvantage when the project being assessed gives higher payoffs later in its life cycle. The payback period can also be used as the basis for accepting or rejecting projects in the case that the payback period is higher or lower than a certain number of years previously defined to differentiate between projects (Al-Ani, 2015). Farm managers using the payback do not usually take other strategic variables into account when they evaluate capital budgeting decision. The strategic variables in question here will include profitability and profit maximization, size of the investment vs the business, the level of uncertainty and risk attached, possible market obstacles, and the timing of the compensation due to management. The technique enables the manager to at least consider risk of investment by examining how long it will take to recover the cost of investment (Al-Ani, 2015). Table 11.2 shows the cash flows of two alternative projects. The two investment projects

Table 11.2. Project A and B scenario. Projects A and B both cost $12,000 and have the following cash flow streams.

Year	0	1	2	3	4
Project A	–$12,000	$4000	$6000	$4000	$2000
Project B	–$12,000	$2000	$4000	$4000	$8000

have the same capital outlay but the difference is in the cash flows and their timing. Investments with a longer payback introduce a higher level of risk as there can be a multitude of factors that may affect the project cashflows in the intervening period. However, the other strategic variables are not ideally catered for. The payback period is especially suitable for short-term focused investors interested in generating liquidity. This links with the pecking order theory where investors are interested in projects that create immediate liquidity.

Project A has a payback of 2.5 years (2 years 6 months), and Project B has a payback period of 3.25 years (3 years 3 months). A farmer may decide that any project he will accept must have a payback period of no more than 3 years. Project A would be accepted, while Project B would be rejected because its payback period is longer than 3 years. Project A allows the farmer to recover $10,000 within two years and the initial investment is recouped within year 3. Project B allows the farmer to recover $10,000 within 3 years and payback occurs three months into the fourth year.

The major weaknesses cited for payback by most authors is its failure to take into consideration cashflows after the payback period and the fact that it ignores the time value of money. It has been suggested that the latter can be resolved by introducing a modified payback period, the discounted payback period (DPB) (Azar and Noueihed, 2014), thereby searching the payback period when the accumulated present value of cashflows covers the initial investment outlay (Yard, 2000). Payback period and the discounted payback period are often used for small investment decisions, for example tuning up a tractor to save fuel (Thomas, 2017). Its use as a single criterion seems to have decreased over time, but it is still commonly used as a

secondary measure (Yard, 2000). As far as large investments are concerned the payback ought to be used as a supplementary criterion to the net present value method.

11.2.4 Discounted cashflow methods

When the farmer evaluates the financial feasibility of an investment, the time value of money is an essential consideration. This becomes critical when a project involves cashflow patterns that extend over a number of years. In order to discount all cashflows, an interest rate must be assumed for the period in question. Frequently the desired rate of return on investments by the farmer is used. Sometimes this is a reflection of the known rate of return, which can be earned on alternative investments like bonds and other money market instruments (Nábrádi and Szöllősi, 2007). Helfert (2001) writes that this rate is mostly based on a company's weighted average cost of capital, which embodies the return expectations of capital structure. In other words the WACC calculation takes into account the interest peculiar to each source of funds as in Table 11.3. According to Nábrádi and Szöllősi (2007) it should be the rate of return an investor normally enjoys from investments of a similar nature and risk.

The rate, weighted average cost of capital (WACC) is calculated as follows.

A farmer's sources of finance are:
Ordinary share capital 2,700,000
8% preference shares 1,350,000
10% debentures 450,000
The shareholders expect a dividend of 7% p.a

Debenture interest is an allowable expense for tax purposes. This results in a reduction

Table 11.3. Calculation of WACC.

	Cost (%)	Amount ($)	Weight	Cost × Weight
Ordinary shares	7	2700	60	420
Preference shares	8	1350	30	240
Debentures	7.5	450	10	75
		4500	100	735

in the tax bill and this must be shown in the calculation of the cost of capital supplied by debentures. Corporation tax is 25%. Cost of debentures therefore is 10% (1−0.25) = 10% × 0.75 =7.5%

The weighted cost of capital is:

$$\text{WACC} = \frac{735}{100} = 7.35\%$$

11.2.5 Discounted payback

The discounted payback period (DPP) is the length of time it takes a project's incremental cashflows discounted at the 'opportunity investment rate' to accumulate to investment outlay (Azar and Noueihed, 2014). The only limitation DPP has, just like the simple payback method, is that it ignores cash flows beyond the DPP, but it does so only after assuring the profitability of the project. Bhandari (1986, cited in Azar and Noueihed, 2014, p. 22) writes that 'DPP is closely related to Net Present Value, Internal Rate of Return, and Profit Index. For small business, survival and efficacy in using capital are more in line with the DPP than NPV and internal rate of return (IRR)'. The discounted payback is therefore a compatible investment analysis tool for communal farmers who usually operate on a micro basis. Not only is the method simple to apply but it takes cognisance of the time value of money notion unlike the simple payback period. The discounted payback method sums up discounted cash inflows until they equal the discounted cash outflows, the result being NPV = 0. DPP approximates the internal rate of return in that IRR is the discount rate that makes the benefits equal to the costs until the NPV is equal to zero. However, the DPP only considers the discounted cashflows needed to make the NPV = 0 while IRR considers all the cashflows. The discounted payback method takes into account the time value of money. Taking project A by the farmer and using 15% as the farmer's preferred cost of capital (WACC) the calculation of the NPV will be done as in Table 11.4.

The discounted payback period is just over 3 years. This compares with the payback period of 2.5 years. Therefore, taking the time value of money into account, it will take 3 years for the project to reach a position where NPV is equal to zero.

Table 11.4. Calculation of NPV.

	Cashflow ($)	PV factor (WACC = 15%)	Present value ($)
Cost	−12,000	1.000	−12,000
Year 1	4000	0.8696	3478
Year 2	6000	0.7561	4537
Year 3	6000	0.6575	3945
Net present value			40

11.2.6 Net present value (NPV)

Net present value is the difference between the present value of all cash inflows and the present value of all cash outflows over the period of the investment (Arnold and Nixon, 2013). Present value refers to today's value of an amount of money that will be received or paid in the future (Cruz and Singerman, 2019). A rational person would prefer money in hand than money to be collected later on. Money in hand can be used to generate more money in business or be put in the bank and earn interest. Future money is less valuable because inflation erodes its buying power (Stelling et al., 2018). So the money value now is called the present value, so we obtain the present value of the future cashflow then decide whether a project is worth investing in (Dai et al., 2022). Net present value, which accounts for the time value of money, is a common metric for examining an investment, and is considered a superior method over other approaches (Thomas, 2017). The present value of the money is calculated using the cost of capital, discount rate which is also called the hurdle rate, and cutoff rate. The NPV analysis accounts for risk by factoring the risk into the discount rate and applying the discount rate to the cashflows to calculate the risk-adjusted performance measures. The cost of capital is the rate of return demanded by those who supplied capital for the investment, i.e., the capital suppliers (Jory et al., 2016). New agricultural technologies offer different benefits, including reduced costs or increased revenue. In order to estimate the net present value, it might be necessary to forecast any increased sales to estimate additional revenue due to adopting a new agricultural technology

(Thomas, 2017). It is also imperative that incremental costs of production are included as well as costs incurred and benefits accrued due to the investment including costs that would be incurred without the investment in the new technology. The simple decision rule is to accept projects with a positive present value and reject those with a negative net present value (Lucey, 1996). However, interpreting net present value is at times difficult, if net present value is positive, it means that the return on the investment is expected to exceed the discount rate. An anticipated follow-up question is what the rate of return is on the investment (Thomas, 2017).

The formula:

$$\text{NPV} = \sum_{t=1}^{n} \frac{C_t}{(1+k)^t} - 1$$

where C_t = net cash flow at time t
k = cost of capital

The cashflow stream for Project C is shown in Table 11.5.

The cost of the investment is $12,000 and this will result in cash flows of $8000 in year one and $6000 in years two and three. The firm's cost of capital is 10%.

The cash receipts will be received in the future, so to find their present values they have to be discounted back. Table 11.6 show how the farmer can determine the net present value of the proposed project.

Table 11.5. Project C cash flow stream.

Year	0	1	2	3
Cashflow	12,000	8000	6000	6000

Table 11.6. Determination of the net present value of the proposed project.

	Cash flow ($)	PV factor (15%)	Present value ($)
Cost	12,000	1	−15,000
Year 1	8000	0.8696	6957
Year 2	6000	0.7561	4537
Year 3	6000	0.6575	3945
NPV			439

11.2.7   Internal rate of return (IRR)

The internal rate of return is the discount rate that sets the NPV to zero and can be viewed as the 'true' return of the project (Arnold and Nixon, 2013). For an investment to be worth putting money into, the rate of return on the money to be invested must be higher than the opportunity cost of capital. The IRR is the maximum opportunity cost of the capital that can be accepted by the investor (Dai et al., 2022). One of the benefits of using the internal rate of return is that there is no need to select a discount rate generally. If the internal rate of return is calculated to be greater than or equal to your minimum required rate of return to make an investment (e.g. discount rate or hurdle rate), then the investment is economic (Thomas, 2017). The simple decision rule is to accept projects with an IRR that's greater than the cost of capital and reject those with an IRR that's less than the cost of capital (Steven, 2019). An IRR that is less than the cost of capital means that the project does not earn its cost of capital hence the grounds for rejection. If the investment's IRR is equal to the cost of capital, the firm should be indifferent about accepting or rejecting the project (Jiang, 2016). IRR in application is one of the more demanding methods, the calculation of which requires certain mathematical knowledge, especially since its calculation is made by trial and error or modern approach using software such as MS Excel (Miletić and Latinac, undated). IRR is also called the marginal efficiency of capital or yield on the investment. In analysing an investment, the IRR is taken to be a measure of the profitability of its expected cashflow before taking into consideration the investment's cost of capital.

The formula:

$$\text{IRR} = \sum_{t=i}^{n} \frac{C_t}{(1+r)^t} - 1 = 0$$

where r = internal rate of return
t = the number of time periods

Say the farmer wants to calculate the IRR of Project X. At a discount rate of 15%, the NPV is $439. To find the IRR we have to reduce the NPV to zero. It is therefore necessary to increase the discount rate to reduce the NPV. In this case the farmer may try 25% as illustrated in Table 11.7.

Table 11.7. Calculation of the IRR.

	Cashflow ($)	PV factor (25%)	Present value ($)
Cost	12,000	1	−15,000
Year 1	8000	0.8	6400
Year 2	6000	0.64	3840
Year 3	6000	0.512	3072
Net present value			−2048

Taking the two NPVs, that is $439 at 10% and −$2048 at 20%, we can also find the IRR using the formula below:

$$IRR = X + \frac{pq*ac}{ad}$$

where X = the rate giving the positive NPV, in this case 10%

pq = the distance between the two rates used to give the NPVs

ac = the positive NPV

ad = the positive NPV + the negative NPV (Randall, 1996)

Therefore:

$$IRR = 10\% + \left(10\% \times \frac{439}{439+2048}\right) = 11.76\%$$

Unfortunately, according to Magni (2010) the IRR has serious flaws, among others:

1. Multiple real-valued IRRs may arise.
2. Complex-valued IRRs may arise.
3. The IRR is, in general, incompatible with the NPV in accept/reject decisions.
4. The IRR ranking is, in general, different from the NPV ranking.
5. The IRR criterion is not applicable with variable costs of capital.
6. It does not measure the return on initial investment.
7. It does not signal the loss of the entire capital.
8. It is not capable of measuring the rate of return of an arbitrage strategy.

The internal rate of return also does not reveal the duration of the investment. These flaws have led practitioners and academics on a quest for a solution to these shortcomings. The

search has resulted in the ushering in of the average internal rate of return (AIRR) method.

11.2.8 The modified internal rate of return (MIRR)

The MIRR of an alternative is the discount rate at which the present value is exactly equal to the future value. The MIRR may or may not be a prominent method used for economic decision making; however, given the prominence of the internal rate of return and the many shortcomings of this metric, it is prudent to discuss the modified internal rate of return (Thomas, 2017). The MIRR is more complex to calculate but the advantage it has is that it avoids many of the demerits of the IRR. Xie and Chen (2021) highlighted the main demerits addressed by the MIRR:

- First the implicit reinvestment rate assumption of IRR is unreasonable. The IRR method assumes that net cashflows can reinvest and reinvestment rate of cashflows is the IRR. The assumption is usually unrealistic. IRR can be higher than the reinvestment rate of return or IRR can be lower than it. In practice, if IRR is higher than the reinvestment rate of return, the assumption is invalid.
- Secondly the calculation of the IRR is difficult. In general, it cannot calculate directly, and often needs an iterative trial-and-error procedure.
- Thirdly there is the problem of multiple or nonexistent internal rates of return for unconventional cashflow alternatives. If sign of cashflows change more than once, an alternative may have several IRRs. The IRR method is only suitable for the conventional cashflow alternatives. When an alternative may have several IRRs, which IRR is reasonable? In this case the IRR method is commonly believed invalid or meaningless. So it is universally regarded as its fatal flaw.

The MIRR is determined by calculating the rate that causes the present value of the terminal value of the project's inflows equivalent to

Table 11.8. Project A and B MIRR.

Project	0	1	2	3
A	−100	20	20	110
B	−100	60	50	30

the present value of the project's cash outflows (Correia *et al.*, 2015). The terminal value is determined by reinvesting the cash flows at the cost of capital or any other specified rate as illustrated in Table 11.8.

The formula:

$$\left(1 + \text{MIRR}\right) = \left[\frac{\sum\limits_{t=0}^{N} \text{CIF}_t * \left(1 + k\right)^{N-t}}{\sum\limits_{t=0}^{N} \frac{\text{COF}_t}{\left(1+k\right)^t}}\right]^{\frac{1}{N}}$$

where COF_t = the firm's cash outflows at time period t

CIF_t = the firm's cash inflows at time period t

k = the firm's available investment rate or rate of return, and

MIRR = the firm's modified rate of return. Such that,

$$\left(1 + \text{MIRR}\right) = \left(\frac{\text{TV}}{\text{PV Costs}}\right)^{\frac{1}{N}}$$

where TV = the terminal value of the firm's benefits. Example worked out by Correia *et al.* (2007).

The internal rates of return for project A are 17 and 21% for project B. However, if future cashflows can be reinvested only at cost of capital, then this results in a reduction in the value of project B. If the cashflows are reinvested at a cost of 12%, then the terminal value of inflows of project A and project B will be $157.49 million and $161.26 million respectively. The MIRR is calculated by applying the internal rate of return method to the terminal values of the inflows, and the internal rates of return of each project are now calculated at 16.3% for project A and 17.3% for project B. This allows the investor farmer to specify the reinvestment rate at the cost of capital and still evaluate projects with the modified internal rate

of return rather than be restricted to the use of the NPV.

11.2.9 The average internal rate of return and the real options analysis

The AIRR and the options analysis are recent methods in capital investment analysis and because of the advanced nature of these two techniques we will not delve deep into their complexities but will attempt to simply introduce you to their method. Nevertheless, what is important to take away from the following paragraphs is that students of agribusiness are aware of the existence of these techniques and may request them of experts they might engage to advise them on project financing.

11.2.10 The average internal rate of return (AIRR)

An alternative method that can be used in farm investment analysis is the average internal return. The average internal rate of return (AIRR), introduced by Magni (2010), fixes many deficiencies associated with the traditional IRR, including apparent inconsistency with NPV (Hazen and Magni, 2021). Therefore AIRR is a new approach for financial modelling. In particular, it enables measuring the economic profitability of any project and assessing shareholders value creation while extracting several of pieces of economic information. The approach as noted above avoids the usual pitfalls associated with IRR and enriches the traditional NPV analysis. According to Marchioni and Magni (2018), it has the following features:

- It guarantees existence and uniqueness of the rate of return.
- It provides both the project rate of return (project AIRR or average return on investment) and the equity rate of return (equity AIRR or average return on equity).
- It leads to economically rational decisions.
- It is consistent with the NPV and decomposes the latter into project scale and the project's economic efficiency.

- It is consistent with the input data and the estimates on incremental revenues and costs.
- It copes easily with time-varying cost of capital.
- It may be applied to financial assets as well as real assets.

The formula for AIRR is as follows.

The project AIRR is defined as the ratio of the total profit to total invested capital (Magni, 2010, eq. 7):

$$i_A = \frac{PV(I/r)}{PV(C/r)} (1 + r).$$

where i_A = average internal rate of return
PV = present value
I = income
C = capital amounts
r = cost of capital.

The AIRR has a straightforward economic interpretation, for it is, literally, a ratio of total profit earned to total capital invested and, at the same time, a (capital weighted) average of the one-period returns on investment (ROI) (Hazen and Magni, 2021). Any sequence of capitals (capital stream) univocally determines a sequence of one-period IRRs (internal return vector) (Magni, 2010). According to Magni (2010) a corresponding arithmetic mean is shown to represent an unfailing economic yield, here named 'average internal rate of return' (AIRR). The AIRR may then be interpreted as the unique real-valued rate of return on the capital invested in the project (Hazen and Magni, 2021).

11.3 Real Options Analysis

Capital budgeting, investment analysis over the years has mostly relied on the payback method and discounted cashflow techniques such as the net present value method and the internal rate of return method discussed above. The DCF approach as illustrated earlier determines the net present value of the project by discounting the future cash flows using a risk-adjusted discount rate, which is usually the weighted average cost of capital (WACC), that is if the current venture and the new project are being undertaken within the same business environment facing the same risks. However, Guthrie

(2009, cited in Mashamba, 2016) posits that these techniques do not incorporate managerial flexibility. This flexibility in capital budgeting and decision making is known as 'real options' (Mashamba, 2016). There has been a growing use of real options analysis as a tool for valuation and strategic decision making in the recent past (Mbuthia, undated). Mason and Merton (1985) cited in Crouhy *et al.* (2019) define project flexibility as a 'description of options made to management as part of the project'. The concept of a real option was developed, at least partly, as a response to the inadequacy of the traditional DCF approaches for the valuation of projects under uncertainty (Zhang, 2011). This flexibility, as noted by Muharam (2010, cited in Mashamba, 2016), comes in different forms mainly: deferral/waiting options, altering options, switching options, growth/expansion options, abandoning or shutting down options. These options present the farmer with more opportunities for building value within investments. According to Evans (undated), three common sources of options are:

1. Timing options: the ability to delay investment in the project.
2. Abandonment options: the ability to abandon or get out of a project that has gone bad.
3. Growth options: the ability of a project to provide long-term growth despite negative values.

For example, a new research programme (research and development) may appear negative but it might eventually lead to new product innovations and market growth which contributes to the sustainability drive. Real options 'in' projects are options created by changing the project variables (de Neufville *et al.*, 2006).

There is also a need to consider the growth options of the project as well, this being targeted towards spreading the benefits further. Option pricing is the additional value that we recognize within a project because it has flexibilities over similar projects. These flexibilities help the farmer to manage capital projects therefore failure to recognize option values can result in an undervaluation of a project's flexibilities over similar projects. According to Amram and Kulatilaka (1998) viewing investment

Table 11.9. Real options.

Present value of cashflows < Salvage value	Abandon the project
Present value of cashflows > Salvage value	Do not abandon the project

opportunities as options on real assets provides valuable insights and may challenge popular beliefs, one of which is the perception that uncertainty is bad. Real option analysis pays careful attention to market opportunities related to the investment at hand (Mbuthia, undated). In a scenario where the farmer is looking at two options, that is expand or abandon the project, the decisions will be made as follows. For an expansion option, the real option theory states that if:

Option value > Current NPV Exercise the option

Option value < Current NPV Do not exercise the option

Thus the farmer under such circumstances will only take up the project expansion option if it will result in an increased total net present value. However, if the project under consideration will result in a net present value lower than the one being currently realized the farmer has no reason to expand the project.

If the farmer has to decide on continuing with the project or abandoning it, the real option theory is shown in Table 11.9.

The calculations involved in the real options analysis are complex and tedious making the method unattractive to investors. However where one can engage a consultant the real options analysis is an alternative way to look at the investment.

11.4 Conclusion

It is paramount for companies rationing their capital to use different models to select projects

(Nyarombe and Gwaro, 2015). What is important is to ensure that an objective decision is made. This may require the application of more than one method. However, the factors to consider in an investment analysis are size, cost, risk and the expected returns. These factors will also assist in determining the methods of analysis to apply. The importance of informed investment decisions is that risk of loss is reduced and the investment is sustainable. Lack of these investment decision making tools will result in investors facing a myriad of uncertainties without any mitigating measures.

We therefore suggest that for the decision to be well informed the investor should incorporate methods that go beyond number crunching and consider other measures that deal with risk and uncertainty. These include sensitivity analysis, which asks a number of 'what if' questions. These questions address possible changes in the quantifiable and non-quantifiable project variables and attempt to forecast outcomes should such changes occur. Scenario analysis according to Brealey *et al.* (2006) is project analysis giving a certain combination of assumptions allowing managers to look at different but consistent combinations of variables. Simulation analysis according to Ross *et al.* (2005) estimates the probabilities of different possible outcomes, for example the Monte Carlo simulation. Lucey (1996) suggests the use of the break-even analysis pre-investment to help set activity targets and check on whether the targets are attainable. In making the final decision the investor must also consider the qualitative side of decision making. There are factors that cannot be quantified within the economic environment, which are however as important in decision making.

References

Al-Ani, M.K. (2015) A strategic framework to use payback period in evaluating the capital budgeting in energy and oil and gas sectors in Oman. *International Journal of Economics and Financial Issues* 5, 469–475.

Alkaraan, F. and Northcott, D. (2006) Strategic capital investment decision-making: a role for emergent analysis tools? *The British Accounting Review* 38(2), 149–173. DOI: 10.1016/j.bar.2005.10.003.

Amram, M. and Kulatilaka, N. (1998) *Real Options: Managing Strategic Investment in an Uncertain World*. Harvard Business School Press.

Arnold, T. and Nixon, T. (2013) Measuring investment value: free cash flow, net present value, and economic value added. In: Baker, H.K. and English, P. (eds) *Capital Budgeting Valuation*. John Wiley & Sons, Hoboken, NJ, USA, pp. 57–77. DOI: 10.1002/9781118258422.

Azar, S.A. and Noueihed, N. (2014) The discounted payback in investment appraisal: a case study. *International Journal of Business Administration* 5(5), 58. DOI: 10.5430/ijba.v5n5p58.

Brealey, R.A., Myers, S.C. and Allen, F. (2006) *Corporate Finance*, 8th edn. McGraw-Hill.

Correia, C., Flynn, D.K., Uliana, E. and Wormald, M. (2007) *Financial Management*, 4th edn. Juta Publishers, Cape Town.

Correia, C., Flynn, D., Uliana, E., Wormald, M. and Dillon, M. (2015) *Financial Management*, 8th edn. Juta Publishers, Cape Town.

Crouhy, M., Galai, D. and Wiener, Z. (2019) *World Scientific Reference on Contingent Claims Analysis in Corporate Finance*, Foundations of CCA and Equity Valuation, Vol. 1. World Scientific Publishing Company. DOI: 10.1142/9857-vol1.

Cruz, J. and Singerman, A. (2019) Understanding investment analysis for farm management. *EDIS* 2019, 4. DOI: 10.32473/edis-fe1060-2019.

Dai, H., Li, N., Wang, Y. and Zhao, X. (2022) The analysis of three main investment criteria: NPV IRR and payback period. In: *2022 7th International Conference on Financial Innovation and Economic Development (ICFIED 2022)*, Zhuhai, China. DOI: 10.2991/aebmr.k.220307.028.

de Neufville, R., Asce, L., Scholtes, S. and Wang, T. (2006) Real options by spreadsheet: parking garage case example. *Journal of Infrastructure Systems* 12(2), 107–111. DOI: 10.1061/(ASCE)1076-0342(2006)12:2(107).

Evans, M.H. (undated) Course 3: capital budgeting analysis 19.

Farooq, A. and Sajid, M. (2015) Factors affecting investment decision making: evidence from equity fund managers and individual investors in Pakistan. *Research Journal of Finance and Accounting* 6, 2015.

Gaffney, J., Challender, M., Califf, K. and Harden, K. (2019) Building bridges between agribusiness innovation and smallholder farmers: a review. *Global Food Security* 20, 60–65. DOI: 10.1016/j.gfs.2018.12.008.

Gurau, M.A. (2012) The use of profitability index in economic evaluation of industrial investment projects. *Proceedings in Manufacturing Systems* 7, 55–58.

Hazen, G. and Magni, C.A. (2021) Average internal rate of return for risky projects. *The Engineering Economist* 66(2), 90–120. DOI: 10.1080/0013791X.2021.1894284.

Helfert, E.R. (2001) *Financial Analysis: Tools and Techniques, A Guide for Managers*. McGraw-Hill.

Jiang, L. (2016) Application analysis on internal rate of return rule for investment decision. In: *Proceedings of the 2016 International Conference on Humanity, Education and Social Science*, 2016 International Conference on Humanity, Education and Social Science, Hong Kong, China. DOI: 10.2991/ichess-16.2016.106.

Jory, S.R., Benamraoui, A., Boojihawon, D.R. and Madichie, N. (2016) Net present value analysis and the wealth creation process: a case illustration. *The Accounting Educators' Journal* 26, 86–89.

Lucey, T. (1996) *Costing*. Bloomsbury.

Magni, C.A. (2010) Average internal rate of return and investment decisions: a new perspective. *The Engineering Economist* 55(2), 150–180. DOI: 10.1080/00137911003791856.

Marchioni, A. and Magni, C.A. (2018) Investment decisions and sensitivity analysis: NPV-consistency of rates of return. *European Journal of Operational Research* 268(1), 361–372. DOI: 10.1016/j.ejor.2018.01.007.

Mashamba, T. (2016) Real options valuation: the modern day technique in capital budgeting and decision-making? *Real Options Valuation* 4, 12.

Mason, S.P. and Merton, R.C. (1985) The Role of Contingent Claims Analysis in Corporate Finance. In: Altman, E.I. and Subrahmanyam, M.G. (eds) *Recent Advances in Corporate Finance*. Richard D. Irwin, Homewood, Ill, pp. 125–144.

Mbuthia, G. (undated) *Potential Applications of Real Option Analysis in the Construction Industry in Developing Countries*. University of Cape Town.

Miletić, M. and Latinac, D. (undated) Internal rate of return method - a commonly used method with few advantages and many disadvantages? In: *Conference Proceedings of the 4th Contemporary Issues in Economy & Technology*, CIET 2020.

Nábrádi, A. and Szőllősi, L. (2007) Key aspects of investment analysis. *Applied Studies in Agribusiness and Commerce* 1, 53–56. DOI: 10.19041/APSTRACT/2007/1/7.

Nyarombe, F. and Gwaro, S. (2015) An investigation of capital budgeting techniques on performance: a survey of selected companies in Eldoret town. *International Journal of Business and Management Invention* 4, 54–70.

Randall, H. (1996) *Advanced Level Accounting*, 3rd edn. Letts Educational, Chiswick, London.

Rappaport, A. (1965) Discounted payback period. *Management Services: A Magazine of Planning Systems, and Controls* 2, 4.

Ross, S.A., Westerfield, R.W. and Jaffe, J. (2005) *Corporate Finance*, International Edition. McGraw-Hill.

Savvides, S. (1994) Risk analysis in investment appraisal. *Project Appraisal* 9(1), 3–18. DOI: 10.1080/02688867.1994.9726923.

Stelling, S., Syah, T.Y.R., Indrawati, R. and Dewanto, D. (2018) Role of payback period, roi, and npv for investment in clinical health business. *International Advanced Research Journal in Science, Engineering and Technology* 5, 78–82.

Steven, G. (2019) Performance operations: investment appraisal methods. *Financial Management (UK)* 48–49.

Thomas, D.S. (2017) *Investment Analysis Methods: A Practitioner's Guide to Understanding the Basic Principles for Investment Decisions in Manufacturing*. National Institute of Standards and Technology. DOI: 10.6028/NIST.AMS.200-5.

Xie, M. and Chen, W.K. (2021) Modified internal rate of return. *Turkish Journal of Computer and Mathematics Education* 12, 4087–4090.

Yard, S. (2000) Developments of the payback method. *International Journal of Production Economics* 67(2), 155–167. DOI: 10.1016/S0925-5273(00)00003-7.

Zhang, Y. (2011) An analysis and comparison of real option approaches for project valuation under uncertainty. Thesis, Master of Commerce, University of Otago, New Zealand.

12 Blockchain Technology, Sustainability and Future of Public Input Distribution in Zimbabwe

Archeford Munyavhi*, Basil Shumbanhete, Tatenda Mapfumo and Luckmore Marodza

Marondera University of Agricultural Sciences and Technology (MUAST), Zimbabwe

Abstract

The aftermath of the agrarian reform in Zimbabwe has been characterized by a drastic decrease in agricultural production, plunging the nation into a state of food insecurity and net food importer. The government has made several efforts to restore the country's former glory through public inputs distribution particularly targeting maize production. These efforts have culminated in the introduction of Operation Maguta, the Presidential Input Scheme, Pfumvudza and the Command Maize Input Scheme with the sole aim of eliminating increased food import expenditure, introducing import substitution in light of comparative advantages, and improving food production as the country moves towards poverty alleviation among farmers and the country at large. However, since the agrarian reform, all the government's efforts in public input distribution have been characterized by lack of trust due to red tape and high levels of centralization, manipulation of financial records and lack of mutual trust among the responsible government officials and the citizens. This is threatening the sustainability of public input distribution as well as the effectiveness of government efforts to alleviate poverty among citizens and national food security. The incidents of corruption and mismanagement of public inputs distribution remain high and the perpetrators go unpunished because of the manipulation of the recording system being used. With the increasing call for agriculture 4.0 to meet the increasing demand for food to feed the growing national population, blockchain technology has been heralded as a tool in providing technological sustainable agricultural solutions for handling public input distribution and eradicating corruption. This chapter proffers vital insights that blockchain technology can reduce corruption in public input distribution by making it easier to track all financial government/public processes. The challenges to be addressed in this chapter are supply chain related in the distribution of agricultural inputs in Zimbabwe, where the inputs that are meant to drive agro-based production and be an impetus for development and poverty alleviation are being diverted to the black market, the proceeds of which are privately used.

Keywords: Sustainability, Public input, Blockchain technology

12.1 Introduction

In June 1992, at the Earth Summit in Rio de Janeiro, Brazil more than 178 countries adopted Agenda 21, a comprehensive plan of action to build a global partnership for sustainable development to improve human lives and protect the environment. The summit led to the elaboration

*Corresponding author: munyavhiarcheford@gmail.com

© CAB International 2023. *Sustainable Agricultural Marketing and Agribusiness Development: An African Perspective* (eds B. Nyagadza and T. Rukasha)
DOI: 10.1079/9781800622548.0012

of eight Millennium Development Goals (MDGs) to reduce poverty by 2015. 2015 was a landmark year for multilateralism and international policy shaping with the adoption of several major agreements including Transforming our World: the 2030 Agenda for Sustainable Development with its 17 Sustainable Development Goals (SDGs) was adopted at the UN Sustainable Development Summit in New York September 2015.

In an effort to address food insecurity bedevilling the previously disadvantaged indigenous African small-scale farmers in the 2000s, the Zimbabwe government launched the agrarian land reform programme (Zikhali, 2008; Chingarande, 2020). The agrarian reform significantly empowered African indigenous small-scale farmers as a reward for their support in a well-fought liberation struggle (Scoones *et al.*, 2011; Moyo, 2011). The aftermath of the agrarian reform resulted in shifts in land ownership with poor resourced African indigenous farmers replacing over 4500 white commercial farmers (Moyo *et al.*, 2013). The agrarian reform led to the key actors in financing development such as the International Monetary Fund (IMF) and the World Bank, which were supporting agricultural production, to cease financial assistance extended to Zimbabwe (Govere *et al.*, 2009). The agricultural production as well as contribution to the fiscus drastically decreased and the nation became a perennial food importer (Mazwi *et al.*, 2019). This posed sustainability threats to national security and food security, which led to deliberate government efforts to support farmers through provision of inputs.

The pursuit of national food security dovetailed into the United Nation's SDGs premised on food security, particularly goal 2 of zero hunger and goal 12 of responsible production and consumption. The Zimbabwe government intervened by supporting the agricultural sector (Moyo and Yeros, 2007). According to FAO (2014), food security is defined as a state where all people, at all times, have physical and economic access to enough safe and nutritious food to meet their dietary needs and food preferences for an active and healthy life. To ensure national food security and boost agricultural production, the government of Zimbabwe came up with public input distribution schemes meant to assist farmers in accessing the

inputs adequate in meeting production levels enough to improve the national food security and food self-sufficiency targets. The schemes include *operation maguta*, the presidential input scheme, *pfumvudza*, farming God's way, and the command input scheme (Mazwi *et al.*, 2019). Of concern, however, is despite the deliberate efforts by government to provide agricultural inputs under all these initiatives, the aftermath in terms of increased productivity, food import elimination, poverty reduction and economic growth is inconsistent (Zikhali, 2008; Mazwi *et al.*, 2019). The nation remains entrapped in debt over the food import bill.

In line with the SDGs, the Zimbabwe government also adopted the National Development Strategy 1 (NDS1), which is a bold transformative measure premised on the attainment of all the SDGs that prioritized ending poverty and ensuring national food security and nutrition by 2030 (FAO, 2014). The NDS1 aims to improve food self-sufficiency and to restore the Southern African Development Community (SADC) regional breadbasket status. It entrusts attainment of these goals through transforming the agricultural sector by adopting modern information and communication tools (ICTs) and public input distribution initiatives such as pfumvudza, the command agriculture scheme and the presidential input scheme (Bhattacharjee and Raj, 2013; Mazwi *et al.*, 2019).

Public inputs are sourced and procured in compliance with the public procurement and public asset authority act of 2003, then distributed to farmers before the onset of planting season based on the agro-ecological region. Public input distribution refers to any agribusiness associated with provision of farming and farming-related inputs for production (Yadav *et al.*, 2019), while agricultural inputs refer to resources or materials used in production or handling of agricultural products. Public input distribution consists of multiple stakeholders and actors along the distribution chain, such as farmers, extension officers, distributors, wholesalers and retailers (Kamilaris *et al.*, 2019). Although government public input distribution is pinned on positive objectives to curb the drastic decrease in agricultural productivity and production as well as efforts to increase import substitution where food is concerned, there have been worrisome concerns emanating from failure to boost the national food production levels. The traditional approaches to public agricultural input distribution over the past two decades have failed to transform agricultural productivity in Zimbabwe.

The current system used in the input distribution matrix does capture all information with regards to public input distribution, but on paper, which may be destroyed. It also makes use of isolated servers and databases of confidential agencies (TNH, 2009). It favours limited access to data for the general public, hence it is amenable to fraud, manipulation, embezzlement, politicization and corruption, resulting in substantial revenue losses as well as problems of product imitation and exchange (Motta et al., 2020). In addition, it is characterized by lack of transparency in which the entire input supply chain bears an impact of producing more artificial food shortages . The agriculture sector in Zimbabwe is known to be the least digitized industry, which did not utilize advantages of the internet, due to non-existence of connectivity.

Meanwhile, a couple of interventionist approaches have been undertaken to deal with unpleasing effects of corruption in government, the private sector and general public aiming to resuscitate the economy in general, and the agricultural sector in particular. Some of the popular interventions include the much publicly celebrated Operation Restore Legacy, in which the top officials in the ruling ZANU-PF party who were alleged to be the serial perpetrators of corruption hence were not addressing corrupt activities in the economy were removed from office. The Operation Restore Legacy provided short-term solutions to national revenue loss, but proved ineffective in the fight against corruption in the long run.

There is also an anti-corruption commission known as Zimbabwe Anti-Corruption Commission (ZACC) (Guma, 2011). However, ZACC has proven to be highly inefficient in addressing the issues of revenue losses in Zimbabwe's economy in general and public input distribution in particular (Global Integrity, 2011). ZACC's main contention in answer to accusations of being ineffective is that they do not have prosecuting authority. In light of these problems within public input distribution systems, there is a need for an efficient and sustainable agricultural input supply system.

This study proposes blockchain technology as a revolutionary and promising solution for the agricultural industry as it makes it possible to address current fraudulent and embezzlement challenges (Hughes et al., 2019; Bumblauskas et al., 2020). Blockchain in general is a circulated database technology that upholds the continuously growing list of data records that are confirmed by all the contributing nodes in the network. These approaches emphasize traceability of financial transactions (Ronaghi, 2021), which can bring transparency to corrupt systems, and verifiability and immutability to transactions in public distribution systems.

12.2 Challenges Facing the Agricultural Supply Chain

High reliability is an essential attribute for a successful supply chain system in today's competitive environment (Jia et al., 2018). Guess et al. (1996) define reliability as the probability that an item will perform a required function under stated environment and operational conditions for a stated period of time. With regards to supply chains, reliability can be defined as the probability that an item or players within the chain will perform a required function under stated environment and operational conditions for a stated period of time. Warburton and Stratton (2002) opine that supply chain reliability is the ability of the supply chain to meet the logistic performance expectations of customers. Supply chains have failed to meet their expectations in recent days rendering them unreliable. In agriculture this could be a difference between a bumper harvest and a loss as seasons (time) are of essence. The Covid-19 pandemic only made the situation worse as the logistics side of the supply chain was heavily impacted by the interventions meant to curb the spread of the virus. Munharo et al. (2021) argue that Covid-19 disrupted medicine supply schedules for prescription refills and routine laboratory testing for patients with chronic noncommunicable and infectious diseases and this may lead to increased odds of death from Covid-19 infections. The same scenario exists in the agricultural input schemes. Such disruptions have resulted in supply chains failing to deliver goods on time for end user use, with some farmers meant to benefit from the schemes having to source inputs late under emergency conditions inevitably at exorbitant prices.

Modern supply chain operability is affected by incomplete information. Bala (2014) argues that: 'Information sharing in a supply chain faces several hurdles. The first and foremost challenge is that of aligning incentives of different partners.' The supply chain is composed of different players and each with their own interest and prospects. Some players may decide to withhold information so as to obtain monopolistic advantage to increase their profits in the process. Even in instances where information sharing would be beneficial to all players, they may decide to play a non-collusive game to maintain power over other players. Such rent-seeking behaviour leads to welfare loss and increase in cost to other players in the supply chain.

For supply chains to effectively function, efficiency is of paramount importance. For agriculture to grow as a business there is a need to inculcate efficiency in the supply chains. FAL (2014) argues that: 'given the regional and global span of supply chains, business is often done in incomplete, backward-looking and reactive frameworks which are neither complementary nor integrated across borders. These create inefficiencies in logistics and supply chain systems and do not proactively contribute towards improving our collective sustainability.' The pursuit of comparative advantages and economies of scale have led to international markets and inevitably longer supply chains. Organizations have to rely on third parties to get their products from the plant to the customer efficiently. Engaging third parties makes business sense as one will focus on the core business activities while passing the non-core activities, such as distributing products, to professionals in the supply chain management business. However, manufacturers must remain awake to the risks created by the use of third parties. The agency–principal relationship and its challenges inevitably have to be addressed as many times the supply chain partners will attempt to fraudulently benefit from transactions they are meant to undertake on behalf of their principals. It is vital for the executive staff to understand that while they are bound by ethical norms and laws their supply chain partners may not be bound by the same rules.

Growth in the global economy comes with increase in demand for goods and services. For these to be delivered from the producer to the consumer on time there is a need for an efficient supply chain that minimizes lead times. In other instances producers also face bottlenecks in the production processes due to delays in the delivery of inputs. This affects production processes, which increases costs for the firms, which are later passed to the final consumer in the form of high prices.

Barriers to international trade have been another impediment affecting global supply chains. This challenge is more pronounced in developing countries that are sluggish in adoption of trade facilitation measures. Sequeira and Djankov (2010) stated that on average it takes three times as many days, nearly twice as many documents and six times as many procedures to trade in many African countries than in high-income economies. Soko (2013) further stressed that, on average, Africa has the longest customs delays in the world. It is estimated that border delays in southern Africa cost the region about US $48 million annually. He went on to say that the average delay it takes in customs for the different regions of the world are 12.1 days for sub-Saharan Africa, 7.2 days in Latin America, 5.5 days in Asia and 3.9 days in Europe. Such delays slow the movement of goods making supply chains inefficient. Makochekanwa (2013) further confirmed this phenomenon by stating that Mauritius was the best performer in this category, with both the exporter and importer requiring 13 days each to conduct all the formalities needed before exporting or importing. Seychelles came closer, with 16 days required before exporting and 17 days when importing. The worst performers with regards to days needed before exporting and importing are: Zimbabwe (which requires 53 days before exporting and 73 days when importing); Angola (48 days before exporting and 45 days when importing); Zambia (44 days before exporting and 56 days when importing) and Democratic Republic of Congo (44 days before exporting and 63 days when importing). This is an indication that in most developing countries, supply chain processes are being lengthened by administrative processes that can be easily avoided.

Lall (1994) shows that a typical trade transaction passes through 27 to 30 parties including brokers, banks, carriers, sureties and freight forwarders. At least 40 documents are needed, not only for government authorities but also for related businesses. Over 200 data elements are typically requested, of which 60–70% are rekeyed at least once, while 15% are retyped 30 times. This need for documentation, which varies from country to country, creates a lot of delays, increases transaction cost and is a major source of corruption. This was confirmed by Makochekanwa (2013) in his study on developing countries where he shows that most developing countries required a lot of documents when exporting and importing. This makes trade of goods and services difficult, creating stalls in supply chains be it to the final consumer or producers.

Modern supply chains have been plagued with costs given their length and the number of players who participate. Each player passes the cost to the next participant, which usually results in high cost to the final consumer. A report by narrates that 'recent supply-chain disruptions have given rise to a variety of business costs, including both increased costs of operations and reputational costs in terms of customer complaints and damage to brand reputation as companies have struggled to maintain supplies of their goods'. Firms have had to bear the brunt of increasing costs that have been worsened by the Covid-19 pandemic. The rise in costs emanates from increasing input prices, higher labour costs, complex logistics that increase transaction costs, and increase in fuel prices. The challenge of transaction costs was confirmed by OECD (2005), which concludes that estimates of trade transaction costs range from 2 to 15% of the total transaction value and that an approximate estimate for world trade is US$550 billion dollars each year. Such costs result in an increase in prices for final users.

Agricultural supply chains especially in developing countries are also affected by lack of accountability. This has been a challenge in Zimbabwe with specific reference to the government programme Command Agriculture. Mutedzi (2023) alleged that over US$3 billion that was set aside to finance the programme had been missing. Such lack of accountability has ripple effects on the functionality of the supply chain. Also the lack of accountability was noted in cases where farmers could divert the inputs received to other alternative uses thus affecting viability of the government programme. Munhende (2022) stated that police in Plumtree have arrested a woman in the border town after she was allegedly found in possession of Command Agriculture inputs she intended to sell on the black market. In the same article, it is further stated that a government minister had been arrested for defrauding the government inputs worth US$73,300. It was further noted that this lack of accountability was leading to low recovery rates.

12.3 Blockchain and Supply Chain

To address some of the issues and challenges raised above there is a need to look towards technology. Technology has become an undeniably integral part of society. The great technological progress, availability of the Internet and smart devices everywhere, and rise in customers' awareness have helped the spread of this revolution (Chen et al., 2022). Blockchain technology has been the most disruptive and transformative of the technologies. Blockchain, the digital record-keeping technology behind Bitcoin and other cryptocurrency networks, is a potential game changer in the services industry where it is playing a major role in the secure distribution and payment for services and accompanying products.

Blockchain technology is the underlying technology of cryptocurrencies (Pal et al., 2021). Although still in its relative infancy, blockchain technology has already triggered much interest, and a lively debate is ongoing about its future progression and the important benefits that it could bring in the context of the transfer of assets within business networks (Cocco et al., 2017). This is highly inevitable considering the pace at which technology is permeating all aspects of life. Although blockchain has mostly been applied in facilitating usage and adoption of cryptocurrencies such as bitcoin, its potential goes beyond cryptocurrencies to various economic and social transactions, thus having the potential to spread to various other business applications as well (Lindman et al.,

2017; Chong *et al.*, 2019; Hughes *et al.*, 2019). The technology has the potential to upend entire industries. Blockchain is a technology that has desirable features of decentralization, autonomy, integrity, immutability, verification, fault-tolerance, anonymity, auditability and transparency (Guo and Yu, 2022).

The challenges within the agricultural input supply system in Zimbabwe can to a greater extent be resolved by employing blockchain technology in undertaking the supply transactions and compiling the requisite paperwork. The immutability characteristic renders the technology fraud-proof, a feature that will benefit the government as this will reduce monitoring costs related to the input distribution schemes. Fraud usually involves the falsifying of transactional documents, changing figures and other relevant information prejudicing the intended beneficiaries in a transaction. Blockchain technology does not allow deletion and modification operations on the transactions and other information stored on its ledger (Guo and Yu, 2022). The technology prevents the falsification of documents as high cryptographic encryption, continuous scrutiny of transactions by network members and distributed transaction data make it highly challenging to manipulate or tamper with any transactions within the system (Beck *et al.*, 2017). This trait of blockchain makes it a very reliable and trustworthy technology that can be adopted to steer innovations and significantly increase productivity in various business domains such as healthcare, supply-chain management, digital art management, entrepreneurship, financial industry, and so on (Pal *et al.*, 2021). Blockchain can greatly improve supply chains by enabling faster and more cost-efficient delivery of products, enhancing products' traceability, improving coordination between partners, and aiding access to financing. The main objectives of the supply chain are listed as cost, quality, speed, dependability, risk reduction, sustainability and flexibility (Kshetri, 2018). Blockchain is not only a new type of internet infrastructure based on distributed applications, but also a new type of supply chain network (Chen *et al.*, 2018).

Blockchain technology provides the integrity and availability that allows participants in the network to write, read and verify transactions recorded. However it does not allow deletion and modification operations on the transactions and other information stored on its ledger

(Guo and Yu, 2022). As a database that offers 'data security, transparency and integrity, anti-tempering and anti-forgery, high efficiency and low cost blockchain can potentially be applied in numerous business activities which involve data exchange and require security' (Nowiński and Kozma, 2017). The application of blockchain technology in supply chain management will save organizations resources currently being channelled towards third-party due diligence and risk mitigation. Blockchain technology can track products and help create trustworthy supply chains and build trust between producers and customers (Xiong *et al.*, 2020). The tipping point for adoption of blockchain technology is expected to be 10% of global gross domestic product (GDP) stored on blockchain ledgers by the expected date of 2027 (Mettler, 2016).

In a supply chain, ownership of products changes many times between players until they are delivered to end users or consumers as stated earlier. This creates and presents opportunities for fraudulent behaviour, falsification of documents, rerouting of goods, changing prices and other undesirable end results. For products with low added value, such as agricultural products or commodities and certain types of mining commodities, supply chains function as an aggregation method by which goods are provided by many small-scale producers to larger-scale supply chain partners for further processing towards a finished product. In Zimbabwe the supply chain used to get produce and commodities to the market is usually the same that conveys inputs to farmers, such that fraud and falsification of information is done by the same middlemen who fleece farmers of inputs as well as their produce.

Granted, sustaining operations across a complex chain of resources, activities and farm holdings can be hard for supply chain partners, especially when a huge number of smallholders are players. Lack of visibility and incentives may cause hardships for sustainability. According to ISO standard 9000, traceability concept means 'the ability to trace the history, application, use and location of an item or its characteristics through recorded identification data'. This lack of visibility may result in donor fatigue and withdrawal of support by the government. In order to illustrate how blockchain might impact the needs of supply chain actors, we quoted

Table 12.1. How blockchain can improve the existing limitations of supply chains. From Litke *et al.* (2019).

Supply chain	Current situation	Blockchain impact
Producer	Lack of ability to protect the origin and quality metrics of products transparently.	Benefits from increased trust by keeping track of the production of raw materials and value chain from producer to the consumers.
Manufacturer	Limited ability to maintain the product to the final destination. Limited capabilities of checking quality measured from raw material.	Added value from shared information system with raw material suppliers and distribution networks.
Distributer	Customer tracking systems with poor collaboration capabilities. Limited certification ability and trust issues.	Ability to have proof of location and conditions certifications registered in the ledger.
Wholesaler	Lack of trust and certification of product's path.	Ability to check the origin of the goods and the transformation/transportation conditions.
Retailer	Lack of trust and certification of product's path. Tracking of products between consumers and wholesalers.	Ability to handle effectively the records of malfunctioning products.
Customers	Lack of trust regarding compliance of the product with respect to origin, quality and compliance of the product to the specified standards and origin.	Full and transparent value of the product origin and its whole journey from the raw material to the final production.

Table 12.1 from literature. It presents Litke *et al.* (2019)'s summary on how blockchain responds to the limitations that the actors of supply chains encounter today.

Despite all these celebrated benefits of blockchain in supply chains, in most developing countries the technological infrastructure is still behind. Connectivity levels are still low. In Zimbabwe the challenges are compounded by lack of knowledge, lack of support infrastructure, lack of will power to implement blockchain.

New problems related to blockchain and other assisting technologies should also be addressed to realize projects in the supply chain domain. Blockchain-based supply chain systems need various new legislative regulations. Current blockchain platforms cannot exactly fit the high level of transaction throughput requirement of supply chain systems. Supply chains combine diverse participants with varying interests. Thus, incentives need to be provided, such as efficiency gains, improved liquidity and data security, to motivate all participants. Security and privacy are other important issues. There are data security concerns with the Internet of

Things (IoT) and a lack of commonly accepted baseline protocol standards for IoT interaction. The current IoT ecosystem is built on a central model in which IoT devices are identified, connected and validated. Thus, there is need for transformation for blockchain adaptation.

In summary, blockchain technology can benefit the supply chain in various ways as it does in many other application areas. Employing blockchain in supply chain processes provides transparent, decentralized, secure, faster and low-cost transactions. By removing unnecessary third parties and covering more daily life processes in digital systems minimizes paperwork. Blockchain establishes trust among trading partners. Making more detailed data available in blockchain improves supply chain monitoring ability and safety. This reduces insurance risks. Smart contracts and automated payments are a gamechanger. They promote efficiency and eliminate bureaucracy especially in insurance and traceability.

This study suggests that blockchain technology certainly has an important role for enhancing and fundamentally transforming supply chains in many industries. It will be expected to create more sustainable solutions for supply chain bottlenecks experienced today in many industries such as logistics, agriculture and automotive. By removing the intermediaries with blockchain-based transformations, the transactions will become faster and more secure thanks to cryptography. Therefore, the infrastructures are evolving along with regulatory changes, technological advancements and new financial mechanisms that will facilitate blockchain based supply-chain management systems.

References

Bala, K. (2014) Supply chain management: some issues and challenges: a review. *International Journal of Current Engineering and Technology* 4, 946–953.

Beck, R., Avital, M., Rossi, M. and Thatcher, J.B. (2017) Blockchain technology in business and information systems research. *Business & Information Systems Engineering* 59, 381–384. DOI: 10.1007/s12599-017-0505-1.

Bhattacharjee, S. and Raj, S. (2013) Youth and ICTs for Agricultural Development. In: Gowda, N.K. and Nataraju, M.S. and Veerabhdraiah (eds) *Youth in Agriculture and Rural Development*. New India Publishing Agency (NIPA), New Delhi, India, pp. 169–188.

Bumblauskas, D., Mann, A., Dugan, B. and Rittmer, J. (2020) A blockchain use case in food distribution: do you know where your food has been? *International Journal of Information Management* 52, 102008. DOI: 10.1016/j.ijinfomgt.2019.09.004.

Chen, G., Xu, B., Lu, M. and Chen, N.S. (2018) Exploring blockchain technology and its potential applications for education. *Smart Learning Environments* 5, 1–10. DOI: 10.1186/s40561-017-0050-x.

Chen, Y., Lu, Y., Bulysheva, L. and Kataev, M.Y. (2022) Applications of blockchain in industry 4.0: a review. *Information Systems Frontiers* 1–15. DOI: 10.1007/s10796-022-10248-7.

Chingarande, D. (2020) Background paper on mainstreaming gender into national adaptation planning and implementation in sub-Saharan Africa.

Chong, A.Y.L., Lim, E.T.K., Hua, X., Zheng, S., Tan, C.-W. *et al.* (2019) Business on chain: a comparative-case study of five blockchain-inspiredbusiness models. *Journal of the Association for Information Systems* 20, 1308–1337. DOI: 10.17705/1jais.00568.

Cocco, L., Pinna, A. and Marchesi, M. (2017) Banking on blockchain: costs savings thanks to the blockchain technology. *Future Internet* 9, 25. DOI: 10.3390/fi9030025.

FAL (2014) Efficiency — a Key Ingredient towards Sustainable Supply Chains.

FAO (2014) *The State of Food Security in the World 2014. Strengthening the Environment for Food Security and Nutrition*. International Fund for Agricultural Development, World Food Program, FAO, Rome.

Global Integrity (2011) *The Global Integrity Report: 2011 Methodology White Paper*. Global Integrity, Washington, DC.

Govere, I., Foti, R., Mutandwa, E., Mashingaidze, A.B. and Bhebhe, E. (2009) Policy perspectives on the role of government in the distribution of agricultural inputs to farmers: lessons from Zimbabwe. *International NGO Journal* 4, 470–479.

Guma, L. (2011) Zimbabwe: anti-corruption commission sworn in. Available at: https://allafrica.com/stories/201109021275.html (accessed 14 July 2023).

Guo, H. and Yu, X. (2022) A survey on blockchain technology and its security. *Blockchain: Research and Applications* 3, 100067.

Hughes, L., Dwivedi, Y.K., Misra, S.K., Rana, N.P., Raghavan, V. *et al.* (2019) Blockchain research, practice and policy: applications, benefits, limitations, emerging research themes and research agenda. *International Journal of Information Management* 49, 114–129. DOI: 10.1016/j.ijinfomgt.2019.02.005.

Jia, X., Lirong, C. and Liudong, X. (2018) New insights into reliability problems for supply chains management based on conventional reliability model. *Eksploatacja i Niezawodność* 20, 3. DOI: 10.17531/ein.2018.3.16.

Kamilaris, A., Fonts, A. and Prenafeta-Boldú, F.X. (2019) The rise of blockchain technology in agriculture and food supply chains. *Trends in Food Science and Technology* 91, 640–652. DOI: 10.1016/j.tifs.2019.07.034.

Kshetri, N. (2018) Blockchain's roles in meeting key supply chain management objectives. *International Journal of Information Management* 39, 80–89. DOI: 10.1016/j.ijinfomgt.2017.12.005.

Lall, S. (1994) *Industrial Policy: the Role of Government in Promoting Industrial and Technological Development*. UNCTAD.

Lindman, J., Tuunainen, V.K. and Rossi, M. (2017) Opportunities and risks of blockchain technologies: a research agenda. In: *Hawaii International Conference on System Sciences*. DOI: 10.24251/HICSS.2017.185.

Litke, A., Anagnostopoulos, D. and Varvarigou, T. (2019) Blockchains for supply chain management: architectural elements and challenges towards a global scale deployment. *Logistics* 3, 5. DOI: 10.3390/logistics3010005.

Makochekanwa, A. (2013) Assessing the impact of trade facilitation on SADC's intra-trade potential. In: *AGRONDER Member's Conference*, p. 19.

Mazwi, F., Chemura, A., Mudimu, G.T. and Chambati, W. (2019) Political economy of command agriculture in Zimbabwe: a state-led contract farming model. *Agrarian South: Journal of Political Economy* 8, 232–257.

Mettler, M. (2016) Blockchain technology in healthcare: the revolution starts here. In: *2016 IEEE 18th International Conference On E-Health Networking, Applications And Services (Healthcom)*, IEEE, pp. 1–3. DOI: 10.1109/HealthCom.2016.7749510.

Motta, G.A., Tekinerdogan, B. and Athanasiadis, I.N. (2020) Blockchain applications in the agri-food domain: the first wave. *Frontiers in Blockchain* 3, 6. DOI: 10.3389/fbloc.2020.00006.

Moyo, S. (2011) Changing agrarian relations after redistributive land reform in Zimbabwe. *Journal of Peasant Studies* 38, 939–966.

Moyo, S., Jha, P. and Yeros, P. (2013) The classical agrarian question: myth, reality and relevance today. *Agrarian South: Journal of Political Economy* 2, 93–119. DOI: 10.1177/2277976013477224.

Munharo, S., Edet, A.A., Friday, A.E., Maradze, T.C. and Lucero-Prisno, E. (2021) Impact of COVID-19 on Supply Chains in Zimbabwe. *Journal of Public Health International* 3(4), 33–37. DOI: 10.14302/issn.2641-4538.jphi-21-3824.

Munhende, L. (2022) Plumtree woman arrested for stealing command agriculture inputs. NewZimbabwe.com. Available at: https://www.newzimbabwe.com/plumtree-woman-arrested-for-stealing-command-agriculture-inputs/ (accessed 10 July 2023).

Mutedzi, P. (2023) Command agriculture: no $3bn disappeared. *The Herald*. Available at: https://www.herald.co.zw/command-agriculture-no-3bn-disappeared/ (accessed 10 July 2023).

Nowiński, W. and Kozma, M. (2017) How can blockchain technology disrupt the existing business models? *Entrepreneurial Business and Economics Review* 5, 173–188.

OECD (2005) The costs and benefits of trade facilitation.

Pal, A., Tiwari, C.K. and Behl, A. (2021) Blockchain technology in financial services: a comprehensive review of the literature. *Journal of Global Operations and Strategic Sourcing* 14, 61–80. DOI: 10.1108/JGOSS-07-2020-0039.

Ronaghi, M.H. (2021) A blockchain maturity model in agricultural supply chain. *Information Processing in Agriculture* 8, 398–408.

Scoones, I., Marongwe, N., Mavedzenge, B., Murimbarimba, F., Mahenehene, J. *et al.* (2011) Zimbabwe's land reform: challenging the myths. *Journal of Peasant Studies* 38, 967–993. DOI: 10.1080/03066150.2011.622042.

Sequeira, S. and Djankov, S. (2010) An empirical study of corruption in ports. Available at: https://econpapers.repec.org/paper/pramprapa/21791.htm

Soko, M. (2013) Why trade Facilitation is key to boosting intra-African trade. Evian Group Policy Brief.

TNH (2009) Zimbabwe: corruption Bedevils farming inputs. Available at: https://reliefweb.int/report/zimbabwe/zimbabwe-corruption-bedevils-farming-inputs

Warburton, R.D. and Stratton, R. (2002) Questioning the relentless shift to offshore manufacturing. *Supply Chain Management: An International Journal* 7, 101–108. DOI: 10.1108/13598540210426659.

Xiong, H., Dalhaus, T., Wang, P. and Huang, J. (2020) Blockchain technology for agriculture: applications and rationale. *Frontiers in Blockchain* 3, 7. DOI: 10.3389/fbloc.2020.00007.

Yadav, A.S., Kumar, S., Kumar, N. and Ram, H. (2019) Pulses production and productivity: status, potential and way forward for enhancing farmers income. *International Journal of Current Microbiology and Applied Sciences* 8, 2315–2322. DOI: 10.20546/ijcmas.2019.804.270.

Zikhali, P. (2008) Fast track land reform and agricultural productivity in Zimbabwe. Available at: https://www.researchgate.net/publication/23530976_Fast_Track_Land_Reform_and_Agricultural_Productivity_in_Zimbabwe

13 Re-envisaging Agricultural Public Policy and Governance in Zimbabwe: Prospects and Incongruities

Noah Ariel Mutongoreni*, Patrick Korera, Moses Jachi, Benard Chisiri, Tafadzwa Machaka, Godwin Shumba and Esther Mafunda
Manicaland State University of Applied Sciences, Mutare, Zimbabwe

Abstract

The agricultural sector has been and is still a source of contestation in Zimbabwe. Critical players in the contest are government, farmers, and civil society and donor agencies among others. What has come out of the contest is a myriad of policies and governance frameworks all of which impacted on agricultural production in general and the agribusiness development landscape in particular. The net effect of the public policies and governance thus had a bearing on the individual farmer, consumers and the agribusiness industry potential in Zimbabwe. This chapter will unravel the prospects and incongruities of the Zimbabwean agricultural public policy and governance framework. In doing so, the chapter adopted a qualitative approach. Pursuantly, a review of relevant literature and relevant government policy documents was undertaken. The research results revealed that the agricultural marketing and agribusiness development arena was replete with agricultural public policies and governance inconsistencies. This had the net effect of nurturing agribusiness development and marketing in Zimbabwe. While government's intervention through policies is intended to promote the sector and create a level playfield as well as regulating the sector, the intended outcomes continue to be elusive. Pursuant to this, it is recommended that government adopts a multi-stakeholder, liberal approach and minimize its tentacles in the sector to ensure sustainable livelihoods, sustained growth and productivity of the sector. Agriculture should be regarded as business which finances itself, and gets and pays taxes regularly to the central government. This business model should assist economic development and emancipation of the Zimbabwean citizenry.

Keywords: Agricultural public policy, Agribusiness, Productivity, Sustainable development, Governance framework

13.1 Introduction

Agriculture forms the backbone of developing economies globally, being the solid foundation anchoring their food security, foreign currency earnings and local regional development (FAO, 2022; Davis *et al.*, 2021). The agricultural sector accounts for up to 4% of the global gross domestic product (GDP), and this figure increases to in excess of 25% in the context of least developed economies (World Bank, 2022). The African continent earns an average of 24% of its annual growth from its farmers' efforts (World Bank, 2022). It is estimated that agriculture and agribusiness combined could command a US$1 trillion presence in Africa's regional economy

*Corresponding author: noah.mutongoreni@staff.msuas.ac.zw

© CAB International 2023. *Sustainable Agricultural Marketing and Agribusiness Development: An African Perspective* (eds B. Nyagadza and T. Rukasha)
DOI: 10.1079/9781800622548.0013

by 2030 (World Bank, 2022). The agricultural sector accounts for between 30 and 40% of GDP in sub-Saharan Africa and employs as much as 90% of the labour force. In Zimbabwe, agriculture contributes 17% of GDP and accounts for 4% of the total export earnings (FAO, 2022). Sustained growth in economies anchored on agriculture is not likely to be achieved without adoption of contemporary agricultural practices (Bautista and Thomas, 1999).

As is the case with many developing economies, Zimbabwe's economy is hinged on agriculture. FAO (2022) opine that the backbone of the country's economy is agriculture and that the population is largely rural and more akin to have a livelihood anchored on agriculture. Runganga and Mhaka (2021) argued that agriculture supplies 60% of the raw materials needed by the industrial sector and also contributes about 40% of the total export earnings. In terms of employment, agricultural-related activities provide employment and income to 60–70% of the population (Runganga and Mhaka, 2021). The healthy perfomance of the agricultural sector is hinged on the multisectoral approach to policy formulation. A nation's institutions and policies act as internal mechanisms that give an incentive for various econmic actors to behave in certain ways (Mutambara, 2016). What this implies is that policies and the governance pathway are critical determinants of a nation's economic outcomes. The right policies are thus very critical for the Zimbabwean agricultural sector.

Historically, Zimbabwe has been known for its strong agriculture-based economy, which earned it the most popularized name phrase 'the breadbasket of Southern Africa' (Mutambara, 2016). This breadbasket of Southern Africa status has since disappeared. The country has however transitioned from being the breadbasket to a basket case (Mungwari, 2017), and has since become a net importer of maize (World Bank, 2017). In the year 2022, for example, Zimbabwe imported about 400,000 tonnes of grain from Zambia and Malawi (FAO, 2022). This is evident of a tale of success gone bad (Mungwari, 2017). The poor economic performance can partly be attributed to the manifestation of existence of internal contradictions that are working against the nation's progress (Mutambara, 2016). This chapter re-envisages the agricultural public policy

and governance prospects and incongruences in the post-independent Zimbabwe. In doing so, key terms central to the chapter are defined, and a conceptual framework to guide the chapter has been developed together with a theoretical framework underpinning the chapter. There is also a discussion on public policy and governance prospects and incongruities, and recommendations are given. The conceptual framework guiding this chapter is illustrated in Fig. 13.1.

What the conceptual framework suggests is that a country that has the highlighted phenomenon will experience the end highlighted in the model. If a conducive environment is created for the means, the possibility of achievement of positive results is very high, all things held equal. Currently, in the sub-Saharan region, Zimbabwe is an example of a country reeling under economic challenges as a result of challenges with dealing with the processes, procedure and policies. This failure to attend to the means has resulted in economic decline (Dansereau and Zamponi, 2005; Moore, 2012; Kanonge and Bussin, 2022). The agribusiness and marketing ecosystem is stunted as a result of, among others, supply chain disruptions emanating from both internal and external policy environments. Poverty and unemployment levels are on the rise. For this phenomenon to be addressed, appropriate agribusiness policies have to be developed. The development of the policy should flow from the stakeholders and not from the state. In other words, the policy process should reflect transparency, accountability and responsiveness through stakeholder engagement or involvement, all these being tenets of governance. Public policies should flow from the rails of good governance (Stoker, 2018; Watson, 2019). Once this happens, there is national development, rising GDP, a vibrant agribusiness ecosystem, sustained agribusiness growth, a healthy agribusiness supply chain, rising employment, mushrooming agribusiness ventures, economic growth and sustained livelihoods.

13.2 Theoretical Framework

The chapter was underpinned largely by the systems and elitist theories as the guiding light.

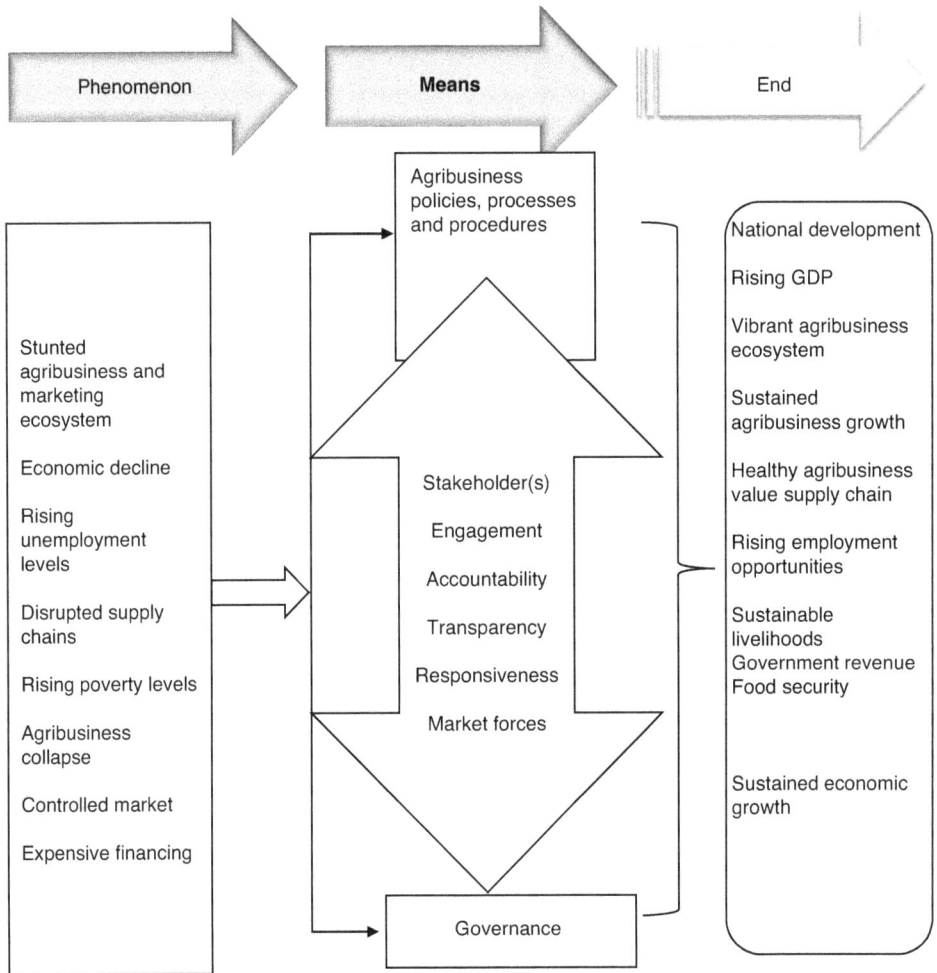

Fig. 13.1. Conceptual framework.

A system in line with the systems theory is an ensemble of components working together harmoniously (Sen, 2012). It also entails groups of objects or elements standing in some characteristic structural relationship to one another and interacting on the basis of certain characteristic processes (Sen, 2012). David Easton in 1953, first articulated the systems theory. The theory suggests that policy emanated from a political system. This political system is made up of the inputs (demands/support) emanating from the public. These demands are processed in the political system and they come out as decisions/ policy constituting the output (Easton, 1964). Inputs are converted by the process of the system into outputs and these have consequences both for the system and for the environment in which the system exists. The institutions found in the economic, social, cultural and international systems that impact on the political process and whose actions are influenced by the political system make up the environment. Using the systems approach, it is assumed that a state of mutual causation exists between public policy and environmental variables (Anyebe, 2018). The output are the works of government and these manifest themselves as decisions or policies. Feedback is the key to check the success of the policies of government. The system is thus a two-way traffic among the rulers and the ruled. The inputs and outputs have a very deep and dynamic inter-relationship. The main directorial belief of this approach is the supposition of equilibrium and that the cooperation of the parts will be necessary for the functioning of the whole system. Once a part of the system is disturbed, the entire system is affected.

The relatively general and abstract nature of systems theory limits its applicability in the study of public policy. Additionally, it does not elaborate much on the methods and processes used in the 'black box' known as the political system to make choices and establish policy (Anderson, 2003). The elite theory on the other hand posits that public policy is mostly a reflection of the interests of the governing elite (Anyebe, 2018). On the same premise Anderson (2003) posits that public policy can be regarded as reflecting the values and preferences of a governing elite. This is opposed to the general belief that pluralism has an inherent mechanism for maintaining equity in the share of power and influence in society. People with power work hard to maintain and improve their social status. There are governing and non-governing members of the elite group. The majority of the population (masses) is destined to be under the authority of the few who have special abilities including skills, affluence, cunning and intelligence. This is how social classes are established. From this vantage point, public policy can be seen as a reflection of the ideals and preferences of the dominant class. Policies or decisions flow downward from the ruling elites to the masses. Contrary to the most common-held view, they do not emanate from masses' demands (Mustafa *et al.*, 2021). The main contention of the elite theory is that the ruling elite, whose preferences are carried out by political officials and agencies, determines public policy rather than the demands and actions of the people or the masses. The elite merely think that they are the only ones with the authority to design and carry out policies that advance the wellbeing of the general populace. As a result, policies pass from the elite to the general public. The elite's policies reflect their principles and they favour maintaining the status quo over drastic change.

In a nutshell, according to Dye and Zeigler (1990), the elite theory asserts that society is split between a small number of powerful people and a large number of powerless people. The majority does not determine societal values; rather, the minority does. The few who rule are not representative of the controlled majority. The elite are disproportionately drawn from society's wealthiest socio-economic groups. To maintain stability and prevent revolution, the ascent of non-elites to elite positions must be gradual and ongoing (Anyebe, 2018). Only non-elites who concur with the fundamental elite consensus are allowed access to governing circles. Elites agree on the fundamental principles guiding the social order and the need to protect it. Public policy reflects the prevalent values of the elite rather than the needs of the majority. Public policy adjustments will be gradual rather than radical. Incremental modifications enable responses to events that endanger a social system with the least amount of system modification or disruption. Apathetic masses have comparatively little direct impact on active elites. Elites have a greater impact on the masses than the populace have on elites (Dye and Zeigler, 1990; Popoola, 2016).

It follows therefore that the elite theory concentrates our attention on the leadership role in policy development and on the fact that a small number of people control a large number of people in every political system. It is challenging to manage the idea that elites dominate and set policy with little input from the general population. The 'establishment rules things', which has been a common complaint in recent years, cannot be shown solely by claims (Anderson, 2003). Elite theory may be more useful for analysing and explaining how policies are formed in some political systems (such as those in developing nations), than in others (such as pluralist democracies in the West) (Anderson, 2003).

13.2.1 Background to the Zimbabwean agriculture ecosystem

Zimbabwe, formerly Zimbabwe-Rhodesia, Rhodesia and Southern Rhodesia was under successive white minority rule from 1890 to

1980 (Chikwanha-Dzenga *et al.*, 1999). During the colonial era, the inhabitants were referred to as Africans, blacks or natives and the colonial settlers as Europeans or whites. The African population were deprived of the best lands and this formed the basis for the liberation struggle, which brought about independence in 1980 (Rakodi, 1996). The black majority were deprived of their land through various laws enacted by the colonial government (Chirisa and Dumba, 2012). Some of the critical legislations enacted were the Land Apportionment Act, 1930 (enacted following the Morris Carter Commission's recommendations on land segregation culminating in the setting aside of 51% of the land to colonial settlers with Africans prohibited from occupying land in European areas), the Native Land Husbandry Act, 1951, and the Land Tenure Act, 1969 (Chirisa and Dumba, 2012). These successive laws had a bearing on the successive agricultural policies enacted post-independence.

The post-colonial government under the leadership of Robert Mugabe was under pressure at independence to address injustices born out of the bifurcated colonial government (Masunungure, 2019). The hitherto interventions among others included the Growth with Equity policy, the Fast Track Land Reform Programme and the Command Agriculture Policy among a host of others. These policies had unprecedented ramifications on Zimbabwe as a nation. The infamous illegal sanctions, political conflict and polarization, international isolation, hyperinflation, rising poverty levels and diseases and the total economic collapse of 2008 can all be traced to land-related policies enacted by government (Bland, 2011). The issue of land and business around agriculture in all its facets is thus an emotive issue in Zimbabwe and a source of contestation. It is against this background, that the agricultural public policy and governance prospects and incongruities in Zimbabwe shall be unravelled.

13.3   Methodology

The study adopted a qualitative research approach. As a consequence, it relied on content analysis. Relevant literature and government policy documents were reviewed.

13.4   Agricultural Public Policy and Governance in Zimbabwe: Prospects and Incongruities

At the advent of independence, Zimbabwe pursued a socialist ideological pathway (Makumbe, 2009). The socialist ideology was in line with the ideals of the liberation struggle, which was centred on the economy and mostly land. To the majority of the black population, it was well received. However, to those who owned capital, this was a cause for concern especially at a time when the world was divided ideologically. Apart from taking a new ideological thrust, the ZANU-PF led government called for a one-party state system. The idea was to ensure that all efforts were channelled towards development. However, it meant the death of dissent voices to policies undertaken by the government as stakeholder participation and engagement became restricted and viewed with suspicion. Policies that had implications on agriculture were expected to be accepted without scrutiny.

Aside from the above, the government adopted a Growth with Equity policy in 1981. This called for popular democratic participation of the public in the ownership and management of the country's natural resources (Makumbe, 2009; Paradza, 2010). In the spheres of agriculture, the government provided assistance to formerly underprivileged rural communities. This encompassed easier access to commercial credit, significantly discounted government agricultural inputs and equipment hire, free agricultural extension services among others (Paradza, 2010). What this meant was that the Growth with Equity policy promoted agriculture as agricultural research, extension and credits were redirected from the large-scale commercial subsector towards the small-scale communal areas where the subsistence-oriented agricultural sector has been traditionally located (Rukovo *et al.*, 1991). These strategies were however undertaken notwithstanding the fact that the nation's economy by then was fragile (Paradza, 2010) and the majority of the people supported did not take farming as a business. What could be deduced is that the independence era ushered in an era of change in the agricultural policy. The intention was to address the previous imbalances created by the miniority governement. The blacks were removed from positions of obstacle to access social authority and responsibility (Wright and Takavarasha, 1989).

The Growth with Equity policy also brought with it the Africanisation of the Public service, which also had an impact on the agricultural sector. The civil service during the colonial rule was dominated by the minority white settlers. The black majority black population occupied lower-level positions. The agriculture value chain was thus firmly under the control of the white minority settlers. Although this policy strategy was positive in terms of including the blacks in the positions of power and control, the majority of them lacked the requisite skills associated with their new positions. Performance setbacks began to be evident (Zinyama *et al.*, 2015). This had a negative impact on the performance of the agriculture sector and the entire civil service.

13.4.1   Old Fast Track Land Reform Programme

The issue of land formed the basis of the liberation struggle that led to the attainment of majority rule in 1980. Land was at the heart of the deliberations of the Lancaster House agreement, which culminated in majority rule. Pursuantly, the post-colonial government embarked on a land redistribution programme. The intention was to alleviate *inter alia* population pressure in the rural areas, increase peasant's access to fertile soils, improve livelihoods of the poor in society, and also bring underutilized or abandoned land to full production (Paradza, 2010). The programme regrettably was abandoned prematurely with less than 7% of the communal area population resettled as the legislation restricted government to acquire land willingly sold by the white farmers. The programme despite lacking funding was too ambitious. In addition, the commercial farmers offered only marginal unproductive land for sale to the government. Corruption was also a major problem. Inputs were not availed to the black farmers in time and on a regular basis. The marketing system did not give incentives to the black farmers to be productive (Muir, 1983). On another note, the Land Reform Programme was not sensitive to climate change as the emphasis was on redistribution of land and provision of input (Kudejira, 2014).

13.4.2   Command Agriculture Programme

The Command Agriculture Programme was introduced during the 2016/17 agricultural summer season. According to the report of the Parliament of Zimbabwe Thematic Committee on Peace and Security on the preparedness of the Grain Marketing Board, the programme was an import substitution-led industrialization concept deliberately meant to empower and incentivize local producers of cereal crops, particularly maize, so as to achieve national food security (GoZ, 2018).

The programme has been run by the military, the Office of the President and Cabinet, and Lands, Agriculture, Fisheries, Water, Climate and Rural Development. It availed inputs, agricultural extension support and loans to farmers. The State acted as the guarantor of the loans. This programme is associated with names such as *Pfumvudza* or *dhiga udye*. Although the programme was well meant, it failed to recognize that not everyone is a farmer and hence intended outcomes were not realized. It had also been noted that the expenditure for the programme was excessive and unsustainable (Pindiriri *et al.*, 2021). The public spending towards Command Agriculture soared, defying any global comparison. In the year 2017 for example, total expenditure on agriculture from the Consolidated Revenue Fund accounted for 5.4% of GDP. It also accounted for 66.7% of agricultural GDP, and nearly a quarter of the national budget (World Bank, 2019). Over and above, the programme lacks transparency and is riddled with corruption in the acquisition and distribution of inputs. The inputs rarely arrive on time and are usually overpriced in Zimbabwean dollars, which is rapidly losing value. Late supply of inputs affected the farmers' farming calendar and output. The programme also created a fertile ground for arbitrage opportunities where it has become more profitable to get the command inputs and sell them in foreign currency on the parallel market than to enter into farming as per the contractual terms. The money would be easy to return due to inflation. However, the default rate has also been very high. Farmers also do side marketing after having received inputs, thus prejudicing the state. The programme has thus become unsustainable for central government (Pazvakavambwa, 2009; Mazwi *et al.*, 2019).

Apart from the above, the programme did not take ecological farming zones into consideration in design and distribution of inputs. The end result was that farmers would get inputs not suited for their respective regions – for instance, fertilizers and maize seeds were supplied to drought-prone Matabeleland regions, which is logically suited for animal husbandry.

13.4.3   Look East Policy

The Look East Policy of the Zimbabwean government is both politically motivated and responds to economic necessities in the absence of donor support from the West. It was also a response to restrictive sanctions imposed by the West following the violent Fast Track Land Reform Programme. The programme entailed closer ties with Indonesia, India, Iran, Russia, Belarus, Malaysia and North Korea. China stood out as the major and pronounced ally and 'all weather friend' (Scoones *et al.*, 2012). Notable positive outcomes of the policy impacting on agriculture included the Harare Water Treatment Project and expansion of the Kariba Hydro Power Station among other major projects (World Bank, 2016). The policy led to a widening of market players in the agribusiness sector. It led to Chinese companies dominating the agribusiness sector as they became monopolistic buyers of products such as tobacco. The pricing system became distorted and this had a negative impact on the farmers, the majority of whom were on the contract farming arrangement.

13.4.4   Contract farming

Contract farming can be understood as a firm lending 'inputs' – such as seed, fertilizer, credit or extension – to a farmer in exchange for exclusive purchasing rights over the specified crop (Prowse, 2012). It is a farming mechanism that is typically seen as useful to help smallholders (Ruml *et al.*, 2021) given that they receive all the basic inputs from the contractor with no payment of cash up front. The contract farming arrangement can help integrate small-scale farmers into modern agricultural value chains, providing them with inputs, technical assistance and assured markets (Minot and Sawyer, 2014). The vast knowledge and expertise that is acquired from contract farming cropping activities can also be transferred to animal husbandry and solve the animal productivity matrix successfully. A lot more positive effects of contract farming on production and household economic welfare and high yields are recorded (Kumar *et al.*, 2019; Vicol *et al.*, 2022). Despite the seemingly lucrative resource provision nature of the contract and higher yield and pronounced profit benefit, many smallholder farmers still regret their decision to participate and thus would prefer to terminate and exit the contract if they could (Ruml and Qaim, 2021). Serious irregularities and incongruities have been raised with respect to the contract farming arrangement, which could be attributing to high dropout rates from contract schemes in other African countries such as Ghana (Ruml and Qaim, 2021).

It is argued that contracts create unequal power relations due to the monopolistic nature of the company (Morrison, 2006; Oya, 2012). The imbalance of power and information between the parties enables agribusiness firms to impose contract terms on small farmers and manipulate quality standards so as to reduce payments to farmers (Little and Watts, 1994). Unequal power relationships in favour of the companies thus have a bearing on the commodity pricing system, which in a majority, if not all of cases, are sponsor controlled and determined. While the market for the commodities is guaranteed, the prices are just too low and do not justify the effort by the concerned farmers (The Zimbabwean Herald, 2016). The contractors determine the price of the products at the end of the farming season. The farmer has no control. This unfavourable pricing system has perpetuated a vicious circle of dependency nets in tobacco and cotton farming in Gokwe, where farmers find it difficult to come out of the net (The Zimbabwean Herald, 2016).

Environmental concerns have also been raised on pesticide and weedicide use that go against environmental management policies. Farmers decry lack of information in the contracts.

Every country can have a different product under contract farming arrangements – India has broilers (Vicol *et al.*, 2022), and Zimbabwe has broilers, tobacco and cotton under contract farming (The Zimbabwean Herald, June 2016). Contract farming has been a central part of Zimbabwean farming business for decades (The Zimbabwean Herald, 2016). There is evidently rare consideration and therefore neglect of staple cereal food crops by the contract farming arrangements (Minot and Sawyer, 2014; Ray *et al.* 2020). In Zimbabwe this has exacerbated hunger and starvation, due to farmer involvement only in cash crops that are supported by contract farming. The introduction of the Command Agriculture Policy in 2006 and in the 2016/17 season was a response to grain food shortages. The overwhelming support and sponsorship for other non-staple cash crops in the contract system in Zimbabwe and other situations comes against the backdrop of maize and other small cereals forming the staple diet in the country. The non-supporting of staple food production should be a cause of food stability worry to Zimbabweans in general, and lawmakers in particular, especially when the majority of the country's Zimbabwean population cannot go to bed without a staple diet cereal supper.

13.4.5 Agricultural mechanization programme

The mechanization programme was meant to provide farming machinery and technologies to the beneficiaries of the Fast Track Land Reform Programme. It was also meant to replace obsolete agriculture equipment. The premise of this programme was that availing land and inputs to the farmers without mechanization support would impact on crop production and food security. While the programme led to the availability of farm equipment, the beneficiaries were the few elites in government. In addition, the equipment sourced were of poor quality and did not have backup spare parts (Obi and

Chisango, 2011). The elite beneficiaries are on record for not returning the loans for the equipment availed. Though the programme benefited a few, the burden was cascaded to every citizen.

13.5 Recommendations

It is recommended that from a policy and governance perspective, government should adopt a multistakeholder thrust, liberal approaches and minimize its tentacles in the agricultural sector to ensure sustainable livelihoods, sustained growth and productivity of the sector. Agriculture should be regarded as business, which finances itself, and gets and pays taxes regularly to the central government. This business model should assist economic development and emancipation of the Zimbabwean citizenry. The role of the government should be reduced to that of a regulator and a provider of a conducive business operational environment. This is critical to avoid the conflict of interest that occurs when an entity becomes both a regulator, buyer and producer.

13.6 Conclusion

The post-independence policy and governance regime in Zimbabwe was born out of a desire to address the imbalances previously created by the colonial regime. The black majority primarily needed land as well as access into the mainstream agricultural business. While today land is in the hands of the majority of the black population, there still exists a myriad of disgruntled stakeholders locally and internationally who, in one way or the other, were affected. As a result, access to meaningful funding, which is an essential condition for successful agribusiness, continues to be elusive. The policy regime in Zimbabwe is moving on an elite rail as it is dictated by the ruling ZANU-PF elites with little or no consultation, engagement or accountability. Lack of a clear governance parameter in the policy process has culminated in negative consequences such as a declining economy, international condemnation and sanctions.

References

Anderson, J.E. (2003) *Public Policy Making: An Introduction*. Mifflin Company.

Anyebe, P.A.A. (2018) An overview of approaches to the study of public policy. *International Journal of Political Science* 4(1), 8–17. DOI: 10.20431/2454-9452.0401002.

Bautista, R.M. and Thomas, M. (1999) Agricultural growth linkages in Zimbabwe: income and equity effects. *Agrekon* 38(S1), 66–77.

Bland, G. (2011) Overcoming a decade of crisis: Zimbabwe's local authorities in transition. *Public Administration and Development* 31, 340–350. DOI: 10.1002/pad.620.

Chikwanha-Dzenga, A.B., Masunungure, E. and Madzingira, N. (1999) Democracy And National Governance In Zimbabwe: A Country Survey Report. Working Paper no 12. Michigan State University, Dept. of Political Science.

Chirisa, I. and Dumba, S. (2012) Spatial planning, legislation and the historical and contemporary challenges in Zimbabwe: a conjectural approach. *Journal of African Studies and Development* 4, 1–13.

Dansereau, S. and Zamponi, M. (2005) *Zimbabwe: The Political Economy of Decline*. Nordic Africa Institute.

Davis, K., Gammelgaard, J., Preissing, J., Gilbert, R. and Ngwenya, H. (2021) *Investing in Farmers: Agriculture Human Capital Investment Strategies*. FAO Investment Centre, Rome. https://doi.org/10.4060/cb7134en

Dye, T.R. and Zeigler, L.H. (1990) *The Irony of Democracy*, 8th edn. Books/Cole, Monterey, Calif.

Easton, D. (1964) An approach to the analysis of political systems. *World Politics* 16, 677–715. DOI: 10.2307/2009452.

FAO (2022) Investing in Farmers: Agriculture Human Capital Investment Strategies.

Goverrnment of Zimbabwe (2018) National agriculture policy framework (2018–2030).

Kanonge, T.T. and Bussin, M.H.R. (2022) Pre-conditions for employee motivation to curb Zimbabwe's academic brain drain. *SA Journal of Human Resource Management/SA* 20 a1819. https://doi. org/10.4102/sajhrm.v20i0.1819

Kudejira, D. (2014) An integrated approach towards moderating the effects of climate change on agriculture: A policy perspective for Zimbabwe. Future Agricultures' Early Career Fellowship Programme.

Kumar, M., Liu, Y., Katul, G.G. and Porporato, A.M. (2019) *Detecting Climate-Stress Induced Forest Mortality Before the Canonical Symptoms Appear*. American Geophysical Union Fall Meeting, San Francisco, CA, pp. 9–13.

Little, P. and Watts, M. (eds) (1994) *Living under contract: contract farming and agrarian transformation in Sub-Saharan Africa*. University of Wisconsin Press, Madison, WI.

Makumbe, J. (2009) *The Impact Of Democracy In Zimbabwe: Assessing Political, Social And Economic Developments Since The Dawn Of Democracy*. Centre for Policy Studies, Johannesburg.

Masunungure, E.V. (2019) The changing role of civil society in Zimbabwe's democratic processes: 2014 and beyond. Available at: https://library.fes.de/pdf-files/bueros/simbabwe/13718.pdf (accessed 11 July 2023).

Mazwi, F., Chemura, A., Mudimu, G.T. and Chambati, W. (2019) Political economy of command agriculture in Zimbabwe: a state-led contract farming model agrarian South. *Journal of Political Economy* 8, 232–257.

Minot, N. and Sawyer, B. (2014) *Contract Farming in Developing Countries: Theory and Experience. Report Prepared for the Investment Climate Unit International Finance Corporation*. IFPRI, Washington, DC.

Moore, D. (2012) Progress, power, and violent accumulation in Zimbabwe. *Journal of Contemporary African Studies* 30, 1–9. DOI: 10.1080/02589001.2012.646748.

Morrison, T.H. (2006) Pursuing rural sustainability at the regional level: key lessons from the literature on institutions, integration, and the environment. *Journal of Planning Literature* 21, 143–152. DOI: 10.1177/0885412206292261.

Muir, K. (1983) Agricultural Marketing in Zimbabwe: Undergraduate Essays. Working Paper 1/83. University of Zimbabwe.

Mungwari, T. (2017) *Representation of Political Conflict in the Zimbabwean Press: The Case of The Herald, The Sunday Mail, Daily News and The Standard, 1999–2016*. Unpublished PhD thesis, University of South Africa, Pretoria.

Mustafa, G., Yaseen, Z. and Arslan, M. (2021) Theoretical approaches to study the public policy: an analysis of the cyclic / stages heuristic model. *PalArch's Journal of Archaelogy* 18, 1307–1321.

Mutambara, J. (2016) *USAID Strategic Economic Research and Analysis – Zimbabwe (SERA) Program: Maize Production and Marketing in Zimbabwe: Policies for a High Growth*. USAID, Zimbabwe.

Obi, A. and Chisango, F.F. (2011) Performance of smallholder Agriculture under limited mechanization and the fast track land reform program in Zimbabwe. *International Food and Agribusiness Management Review* 14, 85–104.

Oya, C. (2012) Contract farming in sub-Saharan Africa: a survey of approaches, debates and issues. *Journal of Agrarian Change* 12, 1–33. DOI: 10.1111/j.1471-0366.2011.00337.x.

Paradza, G. (2010) Reflections on Zimbabwe's Past as Building Blocks for its Future. Centre for Policy Studies.

Pazvakavambwa, S. (2009) Zimbabwe: Achieving household and national food security. A-MDTF. World Bank, Harare.

Pindiriri, C., Chirongwe, G., Nyajena, F.N. and Nkomo, G.N. (2021) *Agricultural free input support schemes, input usage, food insecurity and poverty in rural Zimbabwe.Working Paper Series*. Zimbabwe Economic Policy Analysis and Research (ZIPARU).

Popoola, O.O. (2016) Actors in decision making and policy process. *Global Journal of Interdisciplinary Social Sciences* 5, 47–51.

Prowse, M. (2012) Contract farming in developing countries: a review. In: *A Savoir*, Vol. 12. AFD, Agence française de développement, Paris.

Rakodi, C. (1996) Urban land policy in Zimbabwe. *Journal of Environment and Planning Analysis* 28, 1553–1574. DOI: 10.1068/a281553.

Rukovo, A., Takavarasha, T., Thiele, R. and Wiebelt, M. (1991) The profile of agricultural protection in Zimbabwe. Kiel Working Paper, No. 457.

Ruml, A. and Qaim, M. (2021) Smallholder farmers' dissatisfaction with contract schemes in spite of economic benefits: issues of mistrust and lack of transparency. *The Journal of Development Studies* 57(7), 1106–1119. DOI: 10.1080/00220388.2020.1850699.

Ruml, A., Ragasa, C. and Qaim, M. (2021) Contract farming, contract design and smallholder livelihoods. *Australian Journal of Agricultural and Resource Economics* 66, 24–43. DOI: 10.1111/1467-8489.12462.

Runganga, R. and Mhaka, S. (2021) Impact of agricultural production on economic growth in Zimbabwe. MPRA Paper no.106988.

Scoones, I., Marongwe, N., Mavedzenge, B., Murimbarimba, F., Mahenehene, J, *et al.* (2012) Livelihoods after land reform in Zimbabwe: understanding processes of rural differentiation. *Journal of Agrarian Change* 12, 503–527. DOI: 10.1111/j.1471-0366.2012.00358.x.

Sen, S. (2012) Systems Analysis of David Easton. III Paper V Half 1 Topic 1b.

Stoker, G. (2018) Governance as theory: five propositions. *International Social Science Journal* 68(227–228), 15–24. DOI: 10.1111/issj.12189.

The Zimbabwean Herald (2016) Editorial comment – lets protect our tobacco farmers. Available at: https://www.herald.co.zw/editorial-comment-lets-protect-our-tobacco-farmers/

Vicol, M., Fold, N., Hambloch, C., Narayanan, S. and Pérez Niño, H. (2022) Twenty-five years of living under contract: contract farming and agrarian change in the developing world. *Journal of Agrarian Change* 22(1), 3–18. DOI: 10.1111/joac.12471.

Watson, C. (2019) Governance and the centrality of failure: a comedy of errors. University of Sterling Comedy and social science: New directions in sociology (1). Available at: https://www.stir.ac.uk/research/hub/publication/1430450 (accessed 28 August 2022).

World Bank (2016) Ramani Huria: the Atlas of flood resilience in Dar es Salaam.

World Bank (2017) Enabling the business of agriculture. Available at: http://documents.worldbank.org/curated/en/369051490124575049/Enabling-the-business-of-agriculture-2017 (accessed 11 July 2023).

World Bank (2019) Zimbabwe Public Expenditure Review with a focus on agriculture. DOI: 10.1596/32506.

World Bank (2022) The world bank annual report 2022.

Wright, N. and Takavarasha, T. (1989) The Evolution of Agricultural Pricing' Policies in Zimbabwe: 1970's and 1980's. Working Paper AEE 4/89. University of Zimbabwe.

Zinyama, T., Nhema, A.G. and Mutandwa, H. (2015) Performance management in Zimbabwe: review of current issues. *Journal of Human Resources Management and Labor Studies* 3(2). DOI: 10.15640/jhrmls.v3n2a1.

14 Resuscitation of Neoliberalism in Zimbabwe: Exploring Implications for Agriculture Policy Development

Emmanuel Ndhlovu*

Vaal University of Technology, Vanderbijlpark, South Africa

Abstract

Available studies on Zimbabwe's land discourses and practices mostly focus on issues that require urgent attention and intervention, such as peasant livelihoods, food sovereignty, agrarian productivity, and neoliberal accumulation, among others. As a result, policy issues, though key to land and agrarian practice, do not get the same level of attention. This chapter is part of an emerging strand of literature that argues for the need for a viable agribusiness policy as the centre of a functional agrarian society. Underpinned by a critical document analysis of articles obtained in grey and academic literature, this chapter reflects on Zimbabwe's land discourses and practice in the 'new dispensation' period, evaluates prospects for a viable agribusiness development policy in the context of regrouping neoliberalism, and proposes potential agriculture policy pathways. The study found that shifting standpoints and pro-capital inclinations compromise the government's ability to craft a viable agriculture policy. The chapter proposes the mobilization of political will and the insertion of transformative social policy standpoints in the designing and execution of a viable agriculture policy in Zimbabwe.

Keywords: Agribusiness policy, Land reform, Neoliberalism, Peasants, Zimbabwe

14.1 Introduction

Since independence from British rule in 1980, Zimbabwe has been one of many African countries that focused on agriculture to develop their economies. Today, agriculture dominates the economic practice and activities of the majority of people in the country (Moyo, 2018; Ndhlovu, 2020, 2022a; Mhlanga and Ndhlovu, 2021). The majority of households in the country rely on agriculture for food production, domestic livelihood, job creation, and income generation, for instance (Ndhlovu, 2020). At independence, faced with deep-rooted poverty, formally or informally, the poor, most of whom were located in rural areas, increased their reliance on agriculture for survival. Due to the great potential of agriculture as a livelihood tool in Zimbabwe, the government also focused on this sector to improve the lives of the people (Ndhlovu, 2021a). Agriculture, therefore, received amplified post-independence development support ahead of the other economic sectors. One of the major driving factors for this was affordability in terms of capital availability, but also sectoral expertise in agriculture. At independence, the government lacked financial capital; it therefore opted to commence its development agenda on agriculture essentially because the people

---

*manundhl@gmail.com

© CAB International 2023. *Sustainable Agricultural Marketing and Agribusiness Development: An African Perspective* (eds B. Nyagadza and T. Rukasha)
DOI: 10.1079/9781800622548.0014

already had more indigenous skills in agriculture than in other economic sectors (Ndhlovu, 2021b). The government also adopted pro-poor and pro-peasant land and agricultural policies with a focus on land redistribution and agri-input support. The land redistribution process has been in phases, reaching the highest level in the radicalized redistributions of the 2000s – the Fast Track Land Reform Programme (FTLRP).

With the ascendency of the 'new dispensation' regime to power in November 2017, however, the land and agribusiness policy trajectory is now uncertain and on a precarious edge as it gets submerged under the unfolding neoliberal 'Zimbabwe is open for business' development approach. Once again, neoliberalism, with its proclivities to suppress state sovereignty so as to enable the unrestrained movement of transnational capital, has taken centre stage. A question on the possibility of the emergence of a viable agribusiness policy in the country, thus, emerges. Due to the gravity and urgency of land issues in Zimbabwe, available studies mostly focus on issues that require urgent attention and intervention, such as peasant livelihoods and food sovereignty (Moyo, 2018; Ndhlovu, 2020, 2022a), agrarian productivity (Mazwi et al., 2019), and neoliberal accumulation, among others (Ndhlovu, 2022a). Thus, blind spots remain in relation to the prospects for the emergence of a viable agribusiness development policy. This chapter closes this gap, which had huge policy implications.

This chapter (i) reflects on Zimbabwe's land discourses and practice in the 'new dispensation' period; (ii) evaluates the prospects for a viable agribusiness development policy in the context of regrouping neoliberalism; and (iii) proposes potential agriculture policy pathways. The study is part of the emerging strand of literature seeking to expose regrouping neoliberal capitalism in the country. With the hope to influence agribusiness policy, the study graphically describes the terrible legacy of neoliberalism: aggravated income inequality, weakening worker security, economic insecurity, and the implementation of fiscal austerity policies that leave the poor even more vulnerable (Ndhlovu, 2022a). This is all done in the spirit of broadening existing land and agrarian discourses and practices in Zimbabwe with a view to highlighting potential steps toward the development of a viable agribusiness policy.

The chapter first makes a background discussion of the Zimbabwe land reform policy formulation from the colonial era (Rhodesia) to the present (Zimbabwe). The background discussion is divided into subsections. This is followed by an outline of the research methodology for the study. Thereafter, the results of the study are presented and discussed. Lastly, concluding remarks are made.

14.2   The Evolution of Land Policy in Zimbabwe

This section is in two parts. The first highlights the evolution of land policy during colonialism while the second one flags up land policies since independence.

14.2.1   The politics of land policy formulation in colonial Zimbabwe

Zimbabwe was under British colonial rule from 1893 to 1980. During this period, land and agrarian policies that sustained and bolstered White supremacy dominated both land discourse and practice. Among the many land and agrarian policies adopted was the Rudd Concession (1888) in which King Lobengula of the Matabele was misled to sign a fraudulent document in exchange for a gunboat, 1000 rifles, and a £100 monthly rental (Ndhlovu, 2021a). The Rudd Concession fraudulently gave rights to Rhodes and the British South African Company to occupy the country, prospect for minerals, pass laws, establish a police force, and administer lands (Ndhlovu, 2021a). The Rudd concession, followed by other fraudulent documents including the Lippert Concession (1891), became the basis for racial land and agrarian change policies in Southern Rhodesia.

In 1898, the Native Reserve Order in Council was implemented by the colonial regime to create Native Reserves for Africans and to create more space for occupation by White settlers (Moyana, 1984). The reserves were not only planned in a hurried manner but also ensured that they would be haphazard. They were also made to ensure that Africans were located in low-potential areas so as to make them always available in the labour market. In 1930, the Land Apportionment Act, which eventually sealed the alienation of Africans from land, was adopted. The Act, like its South African cousin, the Native Land Act of 1913, empowered Whites to utterly wrench and alienate Africans from land and confine them to a few areas in which Blacks were allowed to purchase (Purchase Areas (PAs)) (Chitiyo, 2000). The Act shared land as follows: 8.8 million hectares for Native Reserves, 3 million for Native Purchase Areas, 19.9 million hectares for White farming and urban areas, and 7.2 million hectares were unallocated land (Ndhlovu, 2021a). While the Act resulted in overcrowding and overstocking in communal areas (CAs) (Moyana, 1984), for Whites, the Act was described in lurid terms such as the 'Magna Carta of the European', 'the bulwark of White civilisation', and 'the cornerstone of our native policy' (see Ndhlovu, 2021a).

In 1969, the Smith regime enacted the Tribal Trust Land Act, which was to cement the separation of land between Whites and Africans. The Act resulted in the launch of tribal trust lands where traditional leadership structures regained the authority to administer land (Palmer, 1977). In the same year, the Land Tenure Act, which substituted the Land Apportionment Act of 1930, was adopted to categorize lands into European, African, and National lands. Whites who constituted 5% and Africans who constituted 95% got 45,000 acres each, while National land stood at 6500 acres (Palmer, 1977). The Land Tenure Act was replaced by the 1977 Land Tenure Amendment Act, which also led to the adoption of the 1978 Land Tenure Repeal Act. These racial land policies were the centre of political and nationalistic activities that eventually led to independence in 1980. The racial land policies did not, however, cease at independence. The 1979 Lancaster Constitution, which emerged from independence negotiations, sustained racial policies until 1990 (Chitiyo, 2000). The Constitution had some clauses that prevented the newly established government from compulsorily acquiring land. It stipulated that any land redistribution would be in terms of 'willing buyer, willing seller' provisions. Chapter 3, Section 16 of the Constitution required the 'authority [requiring land] to pay promptly adequate compensation for acquisition' and that 'if the acquisition is contested, to apply to the General Division or

some other court before or not later than thirty days after the acquisition for an order confirming the acquisition' (Utete, 2003).

14.2.2   Land policy formulation since independence

When Zimbabwe attained its independence from colonial rule in 1980, only about 6000 white farmers, some few agro-industrial estates (locally and foreign-owned), about 8000 small-scale Black commercial producers, and about 700,000 peasant households occupied most of the good land while the rest of the populace eked for a living on agro-ecologically marginal lands located mostly in the Lowveld where rainfall was inadequate for crop production and temperatures too harsh for animal husbandry (Ndhlovu, 2021a). Thus, the land reform policy adopted in the first decade had three noble and basic components, namely: restitution, tenure security and redistribution (Moyo, 1995). The targeted beneficiaries of the earliest land policy formulation efforts included: the landless poor and those with susceptible dependants such as young children and the elderly of over 55 years; experienced farmers willing to forfeit their land rights in CAs; returning refugees displaced by the 1970–79 liberation war; farmers with Master Farmer certificates; war veterans; and commercial farm workers (Moyo, 1995).

Land policy formulation after independence commenced in September 1980 with the launching of the first phase of land reform: Land Reform Programme, Phase One (1980–1997). The programme involved moving people from CAs into designated lands. The government committed itself to building infrastructure and providing basic services to designated areas so as to reduce the poverty from colonial policies and the war (Moyana, 1984). The government had aimed to acquire 8.3 million hectares from Whites using market-based approaches, and thus, resettle about 162,000 families (GoZ, 1998). However, due to a number of challenges, particularly the Lancaster Constitution, only 43% of the targeted families were resettled (GoZ, 1998). When the Lancaster Constitution expired in 1990, the Land Acquisition Act (1992) was adopted with the view to improve land policy, among other things. Continued poor results in land reform led to the Constitution being amended two times in 1993 (Zimbabwe Amendment Act Nos. 12 and 13). All these efforts, however, made no significant progress in the acquisition of enough land for the needy and the poor. Thus, although the first phase failed to acquire the anticipated land size due to a number of challenges, it is praised for improving population distribution and land use patterns as much larger numbers of people moved into formerly dispersed areas (Moyana, 1984).

The second phase of land reform ran from 1998 to 2002. Due to the slow nature of the previous land reform activities, the second phase was characterized by unplanned grabs of underutilized lands, particularly those in the proximity of communal areas and urban townships (Moyo, 1995). A Land Donor Conference was eventually convened in 1998 in an effort to raise a sum of US$1.9 billion to finance Phase Two of land reform (GoZ, 2001). The intention was to make use of specific criteria to be used for land redistribution. The several specifications included that: land had to be underutilized; the owner of the land was absent or had multiple farms; the farm was oversized and exceeded 1500 hectares; and land was close to communal areas (GoZ, 1998). However, only 0.02% of the required amount was raised at the conference (GoZ, 2001). In the reviewed literature, it is argued that the conference failed due to the failure of the government to clarify to market-oriented donors how the US$1.9 billion would be used (Chitiyo, 2000; Ndhlovu, 2021a). However, even though the Zimbabwean government had clarified how the raised funds would be used, it is unlikely that the donors would have been convinced since the market-oriented approach was preferred.

With poverty intensifying in rural areas; and with about 60% of the total national population categorized as poor in the 1990s (Moyo, 1995), the need for land intensified, leading to various land policy attempts throughout the 1990s and spilling over into the 2000s under the FTLRP. The intention for the FTLRP was to acquire about 3000 farms in a quick manner and release it to people under the A1 (small-sized) model and A2 (commercial farming) fast-track models (Moyo, 2013). The initial criteria for land identification for redistribution under the FTLRP were underutilized land, land that was derelict, foreign-owned land, land owned

by a farmer who owns other farms, or land that was very close to CAs. The state managed to acquire about 1948 farms for redistribution even though the initial criterion was not used (Commercial Farmers' Union, 2001). Before the FTLRP, large-scale commercial farmers owned 11.8 million hectares while the CAs occupied about 16.4 million hectares. However, by 2003, the size of land owned by the large-scale commercial sector was reduced to 12% from 30% while the communal land expanded to 71% from 54% (GOZ, 2003). By the year 2010, the land had been reallocated to over 150,000 urban dwellers, farm workers, peasants in the countryside, and civil servants under the A1 scheme; while an additional 20,000 recipients were allocated A2 farms (Moyo, 2013).

The FTLRP is criticized by some scholars as an economic disaster that ignited food insecurity, joblessness and environmental degradation, attracted economic sanctions and international boycotts, disrupted agro-inputs access and utilization, and, thus, negatively impacted livelihood activities (Zamchiya, 2011). However, in Zimbabwe itself, and also in much of the Southern Africa region and the Global South, the programme is a resounding success which completed the incomplete decolonization taking it into the realm of economic emancipation and delivered the long-awaited land redistributive justice (Mkodzongi, 2013, 2018). With the FTLRP, Zimbabwe managed to initiate a pro-poor agriculture development policy. While other countries on the continent were also making progress with regard to their land reforms, what makes the Zimbabwean case unique is the country's ability to resist fresh land dispossession by foreign capital when most countries on the continent are easily giving up land in exchange for capital. According to Moyo:

> What is unique in the case of Zimbabwe is that it has rowed against the current to meet the rest of Africa halfway, by breaking up the large-scale farming established in the course of the nineteenth-century scramble, broadening the small-scale capitalist sector, which had also been introduced by the colonial regime and preserving some agro-industrial estates.
>
> (Moyo, 2013, p. 341)

Mazwi *et al.* (2018) posit that as renewed land grabs by monopoly-finance capital gathered pace in Africa, Zimbabwe emerged as a unique case of radicalization with its FTLRP, which, instead of losing land to white capital, actually wrenched land from the same capital. The FTLRP disrupted land-related policies that had remained arched toward the realization of the interests of landed capitalists and a few Blacks who accessed land through patronage networks and largely ignored the plight of peasants (Ndhlovu, 2021a). Mkodzongi (2010) argues that considering Zimbabwe's colonial history of land denial by Whites, it was necessary that the new government at independence make use of any means available and affordable to release land to citizens. Mkodzongi (2010) also takes issue with the notion that land had to be bought from the Whites occupying it arguing that they did not buy it in the first place, and therefore were not entitled to any monetary compensation. Instead, Mkodzongi (2010) required Whites to surrender land wilfully so that the land can be redistributed to Blacks. Since Whites acquired the land in question illegally, it would be important that the land be transferred back to Blacks from whom it was illegally taken. The radical, Pan-Africanist, and pro-poor stance in the land policy and pronouncements adopted in Zimbabwe, therefore, particularly in the late 1990s and early 2000s, is widely viewed as having the potential to serve as the basis for the design and implementation of a viable land policy in the country. However, the 'new dispensation' regime that ascended to power in November 2017 seems to be slowly reversing the pro-poor trajectory of land reform policy formulation as discussed in the next section.

14.3   Materials and Methods

This chapter utilizes the critical document analysis methodology, which is augmented by sectorial expertise abstraction. The articles used in the study included public records (mostly government, Zimbabwe African National Union-Patriotic Front (ZANU-PF), and opposition political party documents), personal documents (newspapers and online journals), and academic evidence (publications), which were obtained in both grey and academic literature using

Table 14.1. List of articles used in the study.

Author(s)	Paper type
GoZ, 2018a	Public record
GoZ, 2018b	Public record
GoZ, 2018c	Public record
GoZ, 2019	Public record
GoZ, 2020a	Public record
GoZ, 2020b	Public record
Mandishekwa and Mutenheri, 2020	Academic document
Mhlanga and Ndhlovu, 2021	Academic document
Mazwi and Mudimu, 2019	Academic document
Mazwi et al., 2018	Academic document
Elich, 2020	Personal document
Njini et al., 2019	Personal document
The Sunday Mail, 2019	Personal document
The Zimbabwe Mail, 2019	Personal document
Tome and Mphambela, 2020	Personal document
Freeth, 2020	Personal document

agribusiness policy, land reform, neoliberalism, peasants, and Zimbabwe as key terms.

The articles were selected manually from various data sources including the website of the Government of Zimbabwe (GoZ) and academic databases, such as the Web of Science, Google Scholar and Researchgate, as well as newspapers. This was meant to generate a balanced analysis of land and agrarian change discourses and practices in the country. A total of 16 articles published since the dawn of the 'new dispensation' were selected for analysis (Table 14.1). The inclusion criteria included that the article is related to the 'new dispensation' period and that it was related to land and agriculture discourse and/or practice. Discourse analysis was used to assimilate viewpoints on the prospects of the development of a viable agribusiness policy in the country.

The use of secondary sources without the support of empirical evidence, however, presents a particular limitation for this study. Other researchers should be motivated to critique the results of the study and carry out more detailed empirical research (shown in Table 14.1) to verify the arguments made here. In doing so, it might be important to include both qualitative and quantitative data so as to provide a generalized overview of the debate.

14.4 Results and Discussion

Since its ascendency in 2017, the 'new dispensation' has either introduced or made policy pronouncements that affect agriculture more than any other sector. The pro-finance capital stance now preferred by the regime under its 'Zimbabwe is open for business' development approach represents a departure from Mugabe's Pan-Africanist and pro-poor approach. Mugabe preferred homegrown and inward-looking development policies and only reached out to the East under the 'Look East Policy' when the need arose (Mazwi and Mudimu, 2019). The 'new dispensation' administration pursues the neoliberal approach in which land is commodity, access to which should be subjected to business principles. Through the Agriculture and Food Systems Strategy, the agricultural sector, under the new administration, is to be transformed by 'making farming a business' (GoZ, 2020a, p. 1). The regime prefers that '...the economy [be] ... founded on sound market principles and principles of legal protection that encourage and protect private enterprise and the fruits thereof' (GoZ, 2018a, p. 3). To create '...investor-friendly and sustainable supply-side policies to stimulate production across all sectors' (ZANU-PF, 2018, p. 20), the regime encourages 'foreign investors [to feel] free to invest' in the agricultural sector 'through contract farming arrangements' (GoZ, 2018a, p. 23). The government has also lowered 'the cost of doing business, including trade and labour regulations' (ZANU-PF, 2018, p. 22). Except for diamond and platinum mining, the Indigenisation and Economic Empowerment Act, which prohibited foreign investors to hold more than 49% of businesses in various sectors, has been scrapped (GoZ, 2018b). This has attracted monopoly-finance capital, particularly to the countryside where

it now displaces rural households in the name of investments (Mandishekwa and Mutenheri, 2020; Ndhlovu, 2022b).

The new government shows no interest in developing or maintaining any of the existing agriculture policies (Njini *et al.*, 2019), but rather clarified that it is in pursuit of '... quick-win investment opportunities for realisation of self-sufficiency and food surpluses that will see the re-emergence of Zimbabwe as a major contributor to agricultural production and regional food security in the Southern Africa region and beyond' (GoZ, 2018c, p. 26). Notwithstanding a huge number of people on its land waiting list (Mazwi *et al.*, 2018), the government has stopped land redistribution 'except for public purposes ... in accordance ... with principles of international law, and subject to the prompt payment of adequate and effective compensation' (GoZ, 2018a, p. 5). The abandonment of the Pan-Africanist land policy and the return to the market-based land delivery system is meant to attract 'direct foreign investment while generating employment for Zimbabweans, to facilitate the entry and stay [settler] of key foreign technical and managerial personnel for the purpose of engaging in activities related with foreign investment' (GoZ, 2018a, p. 5). The government views this as part of the broader 'policy reform initiatives of the New Dispensation to stimulate domestic production, exporting, rebuilding and transforming the economy to an Upper Middle Income status by 2030' (GoZ, 2018c, p. 7).

The previous land reform policies adopted since independence, including the FTLRP, were largely in favour of the peasantry (which makes up the majority of farmers in the country (Moyo, 2016)). The pro-Pan-Africanist Mugabe administration had released a state-based permit system for A1 smallholder farmers and a 99 year lease to A2 small-medium capitalist farmers as the suitable system of land tenure as part of an effort to protect smallholder farmers against the land-grabbing tendencies of capital-rich actors through the market. On the contrary, in the new dispensation, in pursuit of vision 2030, '18,000 A2 farmers are [now] going to be transformed to agricultural entrepreneurs and the farms to become enviable businesses by 2025 [while] 360,000 A1 farmers to become viable and formal Small to Medium Enterprises by 2025' (GoZ, 2020b, p. 1b). Farmers had a deadline of 31 January 2021 to 'submit mandatory production returns', failure of which would result in their farms being categorized as being either abandoned, derelict or underutilized, and therefore, requiring reallocation (GoZ, 2020b, p. 2b). This is a clear return to the market-based approach that the country languished under in the 1980s. Before his death, the then Minister of Lands, Agriculture, Water and Rural Resettlement, Perrance Shiri, indicated that: 'The time [would] come when the government [would] really consider taking back all underutilized land and allocate it to other potential users' (Tome and Mphambela, 2020).

In a move that will reverse the much-celebrated FTLRP, on 31 August 2020, the Ministry of Lands, Agriculture, Water, and Rural Resettlement released a joint statement with the Ministry of Finance and Economic Development revealing that former White farmers could now apply for the repossession of their lands and that the government would '... revoke the offer letters of resettled farmers currently occupying those pieces of land and offer them alternative land elsewhere regardless of model' (GoZ, 2020a, p. 3). The intention to repossess land from peasants to redistribute it to capitalists shows the highest-level policy inconsistency, which not only threatens prospects for improved domestic consumption, thereby further weakening livelihoods, but also further affects livelihoods. It is also likely to condemn the country's peasant households to be farm labourers who are exposed to finance-capital exploitation (Freeth, 2020; Ndhlovu, 2022a). In addition, this also exposes the lack of a viable peasant-oriented agribusiness policy in the country.

The new dispensation regime also terminated the Command Agriculture – a state-run contract project which commenced in 2016 to assist both A1 and A2 farmers who were engaged in cereals production for domestic use. The facility provided farmers with inputs as well as a ready market for their produce. The facility prioritized the tenure security and livelihood rights of farmers and the state over the interests of neoliberal monopoly-finance capital (Elich, 2020). The new regime discontinued the facility and posited that it would no longer play any leading role in agriculture financing, but rather it would only provide a framework that encourages financial businesses to engage farmers on a commercial basis. In the Transitional and Stabilisation Plan (TSP), the state posited that '...reliance on government support [by peasants] for... agriculture Production ... [would] be gradually reduced as initiatives to enhance private sector support gather momentum, that way overcoming potential development of voids in capacitating production by the farmer' (GoZ, 2018c, p. 27). In dismissing Command Agriculture, the Minister of Finance Mthuli Ncube indicated the need '... to reorient the financing model for agriculture to crowd in private-sector financing, reduce significantly government footprint in production, and lessen the dependence on the budget' (GoZ, 2019). The Command Agriculture has since been changed to Smart Agriculture so as to fit in the neoliberal capitalistic language, and to allow the same opportunistic financial institutions which are export-oriented to now also engage with domestically oriented farmers.

The land practices and discourses in Zimbabwe under the 'new dispensation' therefore exhibit a lack of land and agriculture policy consistency. This has not only undermined food sovereignty prospects (Ndhlovu, 2022a) but also livelihoods and tenure security (Mazwi *et al.*, 2018; Freeth, 2020; Elich, 2020). Shonhe concludes that:

> ... the ruling capitalists [in Zimbabwe] have become an extension of global capital, if not captured agents of the latest form of imperialism. Opening Zimbabwe for business transfers surplus value through international trade, unequal exchange, and unequal rewards.
>
> (Shonhe, 2020, p. 276)

In order for Zimbabwe to have a functional agriculture system, there is a crucial need to have a clear and grounded agriculture policy. A clear policy would enable the country to command a prosperous agricultural sector, which will enable it to regain its label as 'the bread basket for southern Africa' (Dabale *et al.*, 2014, p. 38).

14.5   Towards a Viable Agriculture Policy

Zimbabwe should begin its economic development agenda in the agriculture sector. This call might not be entirely incorrect considering that the country is currently poor with over 70% of the population classified as poor (Mhlanga and Ndhlovu, 2021). The failure of the Zimbabwe economy shows up in each and every rating of international repute. The country's economy was ranked 159 out of 190 by the World Bank's Doing Business (2018), 144 out of 159 in 2017 by the Fraser and Cato Institutes' Economic Freedom of the World listing, and 124 out of 137 by the World Economic Forum's Global Competitiveness Index (2017–2018) (The Zimbabwe Mail, 2017). Therefore, one of the factors that should promote a focus on agricultural development than the rest of the available economic sectors is the sectoral expertise in agriculture. The government is resource-poor, and therefore should opt to begin its development agenda on agriculture mainly because already its general population has existing indigenous skills in agriculture, and not in the other economic sectors. The development of a sound agriculture policy should therefore be informed by the realization of the fact that Zimbabwe is an agrarian country in which most people depend on land and agriculture for livelihoods and as a major source of income.

A viable agriculture development policy, as proposed by Mkandawire (2001) could be one informed by the transformative social policy as it has the potential to lead to the realization of the nation-building project. The transformative social policy is based on 'the norms of equality and social solidarity... these norms serve many functions: production, protection, reproduction, redistribution, and social cohesion or nation building' (Adesina, 2009). An agriculture policy that is guided by the transformative social policy has the potential to generate production and redistributive and and protective roles which are critical for national development and nation-building (Mkandawire, 2001). This implies that the resultant policy will not only ensure tenure security, but it will also motivate and guarantee increased production by farmers of all categories to increase production (Ndhlovu, 2021b). Adesina (2009) avers that a transformative social policy comprises the enhancement of human welfare and the transformation of social institutions, social relations and the economy. In this view, a viable agribusiness policy would be one which links agriculture policies to social and economic policies, and which recognizes political influences, shifting power relations, the existing diverse levels of resource access, and ideological orientation in development discourse and practice (Mkandawire, 2001). A viable agriculture policy in the country would be one in which citizen participation is placed at the centre. The neoliberal 'Zimbabwe is Open for Business' approach is devoid of a transformative social strategy, and therefore, is not useful for crafting an agriculture policy.

The resuscitation of neoliberal standpoints for the sake of attracting foreign direct investment renders the state incapable of crafting a useful policy for the agriculture sector. Opposition politics in the country also harbours some neoliberal standpoints with regard to land access, ownership, and utilization. As a result, it is difficult to project the trajectory of agriculture policy formulation and implementation in the country. This chapter proposes the mobilization of political will and the insertion of transformative social policy standpoints in the designing and execution of a viable agribusiness policy in Zimbabwe.

14.6   Conclusion

This chapter reflected on Zimbabwe's land discourses and practices in the 'new dispensation' period. The aim was to evaluate the prospects for the emergence of a viable agriculture development policy in the country. The chapter found that contrary to the post-independence period, particularly after the expiration of the Lancaster Constitution in 1990 where pro-poor and African-centred approaches were used in land policy formulation and implementation, the new dispensation regime resuscitated the neoliberal approaches that were abandoned for uselessness in the early 1990s. The new dispensation mirrors the colonial period in terms of instability in land policies. If anything, the pursuit of neoliberal development orientations for the sake of attracting finance capital compromises and underpins the country's ability to craft a viable agribusiness policy that can place the country's smallholder farmer majority at the centre of the much-needed development. The chapter, thus, takes issue with the pro-capital approach and pushes forward the need for a transformative social policy-based framework. The chapter posits that such a framework has the potential to provide guiding principles and practices upon which viable agribusiness policy can be designed in a country that is dominated by peasant households like Zimbabwe. Future research can focus on how political will can be mobilized to enable political leaders to prioritize the interests of citizens during policy design and implementation.

References

Adesina, J.O. (2009) Social policy in sub-Saharan Africa: a glance in the rear-view mirror. *International Journal of Social Welfare* 18, S37–S51. DOI: 10.1111/j.1468-2397.2009.00629.x.

Chitiyo, T.K. (2000) *Land, Violence, and Compensation: Reconceptualising Zimbabwe's Land and War Veterans Debate*. Academy Publishing (Pvt) Ltd, Harare.

Commercial Farmers' Union (2001) *The Commercial Farmer's Union Statement*. Commercial Farmers Union, Harare.

Dabale, W.P., Jagero, N. and Chiringa, C. (2014) Empirical study on the fast track land reform program (FTLRP) and household food security in Zimbabwe. *European Journal of Research and Reflection in Management Sciences* 2, 34–44.

Elich, G. (2020) Mnangagwa's neoliberal assault on the Zimbabwean people. CounterPunchnc.org. Available at: https://www.thezimbabwean.co/2020/02/mnangagwas-neoliberal-assault-on-the-zimbabwean-people/ (accessed 5 August 2020).

Freeth, B. (2020) Farmers oppose compensation deal with Zimbabwe. CajnewsAfrica. Available at: https://www.cajnewsafrica.com/2020/08/11/farmers-oppose-compensation-deal-with-zimbabwe/ (accessed 18 August 2020).

GoZ (1998) *Land Reform and Resettlement Programme Phase II: Policy Framework and Project Document*. Government Printers, Harare.

GoZ (2001) *Zimbabwe's Land Reform Programme*. Government Printers, Harare.

GoZ (2018a) *Investment Guidelines and Opportunities in Zimbabwe*. Government of Zimbabwe, p. 3, sections 1.1 and 1.4.

GoZ (2018b) *Amendments of Zimbabwe's Indigenization and Economic Empowerment Act*. Government of Zimbabwe.

GoZ (2018c) *Transitional Stabilisation Programme: Reforms Agenda, section 362*. Government of Zimbabwe.

GoZ (2019) *Letter of Intent: Attachment 1 – Memorandum of Economic and Financial Policies*.

GoZ (2020a) *Joint Statement by the Minister of Lands, Agriculture, Water, and Rural Resettlement and the Minister of Finance and Economic Development in the wake of the conclusion of on 29 July 2020 of the Global Compensation Deed between the Government of Zimbabwe and the former farm workers*. Government of Zimbabwe.

GoZ (2020b) *Press Statement by the Minister of Lands, Agriculture, Water, and Rural Settlement, Hon, Dr AJ Masuka on Enhancing Production, Productivity and Profitability on A2 and A1 Farms*.

Mandishekwa, R. and Mutenheri, E. (2020) The economic consequences of internal displacement in Zimbabwe. *International Journal of Migration and Border Studies* 6(3), 206. DOI: 10.1504/IJMBS.2020.111440.

Mazwi, F. and Mudimu, G.T. (2019) Why are Zimbabwe's land reforms being reversed? *Economic and Political Weekly* 54, 35.

Mazwi, F., Chemura, A., Mudimu, G. and Chambati, W. (2019) The political economy of command Agriculture in Zimbabwe: a state-led contract farming model. *Agrarian South: Journal of Political Economy* 8, 1–26.

Mazwi, F., Tekwa, N., Chambati, W. and Mudimu, G.T. (2018) *Locating the Position of Peasants under the 'New Dispensation': A Focus on Land Tenure Issues*. Sam Moyo Africa Institute for Agrarian Studies, Harare.

Mhlanga, D. and Ndhlovu, E. (2021) Financialised agrarian primitive accumulation in Zimbabwe. *African Renaissance* 18, 185–207. DOI: 10.31920/2516-5305/2021/18n1a12.

Mkandawire, T. (2001) *Social Policy and Development Programme*. United Nations Research Institute for Social Development,Geneva.

Mkodzongi, G. (2010) Zimbabwe's land reform is common sense. Pambazuka News. Available at: http://www.pambazuka.org/en/category/features/62917

Mkodzongi, G. (2013) Fast tracking land reform and rural livelihoods in Mashonaland west province of Zimbabwe: opportunities and constraints, 2000-2013. Unpublished PhD thesis, University of Edinburgh, UK.

Mkodzongi, G. (2018) Peasant agency in a changing agrarian situation in central Zimbabwe: the case of Mhondoro Ngezi. *Agrarian South: Journal of Political Economy* 7, 188–210.

Moyana, H.V. (1984) *The Political Economy of Land in Zimbabwe*. Mambo Press, Harare.

Moyo, S. (1995) *The Land Question in Zimbabwe*. SAPES Trust, Harare.

Moyo, S. (2013) Land reform and redistribution in Zimbabwe since 1980. In: Moyo, S. and Chambati, W. (eds) *Land and Agrarian Reform in Zimbabwe: Beyond White-Settler Capitalism*. CODESRIA, Dakar. DOI: 10.2307/j.ctvk3gnsn.

Moyo, S. (2016) *Family Farming in Sub-Saharan Africa: Its Contribution to Agriculture, Food Security and Rural Development*. Working paper number 150. Food and Agriculture Organization of the United Nations and the United Nations Development Programme.

Moyo, S. (2018) Third world legacies: debating the African land question with Archie Mafeje. *Agrarian South: Journal of Political Economy* 7, 1–23. DOI: 10.1177/2277976018775361.

Ndhlovu, E. (2020) Decolonisation of development: Samir Amin and the struggle for an alternative development approach in Africa. *The Saharan Journal* 1, 87–111.

Ndhlovu, E. (2021a) Land, agrarian change discourse and practice in Zimbabwe: Examining the contribution of Sam Moyo. PhD thesis, University of South Africa, Pretoria.

Ndhlovu, E. (2021b) The land question in Africa: Sam's perspectives. In: Gumede, V.T. and Shonhe, T. (eds) *Rethinking the Land and Agrarian Questions in Africa*. Africa Century Editions Press. DOI: 10.1007/s11832-016-0776-y.

Ndhlovu, E. (2022a) Changing agrarian discourses and practices and the prospects for food sovereignty in Zimbabwe. In: Mkodzongi, G. (ed.) *The Future of Zimbabwe's Agrarian Sector: Land Issues in a Time of Political Transition*. Routledge, pp. 34–53. DOI: 10.4324/9781003158196.

Ndhlovu, E. (2022b) Political economy of Chisa livelihoods in rural Zimbabwe. In: Helliker, K., Matanzima, J. and Chadambuka, P. (eds) *Livelihoods of Ethnic Minorities in Rural Zimbabwe*. Springer. DOI: 10.1007/978-3-030-94800-9.

Njini, F., Marawanyika, G. and Sguazzin, A. (2019) Zimbabwe to scrap platinum and diamond mine ownership rules. Bloomberg. Available at: https://www.news24.com/Fin24/zimbabwe-to-scrap-platinum-and-diamond-mine-ownership-rules-20190306

Palmer, R. (1977) *Land and Racial Domination in Rhodesia*. Heinemann, London.

Shonhe, T. (2020) Primitive accumulation and Mugabe's extroverted economy: what now under the second republic. In: Ndlovu-Gatsheni, S.J. and Ruhanya, P. (eds) *The History and Political Transition of Zimbabwe: From Mugabe to Mnangagwa*. Palgrave Macmillan, pp. 275–298. DOI: 10.1007/978-3-030-47733-2.

The Sunday Mail (2019) We must be a destination where capital feels safe. Available at: https://www.sundaymail.co.zw/we-must-be-a-destination-where-capital-feels-safe

The Zimbabwe Mail (2017) Four things Mnangagwa can do to revive Zimbabwe's economy. Available at: https://www.thezimbabwemail.com/economic-analysis/four-things-mnangagwa-can-revive-zimbabwes-economy/

The Zimbabwe Mail (2019) 100, some in Mnangagwa Regalia evicted from farm in Bromley. Available at: https://www.thezimbabwemail.com/farming-environment/100-some-in-mnangagwa-regalia-evicted-from-farm-in-bromley/

Tome, M. and Mphambela, L. (2020) *Govt to Repossess Underutilised Farms*. The Herald.

Utete, C.M.B. (2003) *Report of the presidential land review committee under the chairmanship of Dr. Charles MB Utete Vol. 1: main report to his excellency, the president of the Republic of Zimbabwe*. Government Printers, Harare.

Zamchiya, P. (2011) A synopsis of land and agrarian change in Chipinge district, Zimbabwe. *Journal of Peasant Studies* 38, 1093–1122. DOI: 10.1080/03066150.2011.633703.

ZANU-PF (2018) *The People's Manifesto 2018*. ZANU PF.

15 Insights for Sustainable Rural Agribusiness Development Policy

Nyasha Nyakuchena[1]*, Joseph P. Musara[2] and Wellington Bandason[3]

[1]*Marondera University of Agricultural Sciences and Technology (MUAST), Zimbabwe;* [2]*University of the Free State (UFS), South Africa;* [3]*Women's University in Africa (WUA), Zimbabwe*

Abstract

Sustainable management helps decision makers to understand the social–economic–institutional dynamics in agricultural business development systems. The same is true for rural development processes conceptualized and understood from different worldviews. These conceptual, theoretical and practical perspectives are critical in understanding the nuances and intricacies of rural agriculture and how it shapes business developments and livelihoods. The central agribusiness strand on rural agriculture focuses on the human needs approach considering food and nutrition security among rural dwellers. However, the advent of networked supply chains, especially in Southern African food systems, makes it interesting to incorporate sustainable agribusiness management dimensions. Therefore, this chapter unpacks the conceptual, theoretical, and practical perspectives critical in understanding the complexity and diversity of modern rural agriculture in the Southern African landscape with a core focus on agribusiness development pathways. This is critical to reinforce the role of agribusiness policy in sustainable rural development while going beyond the food security pillar and unlocking the income security dimension for creating wealth.

Keywords: Sustainable management, Rural development, Southern Africa

15.1 Introduction

Rural areas in Zimbabwe are administratively all areas outside a city or city boundaries, accounting for an average of 93% of the country's area (FAO, 2021). This traditional dichotomy dividing state territory into urban and rural areas is highly imprecise. It is subject to change as new rural settlements turn into urban areas and rural areas shrink. There are spaces in suburban areas where it is difficult to form a boundary between urban and rural areas (Giuli, 2018). This imprecise definition of local spatial boundaries poses difficulties for theoretical and practical development.

Rural areas are highly diverse as they contain many different forms of land use: agricultural, forest, transport areas, water bodies, commercially unused areas, ecologically valuable territories, rural settlements inhabited by farmers and non-agricultural residents, and an increasing infrastructure base and facilities of public institutions and industrial and service companies (Rukasha *et al.*, 2021). At least functionally, these areas often include small towns, especially towns with urban and

*Corresponding author: gloriousnyakuchena@gmail.com

© CAB International 2023. *Sustainable Agricultural Marketing and Agribusiness Development: An African Perspective* (eds B. Nyagadza and T. Rukasha)
DOI: 10.1079/9781800622548.0015

rural communes. Regions with diverse regional systems are constantly changing, with their nature and functions changing. The weakening of agricultural functions will lead to the abandonment of agriculture, and new production, service and consumption functions other than agriculture will develop. The development of the non-agricultural economy and increasing agricultural productivity are changing the social and occupational structures and space use. Rural spaces are increasingly becoming commodities used by city dwellers, entrepreneurs, tourists, and thus public goods (Shonhe, 2021).

The considerable complexity and diversity of rural areas constitute a coherent theoretical conception of rural area development, enabling efficient and effective planning and development of these areas at national, regional, and local levels. It becomes a problem in any attempt to build a strategy (Shenggen, 2020). Southern Africa suffers from chronic food insecurity and poverty. These problems are likely to continue, especially in rural areas where most people live. Household production is these rural households' primary source of food and income (Shaibu, 2022). Yields of maize, the main food crop in the region, are below estimates achievable given the climate and soil conditions. With limited land available for expansion, the sustainable management aspect of agriculture (i.e. growing more on the same land while maintaining the resource base) ensures that production levels meet future demand. Adopting sustainable management practices in agriculture is highly dependent on the complex diversity of farming systems and household subsystems (Cartmel and Furlong, 2000). As a result, this often leads to unfair gains and further marginalizes groups embedded in the cycle of poverty. The diversity of household production systems makes it challenging to match viable intensive options to farmers' realities. Methods for creating household typologies and modelling the diversity of household responses to different intensification pathways can provide tools for making recommendations based on farmers' options and needs. Stakeholders should therefore embrace the agribusiness approach to rural development and scale up all pillars that enhance sustainability in these fragile communities.

15.2 An Overview of Sustainable Management in Agribusiness

The upward pressure on crucial commodity prices mentioned in the previous section emanates from several underlying factors of varying nature or 'change factors'. These 'driving forces' range from environmental to socio-economic factors, slow-moving to fast-moving factors, and have different impacts on outcomes in the short and long term. Many factors are driving long-term trends in food supply and demand that have also contributed to the tightening of global food markets over the past decade (Virva and Grant, 2021). These trends are driven not only by agriculture and energy but also by environmental and socio-economic changes. Socio-economic changes in population growth and increases in total incomes are among the primary drivers of changing consumer economic behaviour related to the demand for food and energy products (Ravikishore, 2022).

Urbanization, along with these demographic changes, is another factor affecting consumption patterns and changing consumer preferences for food, fibre and energy products. These changes in consumption and consumption preferences lead to increased demand-side stress on food and energy systems. At the same time, other environmental factors may reduce food system supplies due to resource scarcity or deterioration of land and water quality (Campos, 2021). Declining investment in crops and energy technologies could also slow supply growth in the long run, ultimately pushing prices down as demand begins to increase. These provide a range of options for policy makers to consider when deciding how best to address current stresses on food and energy systems or mitigate the severity of such stresses in the future.

15.3 Social–Economic–Institutional–Policy Nexus in Agricultural Business Development Systems

Human systems include social, economic and institutional structures and processes. From an industrial, residential and social perspective, these systems are diverse and dynamic, expressed at the individual level throughout life. They tend to revolve around human goals such as survival, security, happiness, justice and progress. And in this regard, weather and climate often take on secondary importance as sources of benefits or burdens. More important are issues such as access to financial resources, institutional capacity, conflict potential (Abafe, 2021), and stresses such as rapid urbanization, disease, and terrorism. These types of climate change could make a difference, reduce or exacerbate multiple stresses and, in some cases, push multiple stressed human systems past sustainability thresholds in complex interactions with social situations (Ngoc *et al.*, 2021).

For the most part, climate (and climate change) affects human systems in three main ways (Stavros and Dimitrios, 2022). First, it provides a context for climate-sensitive human activities, from agriculture to tourism. Rain-fed rivers, for example, enable irrigation and transportation and can enrich or harm landscapes. Second, climate influences the costs of maintaining a climate-controlled internal environment for human life and activity. Of course, higher temperatures increase cooling costs and reduce heating costs. Third, climate interacts with other kinds of stress on human systems, which may reduce or increase stress. For example, droughts can contribute to rural-to-urban migration, increasing stress on urban infrastructure and socio-economic conditions with population growth. In all these situations, the impact can be positive or negative (Fanchone *et al.*, 2022). However, extreme climate events and other abrupt changes tend to affect human systems because they have less time for adaptation. In contrast, gradual changes reach a threshold at which the effects become noticeable.

15.3.1 Conceptual, theoretical, and practical perspectives of rural agribusiness

15.3.1.1 Agribusiness system concept

Agriculture has evolved into agribusiness and has become a powerful and complicated system that includes everyone involved in getting food and fibre to customers. This system now extends well beyond the farm (Spigel and Harrison, 2017). In addition to farmland, agribusiness refers to individuals and organizations that supply the inputs (such as seed, chemicals and credit) and process the outputs (such as milk, grain and meat). It also encompasses manufacturing food products (such as ice cream, bread and breakfast cereals) and distributing them to customers (such as restaurants and supermarkets). The conventional system has undergone rapid transportation as new industries have emerged and conventional farming activities have expanded and become more specialized. Transportation did not occur overnight (Rukasha *et al.*, 2021). Instead, it developed gradually in response to several circumstances. Understanding how the agribusiness system functions now and how it is likely to change in the future is simpler by knowing how it originated.

15.3.1.2 Background of the agribusiness system

In the majority of industrialized nations, agriculture was the primary industry. Although it was simple for a farmer, output was poor. Even in 1850, the typical American farmer produced enough food to feed the entire population of the states of America. Because of this, most farmers were almost entirely self-sufficient (Campos, 2021). They generated most of the inputs they needed for production, including seed, 'draft horses', feed and basic agricultural machinery. Farm families processed the products they grew to manufacture their food and clothing. All of their output was consumed or used. Households exchanged the meager production that was not used on the farm for cash. The small percentage of the nation's inhabitants that resided in towns and villages were fed and clothed by these supplies.

Several agricultural products made it to the export market and were consumed by customers abroad. There were plenty of inexpensive lands available despite the human resources deficit. Every time a conflict started, farmers noticed a significant rise in the cost of agricultural products. They were obliged to produce more while forced to leave the fields and pay more to employ labourers. This situation caused farmers to become more interested in utilizing freshly created synthetic inputs that reduced labour (Camilla, 2023). Farmers found it more profitable to concentrate on production and bought previously prepared inputs. Others were able to capitalize on this trend and establish a business that catered to the demand for the various inputs needed to produce agricultural output, such as seed, fencing, machinery, and so forth. These businesses expanded into the different agricultural input industry sectors. Input farms, a significant part of agribusiness that offer a variety of technologically based products, produce about 75% of all the inputs used in agricultural production (Giuli, 2018).

Farms that produced commodities and manufactured food underwent a similar transition while the agricultural input sector was growing. In this process, their forms are altered by processing infrastructure to make most commodities, such as wheat, maize, milk, animals, and others, more beneficial and practical for consumers (Ravikishore, 2022). For instance, people would prefer to purchase flour than grind their wheat for pies. These consumers are willing to pay more for the ease of purchasing processed goods (flour) instead of unprocessed agricultural goods (wheat).

At the same time, scientific advancements in food preservation techniques were occurring. Most agricultural products are perishable and only available during the harvest season. Many people attempted to increase product availability through home canning. Advancements in the technology of canning and, later, freezing was triggered by the development of food processing companies (Matovu *et al.*, 2022).

Consumers may now prepare the product any time of the year for less money and labour due to food processing advancements. The process has advanced, and customers prefer finished goods rather than producing their own. Even most farming households nowadays buy food and fibre products rather than processing them themselves (Spigel and Harrison, 2017). The 'processing–manufacturing sector' refers to the businesses that cater to consumers' demands for increased processing and convenience and are a significant component of agribusiness.

For the goods and services they require to generate agricultural commodities, farmers rely on the input industries. Additionally, they depend on third parties to buy raw agricultural commodities, process them, and then distribute them to consumers for final sale (Fanchone *et al.*, 2022). These third parties include commodity processors, food makers, distributors and retailers.

The political structures of the African nations influenced by colonialism have theoretically changed. The new administrations redistributed land during the initial decades of independence, moving it from the majority (black) farmers to the minority (large-scale) (white) farmers (Shaibu, 2022). It was also reported that between the 1990s and 2000s, more than 32 African nations implemented 'land to the tiller' reforms (Shenggen, 2020). For instance, Ethiopia (1975), Zimbabwe (1980, 1990, 1992, 2000), South Africa (1994) and Tanzania (1990) all underwent land reforms, whereas Burkina Faso and Cote d'Ivoire underwent land reforms in the 1990s, and South Sudan underwent land reforms in 2009. This led to the emergence of a new breed of farmers, the smallholder farmers who currently make up the majority of settlers.

15.3.2   Going beyond the human needs approach in rural agriculture: role of sustainable development policy

15.3.2.1   Why sustainable agriculture matters

Most estimates suggest that by 2050 the world's population will reach 9.6 billion (Campos, 2021). This means that there would need to increase food production by 70% to meet the needs of such a large population (Ravikishore, 2022). It is becoming increasingly clear that tough reforms in the agricultural sector are needed to ensure that food systems can meet the challenges of a growing world population. In particular, agriculture needs to transform from the conservative industrial food systems that have shaped food production for centuries towards sustainable farming. For a planet plagued by drought and energy demand, transitioning from traditional industrial food systems to sustainable agriculture holds considerable promise in the long term. While modern agriculture creates many jobs and produces excellent yields in a single harvest season, it also poses devastating problems that require sustainable agricultural practices to solve the mess.

DIVERSITY OF VALUES.   A defining characteristic of industrial agriculture is monoculture, an agricultural system in which farmers plant large land areas with a single crop species (Shenggen, 2020). Excessive reliance on just one plant species can make plants more susceptible to disease, and diseases can spread rapidly from one plant to another, wiping out entire crops.

MAKE PLANTS RESILIENT. Sustainable agriculture is critical in reducing greenhouse gas emissions and conserving energy and water. Sustainable agriculture offers resilience for a planet increasingly vulnerable to climate change as it focuses on growing various crops rather than a single one (Campos, 2021).

STABILIZATION OF FOOD SUPPLY. Consolidation of individual farms into large enterprises, providing economies of scale for individual farmers, is a hallmark of industrial agriculture (Spigel and Harrison, 2017). But developing a large company can be very risky, because if one of the companies runs into a problem, the consequences can have far-reaching implications for food security.

15.3.3 Networking/not working? Understanding the supply chains of modern rural agriculture in Southern African

Digital and analytical techniques offer opportunities to create added value by optimizing agricultural supply chains (Ravikishore, 2022). Companies across many industries have realized that applying digital and analytical technologies to new business models and product offerings can add value. Agricultural stakeholders now realize that these technologies can play a role in optimizing highly complex agricultural supply chains, from farmers to final consumers. By collecting massive amounts of data, leading agricultural players follow companies in other industries by creating digital twins of their physical supply chains. These virtual replicas will allow companies to perform simulations and optimizations, potentially significantly reducing the cost of moving crops through the system (Kuteyi and Winkler, 2022).

This section explains the reasons behind the complexity of agricultural supply networks and how businesses might utilize digital and analytical methods to optimize them. Players might obtain a competitive advantage under challenging market conditions by utilizing technologies like digital twins.

15.3.3.1 Complex supply chains

Supply chain processes are inherently complex across industries because multiple functions interact with potentially different goals and with numerous dependencies between material and information. Fragmented inbound and outbound networks further complicate agricultural supply chains (Rukasha et al., 2021). A typical agricultural supply chain has three steps: from farmer to intermediate silo, from silo to processing plant, and from the processing plant to the customer.

The potential outcomes for each selection make the optimization analysis more challenging. Fragmented supply chains increase the number of potential flows at each stage by two, potentially creating thousands of outcomes. For instance, there can be more than 300 different grain varieties, 300 silos, more than 7000 other storage facilities, and more than 200,000 different transportation possibilities in one organization (Camilla, 2023). The unpredictability of the result makes the issue much more difficult. Two factors are principally responsible for this ambiguity. One is the operational elements, such as the variable yields in each field, and the other is external factors, such as the weather, inputs, farmer expertise, and price changes brought on by imbalances in the world's supply and demand. For instance, an analysis of 10 year production data for sugar uncovered over 150 agricultural scenarios (defined by a range of potential yields per plot) (Ravikishore, 2022).

15.3.3.2 How digital twins can help

Developments in digital and analytical technologies present opportunities for streamlining agricultural supply chains (Stavros and Dimitrios, 2022). The agriculture sector, including information on agronomy, weather, logistics, and market price changes, is gathering more data than ever before. Computing power has increased with data storage capacity and as prices have reduced. Both prescriptive optimization approaches and predictive data science are developed and widely used. Making a digital twin of the physical supply chain from the farm to the end user and using it to undertake virtual simulation and optimization is tempting to employ digital and analytical technology. All supply chain components and their interactions, including procurement, production, inventory points, transportation, warehousing, and the point of sale for finished items (Rukasha *et al.*, 2021) can be included in a digital twin. Depending on the needs of the business, players can modify the mathematical model to add various target functions such as profit, throughput, cycle time, inventory optimization, etc.

The digital twin's strong predictive ability is what makes it valuable. Utilizing artificial-intelligence-based algorithms, it is possible to explore all conceivable planning and scheduling combinations and factors, such as lot size, while performing multivariate function optimization within user-defined limits (Stavros and Dimitrios, 2022). When unanticipated events happen, the planning and scheduling optimizers can be rerun in real-time. Rush orders or different needs, for instance, can be immediately incorporated into the updated plan. Business leaders are prioritizing analytics and digital as their benefit becomes clearer. For instance, more than half of industrial businesses use computer programmes in their everyday business. Despite the potential of this trend, many businesses find it challenging to test digital and analytics successfully. Less than 30% of McKinsey's organizations have successfully switched from a pilot to a fully scaled solution. Most businesses are aware of the technological difficulties in integrating digital twins. This is because of multiple challenges: decision makers developing the solution's main engine, crafting the decision-supporting user interfaces, and integrating the solution into the ICT environment via bi-directional interfaces with current systems. According to Campos (2021), value capture involves three strategies. However, even businesses with excellent technical capabilities can't fully utilize their digital twins. Two additional components are needed for success:

- Sufficient knowledge of the supply chain and industry to determine the objective function of the digital twin. This solution necessitates an end-to-end perspective and a sufficient level of physical and temporal information to support step-by-step accuracy and well-informed decisions regarding the pace of development and execution.
- Change management initiatives for all stakeholders, including agricultural stakeholders and farmers. The farmer must be ready for adjustments to harvest logistics and planning, and these adjustments must make him aware of one potential benefit.

Up to two-thirds of farmers noticed a change in harvest time after one farmer utilized a digital twin to improve his supply chain, and his crop rewards increased by 3–5% (Spigel and Harrison, 2017). Teams working on the agricultural supply chain should reconsider. We should promote the usage of digital twins for planning and exception management rather than depending on frequent schedule adjustments and firefighting.

Farmers now face a more challenging climate, but digital technologies and analytics offer potent tools that unleash new value for those who know how to use them. Achievement rewards are apparent even if the path to adoption is risky. By releasing untapped value from the agricultural supply chain, digital and analytical technologies can give well-positioned industry leaders a new source of competitive advantage.

15.3.4 The role of agribusiness policy in sustainable rural development

The term 'agricultural policy' refers to a group of regulations or rules governing domestic agriculture and importing agricultural goods from abroad (Shaibu, 2022). For the benefit of the agricultural community and the overall national economy, governments often pursue agricultural policies to obtain specific results in the domestic market for agricultural products.

The value chain, which starts with the preparation of primary raw materials for production and concludes with the consumption of the finished product, is taken into account by agricultural policy. The process encompasses all economic activities that happen during these stages. Among them are cultivation, fertilization, crop protection, harvestings, and secondary processing, such as the creation or transformation of goods that enhance the value of agricultural products, installation of facilities for primary processing, distribution, and activities used in higher education (Fanchone *et al.*, 2022). Value chain analysis offers beneficial agricultural strategies that address all primary, secondary and tertiary agricultural sectors (Shonhe, 2021). Policies are crucial in reducing poverty because they set the rules and procedures for enhancing domestic agriculture production.

To support the growth of Somaliland's agricultural sector, the Ministry of Agriculture Development must set policies, rules and regulations governing the importation, supply, usage and quality of all inputs imported for agricultural purposes (Ravikishore, 2022). However, there are still no comprehensive national agriculture policies, laws, or regulations. The Ministry of Agriculture and Development should ensure detailed rules for the sector players' coordination mechanisms, including but not limited to the private sector, international aid groups and other governmental agencies. As a result, once fundamental principles are defined, they will serve as a guide for the Ministry and other sector stakeholders (Virva and Grant, 2021). While guaranteeing that the needs and interests of the general public are adequately considered and protected, it enables stakeholders in the sector to engage in profitable and productive activities.

The agricultural sector in Africa has captured the interest of researchers, policy makers, foreign investors, and the development community over the past ten years. Agriculture is a critical factor in economic growth and is showing promise as a means of reducing poverty and boosting food security. Shaibu (2022) reports that most agricultural research and development initiatives, particularly among smallholder farmers, have concentrated on boosting the production of staple crops (maize, wheat, rice, etc.). The production of staples is merely the tip of the iceberg regarding the vast array of crops and agricultural practices that comprise Africa's food system.

Therefore, research and funding for agricultural growth should go beyond agricultural production and cover every link in the value chain, especially for high-value goods. Development goals do not have the best impact when agricultural production interests are separated from

supporting the expansion and growth of farms and businesses (The Herald, 2021).

Numerous strategy documents on eradicating poverty in Africa stress that growth in the agricultural sector contributes proportionately more to doing so than growth in any other economic sector. Most work and income in sub-Saharan Africa comes from agriculture, especially for the rural poor, where more than 63% of the population resides (Kuteyi and Winkler, 2022). A quarter of Africa's gross domestic product comprises the agricultural sector, which still employs roughly half of the continent's workforce (Virva and Grant, 2021). The agricultural sector must play a crucial role in growth and opportunity across Africa as the global and regional food and agriculture markets expand at an unprecedented rate.

For farm-to-table food, the term agribusiness refers to 'agricultural and all other industries and services that make up the supply chain from the farm through processing, wholesale, retail, and table' (Campos, 2021). It is projected that by 2030, agriculture and agribusiness might contribute nearly US$1 trillion to the regional economies of Africa (Shenggen, 2020). Governments should consequently prioritize agribusiness and the more significant agriculture sector to alter and advance their economies.

15.4   Conclusions and Implications

In situations where environmental and societal inequalities might endanger economic and social stability, sustainable development is about equity in people, their wellbeing and their connections. For example, climate change and supply chain conditions can play a significant role in sustainable development in various contexts because it interacts with economic output and services, human settlements and human cultures. Simply said, depending on geography, economic sector and level of economic and social development already attained, climate change can have a good or detrimental impact on many elements of human progress by increasing the vulnerabilities of impoverished rural communities. Settlements and industries are frequently the focus of policymaking, mitigation efforts and adaptation measures. Thus, it stands to reason that interactions between these two groups will be crucial to many development-oriented approaches to climate change issues within rural agribusiness development.

References

Abafe, E. (2021) Quantitative analysis of farmers' perception of the constraints to sunflower production: a transverse study approach using hierarchical logistic model (HLM). *Sustainability* 13(23), 13331. DOI: 10.3390/su132313331.

Camilla, F.I.P. (2023) When biodiversity preservation meets biotechnology: the challenge of developing synthetic microbiota for resilient sustainable crop production. *Journal of Sustainable Agriculture and Environment* 2, 8. DOI: 10.1002/sae2.12038.

Campos, H. (2021) *The Innovation Revolution in Agriculture A Roadmap to Value Creation: A Roadmap to Value Creation*. DOI: 10.1007/978-3-030-50991-0.

Cartmel, F. and Furlong, A. (2000) *Youth Unemployment in Rural Areas*, Social Research in Transport (SORT) Clearinghouse. Work and Opportunity Series No.18. York Publishing Services Ltd, United Kingdom.

Fanchone, A., Laetitia, N., Dodet, N., Luc, M. and Andrieu, N. (2022) How agro-environmental and climate measures are affecting farming system performances in Guadeloupe: lessons for the design of effective climate change policies. *International Journal of Agricultural Sustainability* 20(7), 1–12. DOI: 10.1080/14735903.2022.2136836.

FAO (2021) A way forward for supporting agricultural innovation in the Lao People's Democratic Republic. Key constraints and opportunities as identified by the agricultural innovation system assessment. Available at: https://www.fao.org/3/cb8057en/cb8057en.pdf (accessed 13 November 2022).

Giuli, K. (2018) Value chain management in Agribusiness. *International Journal of Business and Management* VI, 59–77. DOI: 10.20472/BM.2018.6.2.004.

Kuteyi, D. and Winkler, H. (2022) Logistics challenges in sub-Saharan Africa and opportunities for digitalization. *Sustainability* 14(4), 2399. DOI: 10.3390/su14042399.

Matovu, M., Nankya, R., Lwandasa, H. and Isabirye, B.E. (2022) Heterogeneity in nutritional and biochemical composition of cassava varieties in Uganda. *Journal of Agriculture and Sustainability* 15, 1–36.

Ngoc, V.B., Hung, N.M. and Pham, P.T. (2021) Agricultural restructure policy in Vietnam and practical application for sustainable development in agriculture. *Nanomaterials for Sustainable Development in Agriculture* 2021, 30. DOI: 10.1155/2021/5801913.

Ravikishore, M.P.S. (2022) Frontiers in agricultural marketing: role, challenges and way forward. *Food and Scientific Reports* 3, 34–38. Available at: https://foodandscientificreports.com/details/frontiers-in-agricultural-marketing-role-challenges-and-way-forward.html

Rukasha, T., Nyagadza, B., Pashapa, R., Muposhi, A. and Gikunoo, E. (2021) COVID-19 impact on Zimbabwean agricultural supply chains and markets: a sustainable livelihoods perspective. *Cogent Social Sciences* 7(1), 1928980. DOI: 10.1080/23311886.2021.1928980.

Shaibu, L.K.B. (2022) Stimulating innovations for sustainable agricultural among smallholder farmers: persistence of intervention matters. *Journal of Development Studies* 58(9), 1651–1667. DOI: 10.1080/00220388.2022.2043283.

Shenggen, C.R. (2020) *The Role of Smallholder Farms in Food and Nutrition Security*. Springer, United Kingdom.

Shonhe, T.I.S. (2021) Private and state-led contract farming in Zimbabwe: accumulation, social differentiation and rural politics. *Journal of Agrarian Change* 22(1), 118–138. DOI: 10.1111/joac.12473.

Spigel, B. and Harrison, R. (2017) Towards a process theory of entrepreneurial ecosystems. *Strategic Entrepreneurship Journal* 12(1), 151–168. DOI: 10.1002/sej.1268.

Stavros, K. and Dimitrios, K. (2022) Role of crop-protection technologies in sustainable agricultural productivity and management. *Journal of Agricultural Sciences* 3, 25–29. DOI: 10.3390/land11101680.

The Herald (2021) AMA restructures, seeks sanity in agricultural markets. Available at: https://www.herald.co.zw/ama-restructures-seeks-sanity-in-agric-markets/ (accessed 24 November 2022).

Virva, T. and Grant, D.B. (2021) Exploring supply chain issues affecting food access and security among urban poor in South Africa. *The International Journal of Logistics Management* 33(5), 27–48. DOI: 10.1108/IJLM-01-2021-0007.

16 Procurement Laws in Agribusiness

Charles Tsikada[1]*, Tafadzwa Y. Chiwanza[1], Ernest Mugoni[1], Iscacle Nyanhete[2] and Rumbidzai Pashapa[1]

[1]Marondera University of Agricultural Sciences and Technology (MUAST), Zimbabwe; [2]Midlands State University (MSU), Zimbabwe

Abstract
Governments procure goods and services through public procurement systems. Procurement of goods and services should be conducted in a fair, equitable, transparent, competitive and cost-effective manner. On the other hand, governments around the world use public procurement laws and regulations to implement and control procurement processes in order to achieve part of their national goals. These laws and regulations cover procedures, methods, standards and terms for inviting bids, evaluation, selection, negotiation, awarding and payment of suppliers of goods and services. However, many smallholder farmers fail to meet these procedures, methods, standards and terms because of capacity and resource constraints. There are public food procurement initiatives that are currently being implemented throughout the world in order to help smallholder farmers gain access to markets. However, these initiatives are still relatively new and unstructured. There is still a lack of peer-reviewed research. Additionally, the majority of the available data originates from analyses that mainly rely on qualitative research and case studies. Nonetheless, some key lessons can be drawn from these initiatives. This chapter is organized around the following sections: Section 16.1 defines smallholder farmers from different perspectives; Section 16.2 explains the significance of smallholder farmers; Section 16.3 explains the framework for public food procurement and suggests government-led programmes that might be used to integrate smallholder farmers into the market; and Section 6.4 concludes the chapter, highlighting key issues.

Keywords: Smallholder farmers, Food purchases, Public markets, Supportive public procurement law and regulatory framework

16.1 Defining Smallholder Farmers

The term smallholder farmer is often misunderstood resulting in ambiguity in policy development (Lowder *et al.*, 2016). Smallholder farmers produce food for household consumption and the markets. The subsequent earning acts as a source of income for the household (Carelsen *et al.*, 2021). In most cases, smallholder farmers live in rural areas. The Food and Agriculture Organization came up with a number of indicators to help in defining smallholder farmers (Rossi, 2022). Smallholder farmers are defined in terms of land size, family labour, resources, income, food security and demographics.

16.1.1 Land

Smallholder farmers own or work on less than two hectares of land (Oluwatayo, 2019). However, size threshold differs across countries, regions and

*Corresponding author: tsikadac@gmail.com

socio-economic contexts and stages of development across the country because evolution of the small farm is intrinsically related to the processes of economic development (Rapsomanikis, 2015). The use of land size to define smallholder famers is too simple to capture the complexity of farming systems. For example, a piggery or poultry farm on a two-hectare plot of land is defined as large scale (Rossi, 2022). However, the vast majority of the world's smallholder farmers own or work on less than two hectares of land (Nyambo *et al.*, 2019). In developing countries, most farmers do not have title deeds for the land they work on. Land is also inherited by the household head and divided into smaller parcels that are allocated to household members, usually male members (Ohene-Yankyera, 2004).

16.1.2 Family labour

Most smallholder farmers rely on family labour for production, although hired labour is sometimes used on a seasonal basis (Rapsomanikis, 2015). They normally use outdated production technologies and women play an important role in production (Oluwatayo, 2019). They also lack skills and knowledge to improve their farming activities to enhance their production capacity (Kusnandar *et al.*, 2019).

16.1.3 Demographics

The majority of the smallholder farmers consist of women, children and elderly people. Women provide substantial labour in smallholder farms (Rapsomanikis, 2015). The average age of the house head is 55 years. The majority of these households are headed by females and the average household consists of five individuals (Pienaar and Traub, 2015). Many smallholder farmers only have basic education, primary and secondary school education. Some of these farmers are relatively old, which makes it difficult to use information, communication and technology (Taiy *et al.*, 2017).

16.1.4 Resources

The main objective is usually to cater for the welfare of the family before profit is considered (Mudhara, 2010). Smallholder farmers generally struggle to get their produce to the market at the right time. They suffer from postharvest losses due to poor access to controlled temperature storage facilities, which negatively affects their income (Webber *et al.*, 2013; Tshuma, 2014). They cannot afford to own vehicles that transport their produce to the market. Smallholder farmers do not have postharvest handling equipment and capital to finance their supply chain activities (Bhatia and Janardhana, 2020). For instance, they rely on draught power for tillage, which is provided by animals like cattle, donkeys and horses.

16.1.5 Income

Smallholder farmers have very little material savings, and these savings can easily be wiped out in a single bad harvest (Hystra Hybrid Strategies Consulting, 2015). Some members of their households work off-farm to complement farm income, thereby contributing towards food security. This is a risk management tool that can diversify household income (Rapsomanikis, 2015).

16.1.6 Food security

Food security at household level is when all members of the household have access to safe, sufficient and nutritious food to meet their daily dietary needs all the time (Oluwatayo, 2019). Smallholder farmers grow a variety of crops in order to ensure household food security and as a strategy to stabilize their income and minimize their risks to price fluctuations (Rapsomanikis, 2015). In developing countries, smallholder farmers produce the bulk of the agricultural exports. They supply close to 70% of Africa's food requirements (UNCTAD Secretariat, 2015). Therefore, they also contribute towards the food security of the communities around them. The income they generate from the sale of their produce also enables farmers to purchase other food stuffs that meet the dietary needs for their families. They also concentrate on producing staple food, which increases staple food supply. This leads to a decrease in real prices of staple food, thereby, benefiting the poor and vulnerable members of their community (Oluwatayo, 2019).

In summary, the term smallholder farmer differs in context, by country and agro-ecological zones, and different authors do not agree on one universal definition of smallholder farmer. However, landholding size cannot be used as the only criterion as it can lead to misconceptions as to whether some farmers can be regarded as smallholder farmers or not. Various stakeholders have established definitions either for purely analytical purposes or for the implementation of government programmes. Notwithstanding the wide variations of the definition there are some commonalities. In this chapter, the term smallholder farmer describes a farmer:

- whose farm size is relatively small, depending on the context, country, and ecological region
- who has limited resource to some degree
- who largely depends on family labour
- who provides food security beyond their family, and
- who resides in rural areas.

16.2 Significance of Smallholder Farmers

The United Nations (UN) declared 2014 as the International Year of Family Farming (IYFF) to specifically recognize the important role that smallholder agriculture plays in ensuring food security, reducing poverty and promoting sustainable development (United Nations Conference on Trade and Development (UNCTAD), 2015). In response, the African Union (AU) declared 2014 as Africa Year of Food Security (Graeub *et al.*, 2016). This provided an opportunity to assess how smallholder agriculture is doing globally in terms of food security and sustainable development. Smallholder agriculture has typically been considered a low-production, stagnating industry. Recent studies, however, imply that it has grown to play a substantial role in the world's agricultural value chains. Smallholder farmers are increasingly considered as key solutions to the world's hunger crisis, which has triggered a paradigm shift in

recent policy debates at national, regional and international levels (Fan and Rue, 2020).

According to analysis from the year 2000, there are about 500 million small farms in the world, which produce more than 80% of the world's food (Sabo *et al.*, 2017; Fan and Rue, 2020; Gomez y Paloma *et al.*, 2020). Lowder *et al.* (2014) estimate that 80% of the food produced in Asia and sub-Saharan Africa comes from 12% of the world's farmland which is operated by smallholder farmers. Therefore, smallholder agriculture has higher land productivity compared to large farms, which gives rise to an inverse relationship between farm size and productivity (UNCTAD, 2015). Smallholder farmers contribute to rural economies and play a critical role in natural resource conservation. They use relatively low levels of chemicals as compared to large-scale farmers and are viewed as the guardians of environmental sustainability at community level.

16.3 Public Food Procurement Frameworks for Food Purchases From Smallholder Farmers

The role of the government pertaining to food purchases in promoting social and economic benefits has also gained prominence in recent years (Gaitán-Cremaschi *et al.*, 2022). Generally, governments make large food purchases for public institutions like hospitals, schools, nursing homes and prisons (Gaitán-Cremaschi *et al.*, 2022). In some developing countries, public food procurement has specifically targeted smallholder farmers to promote their economic inclusion and strengthen local food systems (Mensah and Karriem, 2021). Public food procurement can reduce uncertainties and risks associated with market participation by providing an accessible marketing channel and a source of income for smallholder farmers (Miranda, 2018). Access to the market and additional sources of income can generate positive impacts, such as increases in household food security, dietary diversity, investments in production and diversification (Sumberg and Sabates-Wheeler, 2011). Moreover, improvements to smallholder livelihoods can generate positive spillover effects in local economies. In addition, public food procurement can also lead to positive outcomes on health and nutrition, particularly among children and other vulnerable groups (Swensson *et al.*, 2021).

Public food procurement for food assistance has the potential to create an integrated framework that can generate benefits for smallholder livelihoods, food security and nutrition (De Schutter, 2014; Drake and Woolnough, 2016). Government food purchases can target commodities that address the nutritional requirements of vulnerable populations. The food can be procured from smallholder farmers and distributed through different food assistance strategies. There is a growing trend to foster synergies between local food systems, smallholder farmers, and better nutrition through public food procurement.

Different countries in Africa have also piloted initiatives supported by development partners, such as the World Food Programme (WFP) and the UN Food and Agricultural Organization (FAO). Governments of countries, such as Kenya, Ghana, Senegal and Ethiopia, have made substantial progress towards nationally owned programmes. Latin American governments have strengthened their commitment to public food procurement from smallholder farmers by including specific measures in the Community of Latin American Countries Plan for Food and Nutrition Security and Eradication of Hunger 2015. The Committee on World Food Security (CFS) 2015 policy recommendations also contain actions to promote links between smallholder farmers and public food procurement. It is possible to identify some key lessons emerging from current public food procurement initiatives being implemented in different parts of the world.

Public procurement is widely used to advance economic policies, in particular to encourage the development of the small and medium-sized enterprise (SME) sector. Preferential treatment indicates that the proportion of food purchases reserved for smallholder procurement is carried out through either competitive or non-competitive processes. For instance, the WFP has adopted reservation schemes to provide market access to smallholder farmers. A proportion of food purchases has been allocated to targeted farmer organizations and smallholder farmers. This has become the new norm because, traditionally, food was purchased from prequalified large-scale suppliers through competitive tenders. It should be noted that more recently a proportion of the initiative has been reserved for farmer organizations or smallholder farmers. The goal is to increase smallholder farmer engagement in markets and strengthen their productive and marketing capacity. Farmer organizations are selected, according to country-specific criteria developed by the WFP. Farmer organizations can range from local farmer groups and women-only groups to regional federations and farmer unions. However, they must have the minimum capacity to aggregate production and benefit from increases in demand. The sections below describe specific provisions in public procurement and how they have been applied to public food procurement initiatives.

16.3.1 Non-competitive practices

It is not necessary for qualified suppliers to compete on the basis of lowest price and best quality in their bids. Instead, procuring entities can publish a call for bids for the purchase of food, outlining the commodities, quantity, standards for quality and delivery schedule. Smallholder farmers can submit their proposals outlining the commodities and quantities they want to sell. They are expected to obtain an eligibility declaration, which certifies their family farmer status. Prices can be publicized in the public call for food procurement, so that smallholder farmers can decide if the terms are favourable to them. For instance, the national school feeding law in Brazil (Law No. 11947/2009) stipulates that smallholder farmers shall get 30% of the food purchased for school feeding. In this case, a competitive bidding imposed by Brazilian public procurement law is waived. In addition, the WFP uses a non-competitive method through waiving competitive bidding processes when procuring food from smallholder farmers. This can be implemented through direct contracts or forward contracts. In the event of direct contracts, the WFP bargains with farmers to purchase commodities around harvest time. On the other hand, forward contracts can be made with smallholder farmers at planting time for the supply of a commodity in the future at a predetermined quantity and quality for a pre-determined price. As evidenced, governments around the world can use non-competitive practices in order to assist smallholder farmers to access market.

16.3.2 Competitive practices

Smallholder farmers can be given preference for contracts in competitive processes, but they still have to submit bids and compete to obtain the government contract. Lessons can be drawn from Burkina Faso and Rwanda where a certain percentage is set aside by the government for the food reserve to smallholder farmers. In Rwanda, the National Strategic Grain Reserve is using a decree from the government that was issued in 2011, which allots a 40% quota for smallholder farmers (NSGR, 2013). Similar to this, the government of Burkina Faso allocated 30% of purchases to smallholder farmers in 2014. In each case, smallholder farmers must be registered with the appropriate government agencies in order to participate. They are requested to submit bids that include the quantities and prices of the commodities per metric tonne (Amani, 2014). For the WFP, the soft tendering modality is used to conduct competitive processes. Only smallholder farmers who satisfy the requirements are given the opportunity to submit bids and compete for the contract. In addition, to encourage participation from smallholder farmers, the WFP uses less restrictive requirements for tenders, such as supplying small quantities, providing bags with WFP logos, and forgoing bond guarantees.

16.3.3 Subcontracting

Governments do not procure food directly from smallholder farmers, but they require their suppliers to buy a proportion of food from them. This is usually done through traders or processors. For instance, this approach has been commonly used to encourage large government suppliers to subcontract SMEs. The USA has a well-established subcontracting programme for SMEs which applies to all contracts above US$650,000 (International Trade Centre,

2014). Contracts above this threshold require suppliers to submit subcontracting plans for small businesses, including minority- and women-owned businesses. In Thailand, subcontracting schemes have been applied to public food procurement from smallholder farmers in the School Milk Programme. In Ghana, the Ghana School Feeding Programme uses a third-party procurement model with processors in charge of purchasing, preparing and distributing school meals. Processors are allowed to procure 80% of commodities for school feeding from smallholder farmers (Home Grown School Feeding, 2018). The WFP selected traders who are required to purchase from smallholder farmers (WFP, 2016).

16.3.4 Qualification criteria

Qualification criteria are often applied to food purchases conducted by marketing boards or government institutions. Governments have reserved food purchases from smallholder farmers through qualification criteria. Sellers that do not meet specific criteria, for example, smallholder farmers, are eligible to participate in public procurement processes that reserve the entirety of food purchases to one category of supplier. Qualification criteria have been applied to food purchases carried out by governments. In Paraguay, government food purchases have been reserved for smallholder farmers through qualification criteria. However, other sellers can only supply government when procurement from smallholder farmers is not possible (Decree 1056/13). In Brazil, all food purchases are reserved for smallholder farmers. Suppliers who do not meet this criterion are excluded from the procurement process. Brazilian government prioritizes the most vulnerable farmers, including women, land reform settlements, indigenous tribes and slave-descendent communities. Farmers must be certified to be eligible to participate in the programme.

16.3.5 Preferential schemes

Preferential scheme refers to a situation where competitive advantages are given to bidders that meet specific social, economic and environmental criteria. There are two sub-categories of preferential criteria: price preference and procurement award criteria. Bid price preference is the practice of raising the costs of non-preferred suppliers by a predetermined percentage point in order to evaluate their bids. In an open bidding procedure, for instance, a procuring entity might give a target group of suppliers a price preference by raising the bid prices of non-preferred suppliers by 10%, making their bid more expensive. The bid from a targeted supplier, on the other hand, may be discounted by 5%, giving it an edge over other bids. In order to increase competition, a bid from a farmer that is an approved supplier could receive a certain percentage discount. This approach explicitly recognizes that some types of providers will not always be able to compete on price with other suppliers. Consequently, it provides target groups with a price advantage. Only raw agricultural products are covered by it. Purchasing organizations can select the amount of discount they want to offer to regional vendors based on the circumstances.

Preferential treatment has often been used to promote SME access to public procurement markets. According to the World Bank (2016), 10% of countries use bid price preferences for SMEs. SME bids are given precedence in China if their prices are between 6 and 10% lower than those of the competition, whereas in India they are if their prices are up to 15% higher. In Bolivia, public procurement legislation gives a 20% margin of preference to smallholder farmers. In Zimbabwe, a procuring entity may prefer local suppliers when considering bids. There is also the option to offer preference to female-controlled businesses. It is unclear, though, how smallholder farmers might benefit from domestic preference scheme.

16.3.6 Award criteria

Award criteria allows contracting authority to evaluate and compare different bids based on price or on price and quality (Appolloni *et al.*, 2019). Bids that satisfy social, economic, and environmental criteria may be given extra points under this system. Procuring entities grant additional points to bids that exceed the minimal specification when employing award criteria. The bid is frequently evaluated using the relative weighting of each criterion, which represents the proportionate weight given to non-price elements. This method of purchase has been used by the Peruvian school programme and in South Africa to encourage the economic integration of black South Africans (Quinot, 2013; MIDIS, 2013). In the European Union, nearly 90% of contracts for food provision use award criteria in the bid selection process (Caldeira *et al.*, 2017). None the less, research into the application of award criteria to food purchases is still limited.

16.4 Conclusions

In conclusion, participation of smallholder farmers in public food markets is often constrained by many challenges. The most important limitations are those caused by the intense competition involved in public procurement. However, they can be supported through specific public procurement frameworks that remove bureaucratic hurdles, reduce costs, and give them competitive advantages. Initiatives for inclusive public food procurement are necessary policy tools to assist smallholder farmers and their integration into established markets. Public food purchases from smallholder farmers must be closely coordinated. Capacity development is crucial for farmers to participate in the food market. Rules and procedures for public food procurement should take into account the capacity of smallholder farmers. As shown, a variety of strategies can be employed to create a supportive regulatory environment for smallholder farmers through public food procurement. These include, among other things, preferential schemes, non-competitive processes, and award criteria. These must be coordinated to accommodate the needs and abilities of smallholder farmers in order to access the market.

References

Amani, S. (2014) Supporting public procurement from smallholder farmers. P4P Global Learning Series.

Appolloni, A., Coppola, M.A. and Piga, G. (2019) Implementation of green considerations in public procurement: a means to promote sustainable development. In: *Green Public Procurement Strategies for Environmental Sustainability*. IGI Global, pp. 23–44. DOI: 10.4018/978-1-5225-7083-7.

Bhatia, M. and Janardhana, G.M. (2020) Agriculture supply chain management-an operational perspective. *Brazilian Journal of Operations & Production Management* 17(4), 1–18. DOI: 10.14488/BJOPM.2020.043.

Caldeira, S., Bonsmann, S. and Bakogianni, I. (2017) Public procurement of food for health: Technical report on the School setting. Maltese Presidency and the European Commission, Brussels.

Carelsen, C.P.R., Ncube, B. and Fanadzo, M. (2021) Classification and characterisation of smallholder farmers in South Africa: a brief review. *South African Journal of Agricultural Extension* 49(2), 2413–3221. DOI: 10.17159/2413-3221/2021/v49n2a12821.

De Schutter, O. (2014) The power of procurement – public purchasing in the service of realizing the right to food. Briefing Note 08. United Nations Special Rapporteur on the Right to Food, Louvain-la-Neuve, Belgium.

Drake, L.J. and Woolnough, A. (eds) (2016) *Global School Feeding Sourcebook: Lessons From 14 Countries*. Imperial College Press, London. DOI: 10.1142/p1070.

Fan, S. and Rue, C. (2020) The role of smallholder farms in a changing world. In: Gomez y Paloma, S., Riesgo, L. and Louhichi, K. (eds) *The Role of Smallholder Farms in Food and Nutrition Security*. Springer, pp. 13–28. DOI: 10.1007/978-3-030-42148-9.

Gaitán-Cremaschi, D., Klerkx, L., Aguilar-Gallegos, N., Duncan, J., Pizzolón, A. *et al.* (2022) Public food procurement from family farming: a food system and social network perspective. *Food Policy* 111, 102325. DOI: 10.1016/j.foodpol.2022.102325.

Gomez y Paloma, S., Riesgo, L. and Louhichi, K. (2020) *The Role of Smallholder Farms in Food and Nutrition Security*. Springer Nature. DOI: 10.1007/978-3-030-42148-9.

Graeub, B.E., Chappell, M.J., Wittman, H., Ledermann, S., Kerr, R.B. *et al.* (2016) The state of family farms in the world. *World Development* 87, 1–15. DOI: 10.1016/j.worlddev.2015.05.012.

Home Grown School Feeding (2018) *Ghana School Feeding Programme: Way Forward*. Home Grown School Feeding.

Hystra Hybrid Strategies Consulting (2015) Smallholder farmers and business. 15 Pioneering Collaborations for improved productivity and sustainability. Hystra Hybrid Strategies Consulting.

International Trade Centre (2014) Empowering women through public procurement. International Trade Centre, Geneva.

Kusnandar, K., Brazier, F.M. and van Kooten, O. (2019) Empowering change for sustainable agriculture: the need for participation. *International Journal of Agricultural Sustainability* 17(4), 271–286. DOI: 10.1080/14735903.2019.1633899.

Lowder, S., Skoet, J. and Raney, T. (2016) The number, size, and distribution of farms, smallholder farms, and family farms worldwide. *World Development* 87, 1–13. DOI: 10.1016/j.worlddev.2015.10.041.

Lowder, S.K., Skoet, J. and Singh, S. (2014) *What Do We Really Know About the Number and Distribution of Farms and Family Farms in the World?* Background paper for The State of Food and Agriculture.

Mensah, C. and Karriem, A. (2021) Harnessing public food procurement for sustainable rural livelihoods in South Africa through the National School Nutrition Programme: a qualitative assessment of contributions and challenges. *Sustainability* 13(24), 13838. DOI: 10.3390/su132413838.

MIDIS (2013) *Manual de Compras del Modelo de Cogestión para la Atención del Servicio Alimentario del Programa Nacional de Alimentación Escolar Qali Warma*. Food and Agriculture Organization of the United Nations, Rome.

Miranda, A. (2018) *Public food procurement from smallholder farmers: Literature review and best practices*. International Policy Centre for Inclusive Growth (IPC-IG), Brasilia.

Mudhara, M. (2010) *Agrarian Transformation in Smallholder Agriculture in South Africa. A Diagnos of Bottle-necks and Public Options*. University of Kwazulu Natal, South Africa.

NSGR (2013) *Rwanda National Strategic Grain Reserve Operations and Procedures Manual*. Ministry of Agriculture and Animal Resources, Kigali.

Nyambo, D.G., Luhanga, E.T. and Yonah, Z.Q. (2019) A review of characterization approaches for small-holder farmers: towards predictive farm typologies. Hindawi, Ausha, *The Scientific World Journal* 2019, 6121467. DOI: 10.1155/2019/6121467.

Ohene-Yankyera, K. (2004) Determinats of farm size in land-abundant agrarian coomunities of Northern Ghana. *Journal of Science and Technology* 24, 45–53. DOI: 10.4314/just.v24i2.32916.

Oluwatayo, I.B. (2019) Towards assuring food security in South Africa: smallholder farmers as drivers. *AIMS Agriculture and Food* 4, 485–500. DOI: 10.3934/agrfood.2019.2.485.

Pienaar, L. and Traub, L. (2015) *Understanding the Smallholder Farmer in South Africa: Towards a Sustainable Livelihood's Classification.* International Association of Agricultural Economists.

Quinot, G. (2013) Promotion of social policy through public procurement in Africa. In: Arrowsmith, S. and Quinot, G. (eds) *Public Procurement Regulation in Africa.* Cambridge University Press, Cambridge, pp. 370–404. DOI: 10.1017/CBO9781139236058.

Rapsomanikis, G. (2015) The economic lives of smallholder farmers. An analysis based on household data from Nine Countries. Food and Agriculture Organization, Rome.

Rossi, R. (2022) Small farms's role in the European Union. European Parliament Research Services, Strasbourg.

Sabo, B.B., Isah, S.D., Chamo, A.M. and Rabiu, M.A. (2017) Role of smallholder farmers in Nigeria's food security. *Scholarly Journal of Agricultural Science* 7, 1–5.

Sumberg, J. and Sabates-Wheeler, R. (2011) Linking agricultural development to school feeding in sub-Saharan Africa: theoretical perspectives. *Food Policy* 36, 341–349. DOI: 10.1016/j.foodpol.2011.03.001.

Swensson, L.F., Hunter, D., Schneider, S. and Tartanac, F. (2021) Public food procurement as a game changer for food system transformation. *The Lancet Planetary Health* 5, e495–e496. DOI: 10.1016/S2542-5196(21)00176-5.

Taiy, R.J., Onyango, C., Nkurumwa, A. and Ngetich, K. (2017) Socio-economic characteristics of small-holder potato farmers in Mauche ward of Nakura County, Kenya. *Universal Journal of Agricultural Research* 5, 257–266.

Tshuma, M.C. (2014) Understanding small-scale agricultural sector as a precondition for promoting rural development in South Africa. *African Journal of Agricultural Research* 9, 2409–2418. DOI: 10.5897/AJAR12.1631.

UNCTAD (2015) Commodities and development: smallholder farmers and susatianable development. United Nations, Geneva.

UNCTAD Secretariat (2015) The role of smallholder farmers in sustainable Commodities Production and Trade. United Nations Conference on Trade and Development, Geneva.

Webber, C.M., Chigumira, G. and Nyamadzawo, J. (2013) *Building Agricultural Competiveness in Zimbabwe; Lessons from the International Perspective.* Zimbawe Economic Policy Analysis and Research Unit, Harare.

WFP (2016) *Food Quality and Safety P4P Experiences in Systemic Change.* World Food Programme, Rome.

World Bank (2016) *Sustainable Procurement: An Introduction for Practitioners to Sustainable Procurement in World Bank IPF Projects.* World Bank, Washington, DC.

17 Agribusiness Supply Chain Resilience

Iscacle Nyanhete[1]*, Ernest Mugoni[2] and Charles Tsikada[2]

[1]*Midlands State University (MSU), Zimbabwe;* [2]*Marondera University of Agricultural Sciences and Technology (MUAST), Zimbabwe*

Abstract

The agricultural supply chain is inherently cross-industry with multiple functions interacting. These functions have conflicting objectives and it is further complicated by fragmented inbound and outbound networks. This chapter made use of documentary search as a way of collecting data – critical review of academic books, articles and journals, websites, electronic and print media drawing from past and current experiences on agribusiness supply chains and their resilience. Content analysis helped in developing themes linked to agribusiness supply chain resilience. Many supply chain nodes struggled to operate, and remain resilient and sustainable. Resilience and sustainability are closely linked. Supply chain resilience is a critical ingredient to sustainable agriculture in a world that is continuously changing in order to avoid crisis, remain sustainable, maintain essential services during disruptions and quickly recover. In the event of a disruption, the supply should be able to recover from the disruption and continue to operate at the desired level of connectedness and control over structure and function. However, supply chain resilience might not mean going back to the state prior to the disruptions, because doing so might mean building back some vulnerability. On the other hand, the challenge with agricultural products is they are produced through biological processes and these processes cannot be accelerated in response to an upsurge in demand or delayed as the supply chain faces disruptions. The perishable nature of agricultural products also complicates the management of supply chains. Demand, supply, quality and quantity are volatile, and inventory cannot be used as a buffer. The agriculture supply chains have become complex systems of interconnected and interdependent organizations and their supply chain resilience is a collective property that cannot be derived from an individual member of the supply chain. Therefore, the resilience of a supply should be built around all members of the supply chain. This chapter discusses supply chain resilience, disruptions, shocks and stresses of agricultural chains and opportunities that can be explored to build supply chain resilience of the sector. The chapter underscores the understanding of the sustainable agricultural supply chain resilience.

Keywords: Supply chain resilience, Supply chain disruption, Supply chain, Sustainable supply chain management, Sustainable agricultural supply chain

17.1 Introduction and Background

A sustainable agribusiness supply chain focuses on maintaining environment, economic and social stability for long-term growth (Amer *et al.*, 2018). Supply chain resilience is the adaptive capability of a supply chain in the event of a disruption and its ability to recover from the disruption and continue to operate at the desired level of connectedness and control over structure and function (Mandal *et al.*, 2017). However, supply chain resilience does not mean necessarily going back to the state prior to the disruptions, because doing so might mean

*Corresponding author: nyanheteiscacle74@gmail.com

© CAB International 2023. *Sustainable Agricultural Marketing and Agribusiness Development: An African Perspective* (eds B. Nyagadza and T. Rukasha)
DOI: 10.1079/9781800622548.0017

building back some vulnerabilities (Brown *et al.*, 2017).

The Covid-19 pandemic-induced disruptions were exuberated by the erratic handling of the outbreak by countries in the same geographical or economic region, putting into jeopardy supply chain managers' efforts to respond to disruptions and maintain the desired level of connectedness and control (Modgil *et al.*, 2021). The authorities implemented measures to curb the exponential spread of the virus, which included reducing physical contact, restricting movement of people and maintaining physical distance. These measures disrupted agribusiness supply chain processes, since agribusiness processes rely on physical contact between employees in production, distribution, wholesale and retailing (Koncar *et al.*, 2020). Most agricultural products are produced through biological processes. These systems cannot be accelerated to respond to an upsurge in demand or delayed as the supply chain faces disruptions (Hobbs, 2021). For instance, the production of milk, eggs, beef, vegetables, fruits and other agricultural products cannot be delayed or accelerated in response to an upsurge or a dip in demand Therefore, understanding supply chain process flows is critical for building supply chain resilience.

The agribusiness supply chain is also exposed to varying disruptions that complicates supply chain activities, chief among them seasonality of production, which is exacerbated by varied lead times, low standardization of product quality and quantity, trade restrictions and lack of product traceability (Sharma *et al.*, 2020). On the other hand, farm production is a 'push system', which creates an imbalance between demand and supply, and there is no evidence that farmers are working towards systematic attempts to closely link agricultural production to expected demand (Naik and Suresh, 2018).

The agribusiness supply chain in most parts of the world is inherently affected by challenges such as dominance by small or marginal farmers, fragmented supply chains, low value addition, and poor marketing structures (Rukasha *et al.*, 2021). Supply chain management envisages the understanding of the chain as a whole, accounting for all links that coordinate and synchronize the nodes (Leonczuk, 2021). The fragmentation of agribusiness nodes makes it challenging to build supply chain resilience. The network of nodes in a supply chain face different shocks

and stresses. Therefore, the ability to resist node-specific shocks and stresses should be taken into account when building supply chain resilience (Govindan *et al.*, 2017). The pandemic disrupted supply chains through low productivity due to mandatory social distancing measures and quarantining of truck drivers that disrupted road transportation (Oxfam, 2020). Strictly following advanced hygiene protocols also slowed down production activities as more time was dedicated to sanitization, thereby reducing the overall throughput of production processes and supply chain processes (Hobbs, 2021).

Postharvest storage and transportation are the major causes of shocks and stresses in the agribusiness supply chain as more than half of mangoes, bananas, oranges and papaws produced by smallholder farmers and sold by street vendors is lost to postharvest loss due to poor storage and transportation, and other substandard supply chain processes (Coleman, 2022). The perishable nature of agricultural products complicates the management of agricultural supply chains. Demand, supply, quality and quantity are volatile and inventory cannot be used as a buffer (Naik and Suresh, 2018). Many agribusinesses produce large quantities of good-quality harvest, but do not have reliable swift and equitable means to get their produce to the market resulting in qualitative and quantity losses (Mandisvika *et al.*, 2015).

In the past few decades, the primary focus of supply chain management has been to increase efficiency and reliability through globalization, specialization and lean supply chains. However, these practices have made supply chains vulnerable to disruptions (Perera *et al.*, 2017). While interconnectedness and interdependency have enabled agribusinesses to take advantage of resources that span across the supply chain, this has made them vulnerable to upstream shocks and stresses (Ehie and Ferreira, 2019). Some of these shocks and stresses take place thousands of miles upstream yet businesses at the other end of the chain are negatively affected.

17.2 Methodology

The chapter focused on agribusiness supply chain resilience. Documentary search was used as a way of collecting data. This involved critical review of academic books, articles and journals, websites, electronic and print media drawing from past and current experiences on agribusiness supply chains and their resilience. Content analysis helped in developing themes linked to agribusiness supply chain resilience.

17.3   Literature Review and Theoretical Frameworks

17.3.1   Theoretical frameworks

Supply chain resilience has a few theories; however, this research is guided by two dominant theories – complex systems theory and relational theory.

17.3.1.1   Relational theory

The central argument of the relational view is that inter-firm linkage (supplier development) is a source of relational rents, competitive advantage, substantial knowledge and complementary resources and capabilities (Chen *et al.*, 2013).

17.3.1.2   Complex systems theory

The complex adaptive systems theory advocates that complex systems possess collective properties that cannot easily be derived from their individual constituents (Perera *et al.*, 2017). It focuses on the emergence of order in a system that is operating at the edge of chaos due to disruption. The order is achieved through facilitating transformation and remaining functional in a chaotic environment (Tukamuhabwa *et al.*, 2015).

17.3.2   Supply chain vulnerabilities

Vulnerability is the maximum deviation from normal performance after a disruption occurs and it is one of the aspects that measures resilience. Vulnerability can be determined by predisposition to risk, elasticity to withstand shocks, and strength building (Geng *et al.*, 2014). Supply chain vulnerability is not only a result of a turbulent environment but also a result of

the design of the supply chain itself. Therefore, vulnerability mitigation strategies should be embedded within the agribusiness supply chain strategy (Kumiawan *et al.*, 2017).

Many organizations do not include supply chain vulnerability analysis in their overall business strategy. Planning appropriate mitigating strategies, examining sources of the risks, analysing drivers of vulnerability and measuring the consequences reduces the chances of deviating from normal performance (Kumiawan *et al.*, 2017). However, there is a need to balance the building of supply chain resilience and vulnerability. This is because increasing resilience may also increase vulnerability. For instance, improving resilience by investing in a large amount of inventory makes the business vulnerable to obsolescence, damage of inventory, and pilferage (Zhang *et al.*, 2021).

Many organizations regularly access their risks to identify areas of vulnerability, but the assessment tends to be focused on broader regulatory and financial risk rather supply chain vulnerabilities (Christopher, 2018). Unreliable electricity supply, low internet connectivity, and the high cost of data affects the ability of agribusinesses to recover from the disruption, continue to operate at the desired level and adapt to changes brought about by the disruptions (Moyo-Nyede and Ndoma, 2020).

17.3.3   Agribusiness supply chain

In the past few decades, the focus has also been towards optimization through cost minimization, inventory reduction, and the removal of buffers. The pandemic has exposed vulnerability of these processes and practices (Wahi *et al.*, 2021). Lean agribusiness supply chains have improved efficiency, through low levels of inventory holding along the nodes and compressed lead times. However, these efficiencies came along with risks, as they increased the surface area and magnitude of supply chains risks (Alicke *et al.*, 2020).

Supply chain nodes are interconnected and interdependent. Because of this, the effects of disruptions are not isolated by industry or geography. Interconnectedness and interdependence of supply chains are increasingly becoming complex in nature and any disruptions in any geographical location or industry can bring global shocks and stresses (Perera *et al.*, 2017). Supply chain management is not a support function for implementing agribusiness business strategies, but it is a key element of the overall strategy that drives the firm's performance (Mhelembe and Mafini, 2019). However, top-line growth remains the number one priority for agribusiness executives and supply chain resilience issues are at the bottom of their list, thus supply chain resilience cannot keep pace with the agribusiness' top-line growth initiatives (KPMG, 2016).

The quality of agricultural products is significantly affected by the effectiveness of the supply chain management practices, especially during loading, shipment and unloading. One of the major factors that leads to deterioration of produce quality is waiting time, transportation time and overproduction in relation to controlled temperature storage (Baghizadeh *et al.*, 2021). However, most agriculture supply chains are controlled by a large number of unorganized intermediaries which makes it difficult to coordinate time, transportation, and the relation between volume produced and controlled temperature storage, making supply chains inefficient and difficult to build resilience (Patidar *et al.*, 2018).

The agribusiness sector is slow to embrace innovation because there is lack of digital culture and training, which prohibits digitalization of the industry to reach full potential (Trung *et al.*, 2020). Slow uptake of innovation has led to lack of traceability systems in the agribusiness supply chains, which affects the industry's ability to compete in lucrative markets and build supply chain resilience (Burger, 2021). While technology might be available, it is setting out strategies, investing in resilience and organizing the essential partnership in the supply chain that makes the chain resilient. A fundamental reorganization of information streams and agency relationships based on supply chain management principles is required in order to build agribusiness supply chain resilience (Naik and Suresh, 2018).

17.3.4 Resilience

The etymology of resilience is the Latin word *resilire*, which means jumping or springing back

(Tocci, 2020). However, resilience is not only about jumping or springing back to the original position because this would reinforce the conditions that led to the disruption. It rather involves abilities to transform instead of retaining the status quo. Resilience goes beyond mitigating risks as it enables agribusinesses to gain competitive advantage by learning how to deal with disruptions more efficiently than its competitors (Fiksel *et al.*, 2015). Resilience has four main properties: robustness, redundancy, resourcefulness and rapidity (Keating *et al.*, 2017).

17.3.4.1 Robustness

Supply chain robustness as a component of resilience focuses on maintaining normal operation, while compensating for disruptions with minimal impact on the performance of the supply chain (Zhao *et al.*, 2018). Robustness maintains functional operations when a system is subjected to uncertain interference from internal and external events (Geng *et al.*, 2014). The supply chain resists change without modifying its initial settings; therefore, robustness comes from anticipation and readiness (Bauer and Gobi, 2017). The supply chain should be able to sustain a given amount of strain or demand without losing functionality (The World Bank, 2015). The pandemic, persistent droughts, and animal and plant disease outbreaks have tested the robustness of agribusiness supply chains as they have brought a series of shocks that has caused major supply chain disruptions. Government restrictions on movements of animals due to disease outbreaks such as foot-and-mouth, and social distancing are testing the robustness of agribusinesses. Robustness can be improved by planning and designing measures that mitigate or reduce disruptions upstream, thus reducing vulnerability (The World Bank, 2015).

17.3.4.2 Redundancy

Redundancy is the duplication of capacity that can be used during disruptions, thereby increasing flexibility by using spare capacity to manage disruptions (Adobor and McMullen, 2018). Redundancy is a proactive strategy that duplicate resources and possesses alternative opportunities. It is the extent to which alternative element systems are capable of satisfying functional requirements in the event of disruption (Keating *et al.*, 2017). One element of redundancy is the application of larger numbers of elements compared to what is necessary or generally accepted. These elements may comprise additional resources, human resources, space or information to enhance dependability of the agricultural entity's operations (Jacyna-Golda and Lewczuk, 2017). There are a few ways of creating redundancy, including holding safety emergency inventory, spare plant capacity, and using back up facilities (Tan *et al.*, 2019). Redundancy helps the supply chain to prepare, respond and recover from disruptions. However, modern day supply chains are designed to optimize costs and efficiency. Therefore, there is need to strike a balance between building supply chain resilience efficiency and redundancy as previously stated (Bauer and Gobi, 2017).

17.3.4.3 Resourcefulness

Resourcefulness is the ability to identify problems, define priorities, and mobilize resources in response to disruptions (Sambowo and Hidayatno, 2021). It is the ability to generate value in a resource-constrained environment. Thus, it is a creative use of internal mechanisms to create resilience (Barraket *et al.*, 2019). Resourcefulness is about optimizing what the agribusiness already has and making the best use of resources in the organization's possession by maximizing benefits of its resources, abilities and time (Baldoni, 2010). It reflects on the agribusiness's availability of resources that restore functionality of the supply chain and being able to diagnose problems and initiate solutions by identifying and mobilizing resources, information and technology (Tierney and Bruneau, 2007).

Resourcefulness is not only about maximizing resources during the times of disruptions, but it is the proactive ability to do more with the existing resources in order to build resilience (Baldoni, 2010). Since the outbreak of the Covid-19 pandemic, agribusinesses have been resourceful despite being vulnerable financially, facing skills shortage, technological obsolescence, severe legislation requirements and ever-changing customer needs (Munyanyi and Chimwai, 2019). In the midst of a crisis, being

resourceful is a challenge. This is because it is difficult to mobilize overstretched resources and use obsolete technology, with an unskilled workforce, under legislation requirements to respond to Covid-19-induced disruptions.

17.3.4.4 Rapidity

Rapidity is the ability to recover and get back to equilibrium within a short period following a disruption (Mandal *et al.*, 2017). It is the ability to meet priorities and achieve goals to avoid future disruptions (Sambowo and Hidayatno, 2021). Rapidity is meeting priorities and achieving goals in a timely manner in order to contain losses, recover functionality, and avoid future disruptions (Keating *et al.*, 2017). The measures to curb the exponential spread of the virus disrupted supply chain rapidity through low productivity due to mandatory social distancing requirements and hygiene protocols (Oxfam, 2020).

17.3.5 Supply chain resilience

Supply chain resilience is the adaptive capability of a supply chain in the event of a disruption and its ability to recover from the disruption and continue to operate at the desired level of connectedness and control over structure and function (Mandal *et al.*, 2017). On the other hand, building supply chain resilience is costly and agribusinesses cannot easily pass on the cost of building and maintaining supply chain resilience to their consumers. Supply base reduction, just in time and outsourcing, connectedness, and interdependency have become the latest trends in supply chain management. However, these trends have increased the vulnerability of supply chains (Nel *et al.*, 2018). On the other hand, a supply chain with low interdependence and individual firm resilience will not feel the disruptions across the chain (Sa *et al.*, 2019). Bypassing intermediaries and buying directly from manufacturers or producers can also build supply chain resilience. It shortens the supply chain by reducing the nodes, thereby improving upstream and downstream visibility. Buying directly from manufacturers also gives the manufacturer direct access to information about end users and the retailers also get access to information about raw materials. The manufacturer can also quickly notice changes in demand pattern, for example, the demand for certain products may change due to disruptions and the manufacturer can quickly react in response to the demand. During the peak of the Covid-19 pandemic, demand for larger packages of products increased as consumers were buying in bulk to reduce frequency to the shops in order to lower their chances of contracting the virus.

Improving inter-business relationships also builds resilience, as good relationships eradicate silo-mentality, thereby reducing the time and effort needed to react to disruptions. Good inter-business relationships drape around trust and commitment, and trust and commitment foster group solidarity, and strong group solidarity is a good foundation for building supply chain resilience (Nandonde and Kuada, 2018). Supplier development initiatives increase the suppliers' willingness to participate in joint problem-solving during disruptions. It also creates a bond and togetherness during disruptions (Wang et al., 2013). A number of strategies improve supply chain resilience: increasing flexibility, collaboration and agility (Tukamuhabwa et al., 2015).

17.3.5.1 Flexibility

Flexibility is the ability to adapt or react to unforeseen changes in a short space of time with little effort and at the lowest cost (Obayi et al., 2017). Flexibility focuses on the capability to adapt to external changes and adaptability of the system (Geng et al., 2014). Flexibility aids in rapid response and recovery through availability of alternative choices, including alternative suppliers. It might be achieved through employing a multi-skilled workforce, installing multi-purpose machines and creating flexible contractual agreements (Tukamuhabwa et al., 2015).

17.3.5.2 Collaboration

Traditionally, data is considered as proprietary and it is closely held. The digital revolution has triggered a dramatic escalation in the commercial value of data resources (Negra et al., 2020). Water systems, energy systems and food systems are widely researched subjects in different sectors, such as hydrology, energy and food value chains. However, the findings of this research do not move freely across the hydrology, energy and food sectors, yet agriculture processes cut across these sectors and this makes it difficult to provide clear information on how to improve sustainability (Negra et al., 2020).

Collaboration is an important pillar of sustainable agriculture. Therefore, joint partnership with various stages of supply chain optimizes completive advantage and supply chain resilience. Sharing of assets (materials, labour, infrastructure, facilities, equipment and machines) and capabilities (technology, business processes, policies and finances) spreads the benefits across the entire supply chain (Dania et al., 2016). Collaboration creates inter-organizational relationships, which yields relational rent (supply chain resilience) as company's pool their respective resources helping each other to eliminate deficiencies in their individual processes and resources, thereby creating a resilient resources endowment (Dyer and Singh, 1998).

Collaboration aids in mitigating disruptions before they occur by facilitating information sharing. It can also aid recovery by enabling supply chain partners to share resources and provide a coordinated response (Tukamuhabwa et al., 2015). Through collaboration, information can be accessed by supply chain partners as rapidly as possible and this information can be converted in supply chain intelligence (Christopher, 2018). Collaboration promotes a sense of community and encourages supply chain partners to invest in social capital, and social capital acts as a hedge against collective adversities, spurs collective action and helps supply chain partners to stay coherent against disruptions (Mubarik et al., 2021). Sharing bad news with supply chain partners can provide essential information, which can used to build supply chain resilience (Christopher, 2018).

Collaboration is often viewed as a fraught territory, especially beyond the first tier of suppliers. Therefore, getting information on inventory levels, capacity and flexibility from second or third tier suppliers gives the agribusiness a glance into potential sources of vulnerability (Alicke et al., 2020). Sharing information does not only minimize the immediate impact of disruptions, but enables the agribusiness to overcome long-term impacts (Mubarik et al., 2021).

The challenge in sharing information is not about technology, but the willingness to share information among supply chain partners.

17.3.5.3 Adaptability

Adaptability is the capacity to make changes that prevent occurrence of undesired events, while improving functioning or acquiring new skills in order to achieve resilience (Leonczuk, 2021). It is also the ability to adjust the supply chain design to meet structural shifts in the markets and modify the supply network to reflect changes in strategies, technologies and products (Whitten et al., 2010). Adaptability enables the supply chain network to respond to macro-changes in the environment (government policies, political changes and legal changes) by setting up facilities (resources), and adjusting the supply chain network configuration in a flexible way (Dubey et al., 2015). Highly adaptable agribusinesses sense long-term fundamental changes in the supply chain and respond to such changes by flexibly adjusting the configuring the supply chain (Eckstein et al., 2014). Supply contracts have also become complicated and rigid, thereby negatively affecting the ability of agribusinesses to respond to macro-changes and to adjust their supply chains. There is need to revisit supply contract terms and incorporate clauses that make it possible to adapt to changes with minimum costs and time, and with little effort.

17.3.5.4 Agility

Agility is the ability to return to a position of equilibrium after experiencing some form of deviation from expectation (Mathu and Phetla, 2018). It can also be described as the ability to change the overall systems significantly in order to respond to disruptions and move quickly and nimbly to take advantage of the new opportunities (KPMG, 2016). An agile supply chain has the ability to quickly adapt to future changes and implement transformation to support operations in the event of a disruption (Marche et al., 2019). Taking a systematic approach in assessing and deploying a flexible resources pool, accommodating variability of demand and supply, and reconfiguring internal processes to meet business needs are critical to agility (Kilpatrick et al., 2021). Agility is also a powerful tool that fosters supply chain performance and it acts as an antidote to complexity, disruption and risks (Kilpatrick et al., 2021). Agility is a source of competitive advantage and a building block for supply chain resilience. The use of an agile workforce brings competitive leverage. Working with local small and medium-sized suppliers improves the agility of the supply chain, they do not have cumbersome procedures, which makes product or service alteration and modifications of specifications easier (Martindale, 2022).

Consumers are also testing the agility of agriculture supply chains as their requirements are continually changing. While inventory replenishment is driven by current demand, on the other hand agriculture planning is based on forecasted demand (KPMG, 2016). Supply chain agility has two main characteristics: visibility and velocity (Christopher and Peck, 2004).

VISIBILITY. Visibility is the extent to which supply chain partners access and share mutually beneficial information (Sunmola and Apeji, 2020). Visibility enables supply chain partners to see through the entire chain, the environment it is operating in and the key partners, and identify vulnerabilities along the chain and prepare in the event of a disruption (Adobor and McMullen, 2018). High visibility contributes to lower lead time variability, reduced inventory, short lead time, increased fill rates and supply chain operational efficiency (Ahimbisibwe et al., 2016). Modern supply chains should have end-to-end data-driven ability that enables upstream visibility from the origin of raw materials and downstream visibility to the end user. However, the ability to extract value from the data and quickly take action is what builds supply chain resilience (Lakovou and White, 2020).

Many traditional supply chain networks have poor upstream and downstream visibility with little shared information; hence they are prone to mismatches of demand and supply in the event of a disruption (Ahimbisibwe et al., 2016). However, the pandemic, has pushed agribusinesses to consider end-to-end visibility, integration and optimization, and adopt agile planning and forecasting (Wahi et al., 2021). Visibility of upstream and downstream processes provides necessary information to make decisions, make corrections to plans, identify bottlenecks and react to eliminate the bottle-necks (Leonczuk, 2021). The ability to detect, measure and manage upstream risks is a challenge, as many agribusinesses only track risks in the first-tier suppliers and few have visibility beyond the first tier (Kilpatrick *et al.*, 2021).

Visibility of primary data on order processes, inventory status, transportation and distribution reduce complexity of the supply chain and give better control of supply chain operations and resource allocations (Gunasekaran *et al.*, 2015). Visibility is enhanced through developing monitoring programmes such as Business Continuity Planning (BCP) (Adobor and McMullen, 2018). Gathering, interpreting, synthesizing and coordinating information across the supply chain increases visibility and the ability to respond and adapt to changing environments (Ahimbisibwe *et al.*, 2016). However, many agribusinesses have limited internal visibility and it takes weeks or months before an internal disruption becomes visible. By the time it becomes clear to all it might be too late to react effectively.

VELOCITY. Velocity is the speed at which activities within the supply chain are completed and an order moves through the supply chain from order processing to delivery to the customer (Hand, 2021). Velocity is also the acceleration or the ability to ramp up or down quickly (Christopher, 2018). High velocity fosters mutually beneficial relationships through responsiveness to the increase or decrease in the flow of products to the customers (Hand, 2021). However, to increase velocity, the end-to-end pipeline time must be reduced and acceleration is dependent upon supply chain alignment (Ahimbisibwe *et al.*, 2016). High velocity minimizes lead times and streamlines supply chain processes and enables customers to receive their orders with relatively little turnaround time (Hand, 2021).

17.3.6 Social resilience

Community resilience is the collective ability of a geographically defined area to deal with disruptions and efficiently resume the rhythms of daily life (Aldrich, 2012). Agriculture entities support communities that are close to them and they are considered to be part of the local social fabric (Nair, 2017). The entity and the communities around it are inextricably linked. A resilient entity improves the communities' ability to respond to disruptions like the Covid-19 pandemic (Brown *et al.*, 2017). There is a need to review the current structure of flows of agriculture products and design a sustainable supply chain that is more efficient and equitable and that minimizes wastage (Naik and Suresh, 2018). Various agricultural supply chain stakeholders have costs and collect benefits even though it might not be fairly distributed in some instances. Therefore, there is need to spread the benefits along and across the supply chain, in a fair and equitable way (Dania *et al.*, 2016).

17.3.7 Economic resilience

Economic resilience is the ability to absorb losses caused by disruptions (Brown *et al.*, 2017). Resilient agribusinesses do not wait for a disruption in order to act; they build cash reserves, create margin headroom, move into new markets, proactively go on an offensive and reshape their value proposition (Kohli *et al.*, 2020). Building cash reserves enables the business to invest during the disruptions, and having a margin headroom enables it to accommodate margin shrinkages. Armed with cash reserves, the agribusiness can go on the offensive moving into crops, livestock, fruits and vegetables where others are forced to withdraw, and can operate while taking a punch on its margins thereby increasing its market share (Kohli *et al.*, 2020). Resilient agribusinesses transpose product assortment and adjust their pricing in response to the disruption in order to safeguard sales and find new customers. They also strive to maintain a high level of customer service during the disruption through investing in frontline workers' training thus upgrading their talent pool (Kohli *et al.*, 2020). Training frontline staff also creates customer value through improved effectiveness and it also increases customer satisfaction. On the other hand, well-trained frontline workers know how to align products with current customer needs (Phillipart, 2016).

Agricultural inventory management has conflicting objectives – holding low inventory causes economic losses, while holding higher inventory increases holding cost and high risk of perishability of the products. Products fetch low prices during harvest seasons but withholding harvest requires keeping the product in a temperature-controlled environment, which is costly and results in carbon dioxide emissions (Galal and El-Kilany, 2016). However, temperature-controlled storage allows agribusinesses to extend their selling season, thereby enabling them to fetch higher prices during off-peak seasons (Webber et al., 2013).

17.3.8   Organizational resilience

The organizational culture theory advocates that firms' openness enriches key partners' knowledge base, enlarges accessibility to resources, and encourages participation, empowerment and teamwork (Li et al., 2019). Knowledge, access to resources and teamwork are key elements of building supply chain resilience. Organizational resilience is built through human, material, financial and technological resources, planning, allocating resources and going through scenario exercises (pseudo crisis situations). Organizational resilience is the maintenance of positive adjustment under disruptions such that the organization emerges from those disruptions strengthened and more resourceful (Barasa et al., 2018). There are three components to organizational resilience – survival, adaptation and innovation – and these components help the organization to overcome adversity and thrive as they reinvent themselves while continuing to operate under disruptions (Brown et al., 2017).

Information guides and correct responses to disruptions help to separate myths from facts and create situation awareness. In this era of social media fake news is a reality and this is exuberated by the fact that the situation about supply chain disruptions evolves daily. Getting accurate and comprehensive information about a disruption's severity and dynamics is critical (Bene, 2020).

Organizational culture has a huge influence on the adaptive capabilities of an agribusiness. The culture of denying the existence of risks has brought problems to mankind throughout history. Accepting the existence of a risks fosters willingness to own the risks, and the ability to learn from past experience is a vital element of building resilience (Barasa et al., 2018). Many organizations develop a tendency to deny the existence of problems and strategically use their public relations departments to deny the existence of problems in the organization in order to create a good image in the public eye. However, the denial tendency can filter into the organization's fabric (employees and supply chain partners) and they will develop the problem denial tendency. Organizational structure also has a huge influence on the agribusiness' ability to respond to disruptions. Structures that are too rigid and systematic are less able to adapt quickly to a new normal (Brown et al., 2017). Governance processes (rules and processes) that guide operations are critical to organizational resilience. Decentralized organizations characterized with distributed controls that empower operational level managers and supervisors to make decisions are more resilient (Barasa et al., 2018).

17.3.9   Risk culture

Risk culture is a business model that firms can follow in their day-to-day operations. It promotes learning and knowledge sharing (Neboh and Mbhele, 2020). It is setting an intent across the whole organization and supply chain partners through clear communications, and taking action to address risks. It is also about creating a sense of ownership and allowing warning signs of both internal and external shocks to be shared (Alicke et al., 2020). Openness in communications fosters risk awareness and understanding, which in turn creates an ownership environment where members feel responsible for their actions and decisions (Alicke et al., 2020). Openness in communication also promotes collaboration and a sense of community based on risk awareness.

17.3.10   Risk management

Risk management focuses on providing appropriate strategic solutions in the event of disruptions and to mitigate severe effects of the disruption (Kumiawan et al., 2017). Good risk management is the identification of risks and understanding that their increase or decrease has an effect on the business (Baily *et al.*, 2015). Supply chain risks are continuing to shift and multiply; therefore, there is a need to address higher levels of complexity and volatility while simultaneously managing costs and building new capabilities in a digitally enabled ecosystem (Kilpatrick *et al.*, 2021). Low-frequency high-impact events have a ripple effect. The Covid-19 outbreak is a low-frequency high-impact event and it had a ripple effect on global supply chains (Ivanov and Dias, 2020). The Covid-19 pandemic's disruption was not localized despite efforts by government to curb it from spreading and it cascaded downstream disrupting supply chain performance (Ivanov, 2018). The connectedness and interdependency of agribusiness supply chain magnified the effects of the pandemic, and the effects of the pandemic could not be localized in China, it had ripple effects on the supply chain worldwide (Sa *et al.*, 2019). Outsourcing, supplier consolidation, globalization and low-cost-country sourcing have made supply chains longer and vulnerable to disruptions. A disruption that takes place in one place can cause knock-on effects on supply chains that resonate worldwide (Baily *et al.*, 2015).

The possibility of a partial or full shut down of the whole industry sector and economic regions has never been the considered by many business executives since the Cold War (Ivanov and Dias, 2020). Those that carry out risk assessment and due diligence only carry it out on the immediate suppliers, leaving second, third and fourth tier suppliers, thereby creating risk blind spots and these are sources of any shocks and stresses (Lees and Dixon, 2022). Risk managers normally use historical data to develop risk mitigation strategies, however many agribusinesses do not have the historical data on public health disruptions to develop risk mitigation strategies of the Covid-19 pandemic magnitude (Wakasala, 2020).

There has been a shift from the traditional risk management perspective, which is more reactive to disruptions, to a more proactive approach of building supply chain resilience (Katsaliaki *et al.*, 2021). Traditional risk assessment strategies cannot deal with unforeseeable events like the outbreak of the Covid-19 pandemic. However, adaptable supply chain resilience strategies make a difference between

survival and success (Gunasekaran *et al.*, 2015). Rather than focusing on vulnerabilities and protection, agribusinesses should develop an orientation toward building supply chain resilience and focus should be on building a favourable post-disruption position thus gaining competitive advantage (Forbes and Wilson, 2018).

17.3.11 Supply chain design

The design of the supply chain should consider underlying vulnerabilities and the level of future exposure or susceptibility to unforeseen disruptions (Alicke *et al.*, 2020). Identifying the right partners plays a major role in developing supply chain resilience and an optimally configured chain reduces uncertainties and complexities which are major sources of future disruptions (Gunasekaran *et al.*, 2015). Aligning the interests of supply chain members enables members to be flexible as misaligned interests can cause chaos and it affects members' response to disruption (Ahimbisibwe *et al.*, 2016). An effective supply chain design is planned, purposive and it takes into consideration influencers, design decisions and building blocks (Melnyk *et al.*, 2014). Influencers are business and the political environment; these are a major source of shocks and stresses in agribusiness. The building blocks include inventory, transportation, capacity and technology (Melnyk *et al.*, 2014). Huge amounts of money are invested in building blocks in order to improve performance; however some of the investment in building blocks is done without factoring in supply chain resilience (Govindan *et al.*, 2017). Many organizations have not yet adopted resilience into their structural standing and design of supply chain (Neboh and Mbhele, 2020). They skip detailed mapping of second and third tier suppliers in their supply chain design stage thereby failing to identify relationships and nodes of interconnectivity, despite relationships and nodes of interconnectivity being potential sources of vulnerabilities in the supply chain.

Supply chain design needs to be set in a way that that improves its resilience and agility, and the notion of putting emphasis on efficiency at the expense of resilience can no longer be tolerated after the shocks and stresses brought about by the Covid-19 pandemic (Lakovou and White, 2020). In order to design a resilient supply chain, agribusiness executives need to accept the fact that unpredictability and uncertainty of disruptions are realities (Adobor and McMullen, 2018).

17.4   Conclusion and Recommendations

The recent events in the agricultural sector brought multifaceted vulnerabilities in supply chains. Therefore, maintaining environmental, economic and social stability for long-term growth is a key to success. Resilience is critical to an agricultural organization's survival, especially as the economic and political environment is not stable. Supply chain disruptions will continue to bring shocks and stresses and new sources of disruptions are constantly emerging (O'Marah, 2017). Disruptions are cyclic in nature and recovery should include readiness for the next event, and they affect consumption and buying patterns (Forbes and Wilson, 2018). Supply chain resilience is a critical ingredient to business survival and building supply chain resilience ensures business survival of agricultural entities in the face of disruptions.

References

Adobor, H. and McMullen, R.S. (2018) Supply chain resilience a dynamic and multidimensional approach. *The International Journal of Logistics Management* 29(4), 1451–1471. DOI: 10.1108/IJLM-04-2017-0093.

Ahimbisibwe, A., Ssebulime, R., Tumuhairwe, R. and Tusiime, W. (2016) Supply chain visibility, supply velocity, supply chain alignment and humanitarian supply chain relief agility. *European Journal of Logistics, Purchasing and Supply Chain Management* 4, 34–64.

Aldrich, D.P. (2012) *Building Resilience: Social Capital in Post-Disaster Recovery*. University of Chicago Press, Chicago. DOI: 10.7208/chicago/9780226012896.001.0001.

Alicke, K., Barriball, E., Lund, S. and Swan, D. (2020) *Is Your Supply Chain Risk Blind – Or Risk Resilient*. McKinsey and Company, Washington DC.

Amer, H.M., Galal, N.M. and E.l-Kilany, K.S. (2018) *A Simulation Study of Sustainable Agri-Food Supply Chain*. Industrial Engineering and Operations Management International Society, Paris.

Baghizadeh, K., Cheikhrouhou, N., Govindan, K. and Ziyarati, M. (2021) Sustainable agriculture supply chain network design considering water-energy-food nexus using queuing system. A hybrid robust possibilistic programming. *Natural Resource Modeling* 35, 1–39. DOI: 10.1111/nrm.12337.

Baily, P., Farmer, D., Crocker, B., Jessop, D. and Jones, D. (2015) *Procurement Principles and Management*, 11 edn. Pearson, Harlow.

Baldoni, J. (2010) The importance of resourcefulness. *Harvard Business Review* 1.

Barasa, E., Mbau, R. and Gilson, L. (2018) What is resilience and how can it be nurtured? A systematic review of empirical literature on organizational resilience. *International Journal of Health Policy and Management* 7(6), 491–503. DOI: 10.15171/ijhpm.2018.06.

Barraket, J., Eversole, R., Luke, B. and Barth, S. (2019) Resourcefulness of locally-oriented social enterprises: Implications for rural community development. *Journal of Rural Studies* 70, 188–197. DOI: 10.1016/j.jrurstud.2017.12.031.

Bauer, D. and Gobi, M. (2017) Flexibilty measurement issues in supply chain management. *Journal of Applied Leadership and Management* 5, 1–14.

Bene, C. (2020) Resilience of food systems and links to food security: a review of some important concepts in the context of Covid-19 and other shocks. *Food Security* 12(4), 805–822. DOI: 10.1007/s12571-020-01076-1.

Brown, N.A., Rovins, J.E., Feldmann-Jensen, S., Orchiston, C. and Johnston, D. (2017) Exploring disaster resilience within the hotel sector: a systematic review of literature. *International Journal of Disaster Risk Reduction* 22, 362–370. DOI: 10.1016/j.ijdrr.2017.02.005.

Burger, S. (2021) *Blockchain-Based E-Livestock Supply Chain Tracebility System Launched in Zimbabwe*. Creamer Media.

Chen, D.R., Preston, D.S. and Xia, W. (2013) Enhancing hospital supply chain performance: a relational view and empirical test. *Journal of Operations Management* 31(6), 391–408. DOI: 10.1016/j.jom.2013.07.012.

Christopher, M. and Peck, H. (2004) Building the resilient supply chain. *International Journal of Logistics Management* 6, 197–210. DOI: 10.1080/13675560310001627016.

Christopher, M. (2018) *Logistics and Supply Chain Management*. Pearson, Harlow.

Coleman, A. (2022) Postharvest food loose pose a serious threat to food security. *Farmers Weekly* 6.

Dania, W.P.D, Xing, K., Amer, Y., *et al.* (2016) Collaboration and sustainable agri-food suply chain: a literature review. *MATEC Web of Conferences* 58, 1–10. DOI: 10.1051/matecconf/20165802004.

Dubey, R., Singh, T. and Gupta, O.K. (2015) Impact of agility, adaptability and alignment on humanitarian logistics performance: mediating effect of leadership. *Global Business Review* 16(5), 812–831. DOI: 10.1177/0972150915591463.

Dyer, J.H. and Singh, H. (1998) Relational view, cooperative strategy and sources of international competitive advantage. *Academy of Management* 23(4), 23–35. DOI: 10.2307/259056.

Eckstein, D., Goellner, M., Blome, C. and Henke, M. (2014) The performance impact of supply chain agility and supply chain adaptability. The moderating effect of product complexity. *The International Journal of Production Research* 53(10), 1–14. DOI: 10.1080/00207543.2014.970707.

Ehie, I. and Ferreira, L.M.D.F. (2019) Conceptual development of supply chain digitalization framework. *Science Direct* 52(13), 2339–2342. DOI: 10.1016/j.ifacol.2019.11.555.

Fiksel, J., Polyviou, M., Croxton, K.L. and Pettit, T.J. (2015) *From Risk To Resilience, Learning To Deal With Disruption*. Massachusetts Institute of Technology, Massachussets.

Forbes, S.L. and Wilson, M.M.J. (2018) Resilience and response of wine supply chains to disaster: the Christchurch earthquake sequence. *The International Review of Retail, Distribution and Consumer Research* 28(5), 1–18. DOI: 10.1080/09593969.2018.1500931.

Galal, N.M. and El-Kilany, K.S. (2016) Sustainable agri-food supply chain with uncertain demand and lead time. *International Journal of Simulation Modelling* 15(3), 485–496. DOI: 10.2507/IJSIMM15(3)8.350.

Geng, G., Xiao, R. and Xu, X. (2014) Research on MAS-based supply chain resilience and its self-organized criticality. *Discrete Dynamics in Nature and Society* 2014, 1–14. DOI: 10.1155/2014/621341.

Govindan, K., Fattahi, M. and Keyvanshokooh, E. (2017) Supply chain network design under uncertainty: a comprehensive review and future research directions. *European Journal of Operational Research* 263(1), 108–141. DOI: 10.1016/j.ejor.2017.04.009.

Gunasekaran, A., Subramanian, N. and Rahman, S. (2015) Supply chain resilience: role of complexities and strategies. *International Journal of Production Research* 53(22), 6809–6819. DOI: 10.1080/00207543.2015.1093667.

Hand, R. (2021) Supply chain velocity in today's world. Available at: https://www.shipbob.com/supply-chain-velocity/#:~:text=supply%20chain%20velocity%20refers%20to (accessed 4 September 2023).

Hobbs, J.E. (2021) Food supply chain resilience and the Covid-19 pandemic. What have we learned. *Canadian Journal of Agricultural Economics* 69(2), 189–196. DOI: 10.1111/cjag.12279.

Ivanov, D. (2018) Revealing interfaces of supply chain resilience and sustainability. A simulation study. *International Journal of Production* 56(10), 3507–3523. DOI: 10.1080/00207543.2017.1343507.

Ivanov, D. and Dias, A. (2020) Coronavirus (Covid-19) SARS-C-0V-02) and supply chain resilience: A research note. *International Journal of Integrated Supply Management* 13(1), 90–102. DOI: 10.1504/IJISM.2020.107780.

Jacyna-Golda, I. and Lewczuk, K. (2017) The method of estimating dependability of supply chain elements on the base of technical and organizational redundancy of process. *Eksploatacja i Niezawodnosc-Maintenance and Reliability* 19(3), 382–392. DOI: 10.17531/ein.2017.3.9.

Katsaliaki, K., Galetsi, P. and Kumar, S. (2021) Supply chain disruptions and resilience: a major review and future research agenda. *Annals of Operations Research* 319(1), 965–1002. DOI: 10.1007/s10479-020-03912-1.

Keating, A., Campbell, K., Szoenyi, M., McQuistan, C., Nash, D. *et al.* (2017) Development and testing of a community flood resilience measurement tool. *Natural Hazards and Earth System Sciences* 17(1), 77–101. DOI: 10.5194/nhess-17-77-2017.

Kilpatrick, J., Brown, J., Flynn, R., Addicoat, A. and Mitchell, P. (2021) *Deloitte Global 2021 Chief Procurement Officer Survey; Using Agility to adress Changing Procurement Officer Priorities*. Deloitte.

Kohli, S., Landry, N. and Noble, S. (2020) *How Retailers Can Build Resilience Ahead of a Recession: Recent History Shows How Retailer Scan Prepare For a Down Turn to Create Competitive Advantage.* McKinsey and Company.

Koncar, J., Grubor, A., Maric, R., Vucenovic, S. and Vukmirovic, G. (2020) Setbacks to IoT implementation in the function of FMCG supply chain sustainability during Covid-19 Pandemic. *Sustainability* 12(18), 1–21. DOI: 10.3390/su12187391.

KPMG (2016) *Demand-Driven Supply Chain 2.0 A Direct Link To Profitability.* KPMG, Amstelveen.

Kumiawan, R., Zailani, S.H., Iranmanesh, M. and Rajagopal, P. (2017) The effects of vulnerability mitigation strategies on supply chain effectiveness: risk culture as moderator. *Supply Chain Management; An International Journal* 22(1), 1–15. DOI: 10.1108/SCM-12-2015-0482.

Lakovou, E. and White, C.C. (2020) How to build more secure, resilient, next-gen US supply chains. Brookings, p. 05.

Lees, N. and Dixon, A. (2022) Legal mapping for resilience. *Supply Management*, 15 January, p. 20.

Leonczuk, D. (2021) Factors affecting the level of supply chain performance and its dimensions in the context of supply chain adaptability. *Scientific Journal of Logistics* 17(2), 253–269. DOI: 10.17270/J. LOG.2021.584.

Li, C., Wong, C.W.Y., Yang, C., Sang, K. and Lirn, T. (2019) Value of supply chain resilience: roles of culture, flexibility, and integration. *International Journal of Physical Distribution & Logistics* 50(1), 80–100. DOI: 10.1108/IJPDLM-02-2019-0041.

Mandal, S., Bhattacharya, S., Korasiga, V.R. and Sarathy, R. (2017) The dominant influence of logistics capabilities on integration: emprirical evidence from supply chain resilience. *International Journal of Disaster Resilience in the Built Environment* 8, 357–374.

Mandisvika, G., Chirisa, I. and Bandauka, E. (2015) Post-harvest issues: rethinking technology for value-addition in food security and food sovereignty in Zimbabwe. *Advances in Food Technology and Nutritional Sciences* Special Edition 1, 29–37. DOI: 10.17140/AFTNSOJ-SE-1-105.

Marche, B., Boly, V., Morel, L., Mayer, F. and Ortt, R. (2019) Agility and product supply chain design: the case of the SWATCH. *Journal of Innovation Economics & Management* 28(1), 79–109. DOI: 10.3917/jie.028.0079.

Martindale, N. (2022) Work in partnership with small companies. *Supply Management* 27.

Mathu, K. and Phetla, S. (2018) Integration enhance the response of fast-moving consumer goods manufacturers and retailers to customer's requirements. *South African Journal of Business Management* 49, 1–8. DOI: 10.4102/sajbm.v49i1.192.

Melnyk, S.A., Narasimhan, R. and Campos, H.A.D. (2014) Supply chain design: issues, challenges, frameworks and solutions. *International Journal of Production Research* 52(7), 1887–1896. DOI: 10.1080/00207543.2013.787175.

Mhlembe, K. and Mafini, C. (2019) Modelling the link between supply chain risk, flexibility and performance in the public sector. *South African Journal of Economic and Management Sciences* 22(1), 1–12. DOI: 10.4102/sajems.v22i1.2368.

Modgil, S., Gupta, S., Stekelorum, R. and Laquir, I. (2021) AI technologies and their impact on supply chain resilience during Covid-19. *International Journal of Physical Distribution & Logistics Management* (in press). DOI: 10.1108/IJPDLM-12-2020-0434.

Moyo-Nyede, S. and Ndoma, S. (2020) Limited internet access in Zimbabwe a major hurdle for remote learning during pandemic. *Afro Barometer* 1–11.

Mubarik, M.S., Bontis, N., Mubarik, M. and Mahmood, T. (2021) Intellectual capital and supply chain resilience. *Journal of Intellectual Capital* 23(3), 1469–1930. DOI: 10.1108/JIC-06-2020-0206.

Munyanyi, W. and Chimwai, L. (2019) Risk attitude, risk perception and risk management strategies adoption in Zimbabwean Small and Medium Enterprises. *Journal of Management and Economics Studies* 1(2), 53–68. DOI: 10.26677/TR1010.2019.73.

Naik, G. and Suresh, D.N. (2018) Challenges of creating sustainable agri-retail supply chains. *IIMB Management Review* 30(3), 272–282. DOI: 10.1016/j.iimb.2018.04.001.

Nair, R.D. (2017) The internationalisation of supermarkets and the nature of competitive rivalry in retailing in Southern Africa. *Development Southern Africa* 35(3), 315–333. DOI: 10.1080/0376835X.2017.1390440.

Nandonde, F.A. and Kuada, J. (2018) Perspectives of retailers and local food suppliers on the evolution of modern retail in Africa. *British Food Journal* 120(2), 340–354. DOI: 10.1108/BFJ-02-2017-0094.

Neboh, N.D. and Mbhele, T.P. (2020) Supply chain resilience and design in retail supermarkets. *Journal of Contemporary Management* 17(2), 51–73. DOI: 10.35683/jcm19086.64.

Negra, C., Remans, R., Attwood, S., Jones, S., Werneck, F. *et al*. (2020) Sustainable agri-food investments require multi-sector co-development of decision tools. *Ecological Indicators* 110, 105851. DOI: 10.1016/j.ecolind.2019.105851.

Nel, J., Goede, E.D. and Niemann, W. (2018) Supply chain disruptions: insights from South African third-party logistics service providers and clients. *Journal of Transport and Supply Chain Management* 12, 77–85. DOI: 10.4102/jtscm.v12i0.377.

Obayi, R., Koh, S.C., Oglethorne, D. and Ebrahimi, S.M. (2017) Improving retail supply flexibility using buyer-supplier relational capabilities. *International Journal of Operations & Production Management* 137(3), 343–362. DOI: 10.1108/IJOPM-12-2015-0775.

O'Marah, K. (2017) Supply chain risk 2020, new worries. *Forbes* 4.

Oxfam (2020) *From Risk To Resilience: A Good Practice Guide For Food Retailers Addressing Human Rights In Their Supply Chains*.

Patidar, R., Agrawal, S. and Pratap, S. (2018) Development of novel strategies for designing sustainable Indian agrifresh food supply chain. *Indian Society of Sciences* 43, 1–16. DOI: 10.1007/s12046-018-0927-6.

Perera, S., Bell, M.H. and Bliemer, M.C.J. (2017) Network science approach to modelling the topology and robustness of supply chain networks: a review and perspective. *Applied Network Science* 2(1), 33. DOI: 10.1007/s41109-017-0053-0.

Phillipart, M. (2016) The procurement dilema-short-term savings or long-term shareholders' value. *Journal of Business Strategy* 37(6), 10–17. DOI: 10.1108/JBS-11-2015-0114.

Rukasha, T., Nyagadza, B., Pashapa, R. and Muposhi, A. (2021) Covid-19 impact on Zimbabwean agricultural supply chain and markets: a sustainable livelihood perspective. *Cogent Social Sciences* 7, 1. DOI: 10.1080/23311886.2021.1928980.

Sa, D.M.M., Miguel, P.L.D.S., Brito, R.P.D. and Pereira, S.C.F. (2019) Supply chain resilience: the whole is not the sum of the parts. *International Journal of Operations & Production Management* 40, 92–115. DOI: 10.1108/IJOPM-09-2017-0510.

Sambowo, A.L. and Hidayatno, A. (2021) Resilience index development for manufacturing industry based on robustness, resourcefulness, redundancy and rapidity. *International Journal of Technology* 12, 1177–1186. DOI: 10.14716/ijtech.v12i6.5229.

Sharma, R., Shishodia, A., Kamble, S., Gunasekaran, A. and Belhsadi, A. (2020) Agriculture supply chain risks and Covid-19; mitigation strategies and implications for the practioners. *International Journal of Logistics Research and Applications* 1–27. DOI: 10.1080/13675567.2020.1830049.

Sunmola, F.T. and Apeji, U.D. (2020) *Blockchain Characteristics For Sustainable Supply Chain Visibility*. Industrial Engineering and Operations Management Society, Detroit.

Tan, W.J., Zhang, A.N. and Cai, W. (2019) A graph-based model to measure structural redundancy for supply chain reslience. *International Journal of Production Research* 57, 6385–6404. DOI: 10.1080/00207543.2019.1566666.

The World Bank (2015) *Disaster Risk Management in the Transport Sector: A Review of Concepts and International Case Studies*. The World Bank, New York.

Tierney, K. and Bruneau, M. (2007) Conceptualising and measuring resilience; A key loss restoration. *TR News* 14–17.

Tocci, N. (2020) Resilience and the role of European in the world. *Contemporary Security Policy* 41, 176–194. DOI: 10.1080/13523260.2019.1640342.

Trung, K.D., Kotivirts, J., Norell, V. and Gammelgaard, B. (2020) Sustainable business model innovation in the last mile logistics. Available at: https://www.i-smile.fi/2021/02/10/sustainable-business-model-innovation-in-last-mile-logistics/ (accessed 27 July 2022).

Tukamuhabwa, B.R., Stevenson, M., Busby, J. and Bell, M.Z. (2015) Supply chain definition, review and theoretical foundations for further study. *International Journal of Production Research* 53, 1–32. DOI: 10.1080/00207543.2015.1037934.

Wahi, R., Doctor, P., Chenoy, D. and Gopalan, M. (2021) *Resilience in the FMCG and Retail Sector*. Deloitte-FICCI, New Delhi.

Wakasala, B. (2020) *Supply Chain Resilience and Performance of Supermarket in Nairobi County, Kenya*. University of Nairobi Research Archive.

Wang, Q., Li, J.J., Ross, W.T. and Craighead, C.W. (2013) The interplay of drivers of different of opprtunism in buyer-supplier relationships. *Journal of Academy of Marketing Science* 41, 111–131. DOI: 10.1007/s11747-012-0310-9.

Webber, C.M., Chigumira, G. and Nyamadzawo, J. (2013) *Building Agricultural Competitiveness in Zimbabwe: Lessons from the International Perspective.* Zimbabwe Economic Policy Analysis and Research Unit, Harare.

Whitten, G.D., Green, K.W. and Zelbst, J. (2010) Triple-A supply chain performance. *International Journal of Operations and Production Management* 32, 28–48. DOI: 10.1108/01443571211195727.

Zhang, H., Jia, F. and You, T. (2021) Striking a balance between supply chain resilience and supply chain vulnerability in the cross-border e-commerce supply chain. *International Journal of Logistics Research* 26, 1–25. DOI: 10.1080/13675567.2021.1948978.

Zhao, K., Scheibe, K.P., Blackhurst, J. and Kumar, A. (2018) *Supply Chain Network Robustness Against Disruptions: Topological Analysis, Measurement and Optimisation.* Iowa State University Digital Respiratory. DOI: 10.1109/TEM.2018.2808331.

18 Sustainable Supply Chains in the Agricultural Sector

Ernest Mugoni[1]*, Iscacle Nyanhete[2], Charles Tsikada[1], Rumbidzai Pashapa[1] and Thelma Lilian Munodawafah[1]

[1]*Marondera University of Agricultural Sciences and Technology (MUAST), Zimbabwe;* [2]*Midlands State University (MSU), Zimbabwe*

Abstract

A sustainable agribusiness supply chain is strategic, transparent, and integrates and achieves the agribusiness' social, environmental and economic goals while systematically coordinating business processes to improve long-term goals. Agribusiness supply chains are increasingly under pressure to manage declining land sizes, growing consumer demands and the ever-increasing population. This chapter employed documentary evidence to appreciate the emerging issues in sustainable agricultural supply chains. The study used journals, organizational reports, websites, and reports from the print and electronic media for the purposes of gathering data. Use of content analysis helped in the development of themes under sustainable agricultural supply chain. The chapter unearthed emerging issues in sustainable agricultural supply chains. However, many agribusiness business models are primarily driven by profit and the world's economic system contributes significantly to social inequality and environmental damage. Sustainability is often confused with what is called 'corporate sustainability', which is a term used to refer to the long-term economic success of the business with little consideration of the environment and social factors. Sustainability supply chain challenges include huge challenges of scope 3 emissions reduction, transparency and traceability of the supply chain and new data in the supply chain. Furthermore, sustainable agricultural supply chains need to be reviewed timeously to make sure that they are responsive to the emerging risks and trends relevant to the prevailing circumstances. The transition to a sustainable agricultural supply chain demands all hands on deck as there is a growing need for regenerative agriculture as well as investments in resilient supply chains.

Keywords: Sustainability, Supply chain management, Sustainable supply chain management, Sustainable agricultural supply chain

18.1 Introduction and Background

A sustainable agribusiness supply chain is strategic, transparent, and integrates and achieves the agribusiness' social, environmental and economic goals while systematically coordinating business processes to improve long-term goals (Carter and Rogers, 2008). It is also about achieving a balance between economic growth, environmental issues and social development (Kamble *et al.*, 2020). Agribusiness supply chains are increasingly under pressure to manage declining land sizes, growing consumer demands and the ever-increasing population (Baghizadeh *et al.*, 2021). The ever-intensifying ecological and social stresses underscore the

*Corresponding author: ernmugoni@gmail.com

necessity for a balanced agribusiness approach based on protection of natural resources and improvement of livelihoods and social wellbeing of people while ensuring the profitability of the business (FAO, 2014). However, many agribusinesses are primarily driven by profit and the world's economic system contributes significantly to social inequality and environmental damage (Institute for Human Rights and Business, 2015).

Many authors concur that sustainability at a broader level has three components – the environment (planet), the society (people), and economic performance (profit) – and this perspective corresponds with the triple bottom line concept (Carter and Rogers, 2008). Triple bottom line draws attention to the boardroom, environmental and social consequences of the economic decisions of board members (Arora *et al.*, 2016). It highlights the impact of the agribusiness decisions and processes on the environment, people and financial capital (Wiedmann and Lenzen, 2006).

18.2   Triple Bottom Line

Economic systems exist to serve human wellbeing, and human and economic wellbeing are inextricably linked to the environment wellbeing as they both rely on the resources extracted from the environment (Hammer and Pivo, 2016). The best way to measure sustainability of agribusinesses is to measure their success along the three trajectories called the triple bottom line, and using it to measure success encourages agribusiness executives to adopt sustainable business practices (Arora *et al.*, 2016).

18.2.1   Planet

Planet relates to the organization's efforts to minimize environmental damage through the efficient use of energy and waste reduction (Correia, 2019). It is about assessing the environmental cost of the agribusiness processes from acquiring materials to the eventual disposal of agribusiness products by the end users (Onyali, 2014). Transport links the nodes in the agriculture supply chain and refrigeration

preserves the quality of the products. However, transport and refrigeration rely on fossil fuels to power them resulting in emission of various gases that are detrimental to the environment (Colley *et al.*, 2011). Holding low levels of inventory in order to reduce energy consumption can lead to economic losses emanating from stockouts and unsatisfied customer needs and holding high levels of inventory does not only increase holding costs, it increases the need for energy to meet controlled temperature requirements, which results in more greenhouse emissions from the cooling plant (Galal and El-Kilany, 2016).

18.2.2   People

Agribusinesses are active members of society. They create economic value, generate jobs and they pay taxes that contribute to societal values (Mark-Herbert *et al.*, 2018). Sustainable supply chains migrate from a narrow focus of merely ensuring that food supplies are available to ensuring food is accessible at household level. Food access often depends on individual income; therefore, agribusinesses can enhance food access through managing cost along the chain, thereby ensuring stable prices, improved farmer income and sustainable competitive advantage along the nodes (Aji *et al.*, 2020). Global food security, access to sufficient, safe and nutritious food, has steadily declined since 2015 and this has been exuberated by Covid-19; relying on market mechanisms to deliver the required quantities of food at affordable prices is a challenge (Teeuwen *et al.*, 2022). Even in bountiful supply of food many people may still go hungry because they do not have physical access to purchase their dietary needs that ensure a healthy life (Aji *et al.*, 2020; Leurs *et al.*, 2008; Teherani *et al.*, 2023). The social dimension of sustainable agribusiness covers the impact of business decisions on working conditions, community development, consumer health and safety and labour rights (Dania *et al.*, 2016; Leurs *et al.*, 2008). Using the available resources efficiently improves the wellbeing of customers and those that work along the supply chain nodes while maintaining high levels of profitability (Amer *et al.*, 2018; Leurs *et al.*, 2008; Teherani *et al.*,

2023). Agribusiness employees work under the most challenging conditions – they work under pressure from the business owners, they suffer adverse weather conditions, they are generally underpaid, they are generally illiterate and they lack information about work-related health and safety (Sinha, 2010). Improving working conditions of employees along the chain, and ensuring that business owners respect workers' rights and remunerate them accordingly contributes to sustainable agribusiness supply chains.

18.2.3   Profit

Agribusinesses are not adjunct to aid. Economic activities are not easily directed where the need is great and business will not always drive growth for the whole society (Institute for Human Rights and Business, 2015). While economic, environmental and social factors are equally weighted in terms of significance, in many organizations environmental and social factors are not typically considered on a par with economic factors (Arora *et al.*, 2016). A sustainable agricultural supply chain focuses on maintaining environmental, economic and social stability for long-term growth (Amer *et al.*, 2018). However, sustainability is often confused with what is called 'corporate sustainability', which is a term used to refer to the long-term economic success of the business with little consideration of the environment and social factors (Institute for Human Rights and Business, 2015). Reducing logistical costs is key to improving agribusiness sustainability, reducing farm waste improves food security, good storage facilities help preserve quality and ensures consumers enjoy fresh produce, which improves their wellbeing, and fast and energy-efficient transport reduces emissions (Webber *et al.*, 2013). In recent years the primary focus of supply chain management has been to increase efficiency and reliability through globalization, specialization and lean supply chain procedures (Perera *et al.*, 2017). Lean and global supply chains have improved supply chain efficiency, through low levels of inventory holding along the nodes and compressed lead times (Alicke *et al.*, 2020). However, some of these efficiencies do not protect natural

resources orimprove livelihoods and social wellbeing of people.

18.3   Methodology

This chapter employed documentary evidence to appreciate the emerging issues in sustainable agricultural supply chains. The study used journals, organizational reports, websites, and reports from the print and electronic media for the purposes of gathering data. It further made use of critical review of academic journals and other various articles with the analysis performed through content analysis. Use of content analysis helped in the development of themes under sustainable agricultural supply chains.

18.4   Literature Review and Theoretical Framework

18.4.1   Theoretical framework

This chapter is guided by the sustainability theory supported by the stakeholder theory and the natural-resource-based view theory.

18.4.2   Sustainability theory

The theory entails meeting current needs without compromising the needs of the future generations. The concept of sustainability is further developed and referred to as the triple bottom line (TBL). Slaper and Hall (2011, p.4) define TBL as '...an accounting framework that incorporates three dimensions of performance: social, environmental and financial'. TBL focuses on people, planet and profits, which is commonly known as the three Ps. The concept of TBL was pioneered by Elkington (2004) and has gained prominence around the globe in corporate reports as evidenced by how companies are taking cognisance of environmental, economic and social concerns (Elkington, 2004; Wu, 2013). TBL helps to harmonize the three Ps (ecological, economic and social) of sustainability in an endeavour to attain the basics of all three forms of sustainability especially in discussions as well as advocates that have something to do with sustainability (Elkington, 2004; McKenzie, 2004).

18.4.3 Stakeholder theory

This theory is premised on the notion that a decision must take into consideration any group or individuals who can affect or are affected by the organization objectives.

18.4.4 Natural-resource-based view

This theory supports the notion that sustainability improves the firm's performance. The theory holds that organizational supply chain performance is determined by the manner in which companies deploy, manage and position their internal resources and capabilities.

18.4.5 Sustainable agricultural supply chains

The United Nations set the Sustainable Development Goals (SDGs) to attain sustainability in various areas such as consumption, production, nutrition and agriculture, environmental protections as well as working conditions. These SDGs call for the promotion of an integrated partnership-driven approach with the involvement of multiple actors, which include corporations (Rudloff and Wieck, 2020). This sparked the debate on the involvement of sustainable supply chain approaches in agriculture. The ultimate consumers are seen to be responsible for ensuring and upholding both human rights and sustainability standards throughout the supply chain (Rudloff and Wieck, 2020). Such supply chains involve multiple actors that are located in different countries as well as the different stages such as production, processing, transport and distribution. Multiple actors within the supply chain must be focused predominantly on human rights as well as sustainability through due diligence. The influence on agricultural supply chains is made through overcoming specific problems on human rights

and sustainability, which calls for all approaches to be synchronized and combined (Rudloff and Wieck, 2020). Agricultural supply chains include small numbers of actors that need to be tracked, disciplined and monitored. Agricultural supply chains include storage, pre- and post-distribution, transportation, procurement, processing, production, farming among others. In the case of short supply chains, the ultimate consumers need to enforce supply chain obligations as actors are concentrated in few countries (Rudloff and Wieck, 2020). Enforcing and implementing supply chain rules in developing economies is difficult because it involves dealing with a large number of small producers that are geographically dispersed as the company at the end of the chain has to incur or shoulder the cost (Rudloff and Wieck, 2020).

Drawing some insights from Seuring and Müller (2008), supply chains of agricultural products have to be considered from the initial stage of the supply chain. Small-scale farmers and consumer associations play a critical role for ensuring fair play in trade in the enhancement of sustainable supply chain management in international trade (global market) (Auroi, 2003; OECD–FAO, 2019; OECD–FAO, 2018; UNCTAD, 2020). Attainment of sustainability in the context of agriculture includes meeting challenges such as environmental (the challenge that is ecological in an endeavour to promote good environmental practices), social (this refers to a challenge of improving the conditions of living as well as the economic opportunities for people in the rural areas) and economic (profit oriented challenge of strengthening the viability and the competitiveness within the agricultural sector). To address the three challenges there is a need for deliberate policies and regulations like the Common Agricultural Policy (CAP) (Mayer, 2022). Such programmes seek to enhance the sustainability of the agricultural sector in order to meet the interests of both the governments and the markets, maximizing the supply chain potential to contribute to sustainable agriculture while ensuring the creation of fairer costs of distribution and benefits (Veeman, 2004; Brigstoke, 2004; Peeters, 2012; OECD–FAO, 2019; OECD–FAO, 2018; UNCTAD, 2020). There is a positive impact for the Sustainable Agriculture Initiative (SAI), which prompted a number of companies such as Nestle, Danone Group as well as Unilever to adopt the sustainable agricultural supply chain concept in their business. This initiative led to increased knowledge sharing, environmental awareness, tracing as well as monitoring of practices through the whole supply chain and promoting sustainable agriculture (Jöhr, 2004). To reach sustainability there is need to have mutual awareness (Grimsdell, 1996).

Sustainable agricultural practices help to minimize the environmental impact as well as provide reassurance to the public via countermeasures both in-site and off-site in the agricultural supply chain (Syahruddin and Kalchschmidt, 2011; OECD–FAO, 2019; UNCTAD, 2020). Good agriculture practices as well as optimization helps in the attainment of sustainability in the agricultural supply chain (Pancino *et al.*, 2019). Sustainable agricultural supply chains should also include the use of renewable energy to industries, which helps reduce the industry impact on the ozone depletion. Sustainable agricultural supply chains have a greater economic impact through the provision of rural employment opportunities as compared to the conventional agricultural supply chain (Syahruddin and Kalchschmidt, 2012; OECD–FAO, 2018; UNCTAD, 2020). Technological application has to play a central role in the development of sustainable agricultural supply chains. Absence or lack of technology implementation in the agricultural sector hinders the sustainability of agricultural practices (Cleaver and Schreiber, 1994). Agricultural production and its supply chain operations calls for the advancement of information, sensing, automation as well as control technologies as argued by Sigrimis *et al.* (2001). Such implementation of information and communication technologies plays a critical role in the attainment of a sustainable agricultural supply chain as it presents opportunities for institutionalizing knowledge management in agricultural development (Rao, 2007).

18.5   Sustainability Supply Chain Challenges

There are a number of challenges faced by supply chain actors in their endeavour to make their supply chains sustainable. Such challenges include the following:

18.5.1   Huge challenge of scope 3 emissions reduction

In the past the focus was on reducing greenhouse gas (GHG) emissions that are directly linked to companies' assets and factories. In today's world there is a shift in effort and commitment towards the scope 3 emissions that are generated within the supply chain. A major challenge is on management of data and the scale necessary for the development of a greenhouse gas baseline as well as showing meaningful reduction through the use of regenerative agriculture coupled with on-farm improvements (FAO, 2017). To understand which practices can drive sustainable impact, industry alignment and collaboration are critical ingredients. This is faced with the challenge on ways of financing, measuring, certification and reporting of these (FAO, 2017).

18.5.2   Transparency and traceability of the supply chain

The consumers of today are becoming highly enlightened in terms of understanding sustainability, which goes beyond the popular topics such as human rights and deforestation to the inclusion of complexities of migrant labour, water use, biodiversity and greenhouse gas emissions. This calls for supply chain actors to invest across their supply chains, which in the longterm allows supply chain traceability and transparency giving leverage to both consumers and companies to learn much about the risks and rewards of their investments (FAO, 2017). The challenge will be on balancing the business case for transparency and traceability in the shortterm against the additional costs, data protection-related issues and the possibility of reputational downsides.

18.5.3   New data in the supply chain

The supply chain actors are now gaining access to a whole host of new data sets. Availability of such data brings in topical debate on protection of data as well as pre-competitive collaboration (FAO, 2017). It also brings the issue of how and which systems can be used to perform data mining at farm level.

18.6 Shifting to Sustainable Agricultural Supply Chains

Sustainable agriculture is a productive, competitive and efficient way to produce safe agricultural products, while at the same time protecting and improving the natural environment and social/economic conditions of local communities. Sustainable agricultural supply chains should take the following issues into consideration: conditions, risks and challenges, natural systems as well as the regulations across the supply chain. The three pillars of sustainability, which include environment, social and economic need to be considered as a complete package by members of the supply chain. Sustainable agricultural supply chains need to be reviewed timeously to make sure that they are responsive to the emerging risks and trends relevant to the prevailing circumstances (UNCTAD, 2020).

To attain sustainable agricultural supply chains, the supply chain partners can follow the OECD-FAO guidance and recommendations. This emphasises the need for companies to strengthen their approach to due diligence as translating their policy commitments into practical actions. The OECD-FAO guidance calls for closer collaboration, effective risk management, scheme certification and alignment with key stakeholders in the agricultural supply chain. Members of the supply chain need to collect, disseminate and report quality comparable data across the agricultural supply chain (OECD–FAO, 2019).

Maximizing sustainability in developing countries presents an opportunity for the enhancement of trade and development approaches so as to be equitably integrated into the global economy. International trade allows bilateral and multilateral coordination to take place signalling meaningful sustainability concepts as a global effort (UNCTAD, 2020). The transition to a sustainable agricultural supply chain demands all hands on deck as there is a growing need for regenerative agriculture as well as investments in resilient supply chains (FAO, 2017). Noticeable strides towards advancing more sustainable and ethical global agricultural supply chains have been made by multinational agribusinesses. This is through the prioritization of strategic collaboration of supply chain players across industries, including cross-sector players as well as local partners, in an endeavour to overcome obstacles to sustainable impact in achieving lasting meaningful results (FAO, 2017). In this case there can be convening of stakeholders and building of partnerships for diverse and critical supply chain players for action (Orr and Jadhav, 2018). Each partnership should have mechanisms that can be deployed to attain sustainable impact including putting in place incentive programmes to enhance and facilitate the much-needed change in behaviour. The members need to be represented with each having a clearly defined role (Orr and Jadhav, 2018). This helps leading companies to partner within their industry as well as helping farmers to build stronger sustainable agricultural supply chains.

The use of technology in agricultural supply chains can help attain sustainable agricultural supply chains. Technologies that can be employed include blockchain and artificial intelligence, which enables supply chain actors to enjoy transparency and traceability, optimizing agricultural supply chains (Gupta, 2022). Modern technologies play a critical role in maximizing the complexity of agricultural supply chains. Such information-driven and integrated supply chains enable agricultural actors to reduce inventory costs, value addition on products, resources extension and acceleration of time to market while ensuring customer retention (Gupta, 2022). The use of blockchain can enable farmers to locate alternative markets that can offer better deals, and establish credibility with other institutions like financial ones allowing access to carbon credits coupled with better contractual terms (FAO, 2017).

Sustainable agricultural supply chains can be achieved through advocacy and innovation from the cross-sector supply chain players. These can be innovations that drive action while defining new standards that suit the entire industry. These can be innovative farming practices across the industry with the involvement of other actors such as governments, funders and researchers (FAO, 2017).

Such practices have to cut across the supply chain from both developing and developed economies. Sustainable agricultural supply chains can also be attained through increasing alignment of climate change adaptation and mitigation. Supply chain actors have to build business cases that favour sustainability investments considering a better understanding of the cost of climate risk (Soltanian *et al.*, 2022). This should encourage a circular economy through reduction of food waste, prevention of pollution and control inclusive of protection and restoration of ecosystems and biodiversity (Orr and Jadhav, 2018; Soltanian *et al.*, 2022). Agricultural production should include ways that lead to reduced usage of pesticides and management of risks emanating from the use of such (Mayer, 2022; Soltanian *et al.*, 2022).

18.7 Conclusions and Recommendations

The chapter examined the emerging issues in sustainable agricultural supply chains. Emerging issues in sustainable agricultural supply chains include involvement of multiple actors, which need to be tracked, disciplined and monitored. Enforcing and implementing supply chain rules in developing economies is difficult because it involves dealing with a large number of small producers that are geographically dispersed as the company at the end of the chain has to incur or shoulder the cost. Sustainable agricultural practices help to minimize the environmental impact as well as provide reassurance to the public via countermeasures both in-site and off-site in the agricultural supply chain. Sustainability supply chain challenges include the huge challenges of scope three emissions reduction, transparency and traceability of the supply chain and new data in the supply chain. Furthermore, sustainable agricultural supply chains need to be reviewed timeously to make sure that they are responsive to the emerging risks and trends relevant to the prevailing circumstances. The transition to a sustainable agricultural supply chain demands all hands on deck as there is a growing need for regenerative agriculture as well as investments in resilient supply chains. Agribusinesses provide employment and food security, and contribute to economic growth, and they also provide raw materials to other sectors. The role of agribusinesses is much more than creating jobs, paying taxes and developing technology, they have to determine the nature and purpose of business in a world that has delivered wealth alongside inequality and prosperity alongside environmental damage. The perishable nature of agricultural products complicates the management of agricultural supply chains – demand, supply, quality and quantity are volatile and inventory cannot be used as buffer. Despite receiving a lot of attention, postharvest losses continue unabated in developing countries due to poor coordination, lack of follow-ups and failure to invest in training, which stalls progress.

References

Aji, J.M.M., Rondhi, M. and Addy, H.S. (2020) Linking supply chain management and food security: a concept of building sustainable competitive advantage of agribusiness in developing economies. *E3S Web of Conferences* 142, 1–11. DOI: 10.1051/e3sconf/202014206005.

Alicke, K., Barriball, E., Lund, S. and Swan, D. (2020) *Is Your Supply Chain Risk Blind-Or Risk Resilient*. McKinsey and Company, Washington, DC.

Amer, H.M., Galal, N.M. and El-Kilany, K.S. (2018) *A Simulation Study of Sustainable Agri-Foods Supply Chain*. Industrial Engineering and Operations Management International Society, Paris.

Arora, P., Peterson, N.D., Bert, F. and Podesta, G.P. (2016) Managing the triple bottom line for sustainability: a case study of Argentine agribusinesses. *Sustainability, Science, Practice and Policy* 12(1), 1–16. DOI: 10.1080/15487733.2016.11908154.

Auroi, C. (2003) Improving sustainable chain management through fair trade. *Greener Management International* 43, 25–35. DOI: 10.9774/GLEAF.3062.2003.au.00005.

Baghizadeh, K., Cheikhrouhou, N., Govindan, K. and Ziyarati, M. (2021) Sustainable agriculture supply chain network design considering water-energy-food nexus using queuing system: a hybrid robust possibilistic programming. *Natural Resource Modeling* 35(1), 1–39. DOI: 10.1111/nrm.12337.

Brigstoke, T. (2004) The future strategy for dairy farming in UK. *Journal of the Royal Agricultural Society of England* 165, 1–12.

Carter, C.R. and Rogers, D.S. (2008) A framework of sustainable supply chain management: moving toward new theory. *International Journal of Physical Distribution & Logistics Management* 38(5), 360–387. DOI: 10.1108/09600030810882816.

Cleaver, K.M. and Schreiber, G.A. (1994) *Reversing the Spiral: The Population, Agriculture, and Environment Nexus in Sub-Saharan Africa*. World Bank.

Colley, D., Howard, M. and Winter, M. (2011) Food miles: time for a re-think? *British Food Journal* 113(7), 919–934. DOI: 10.1108/00070701111148432.

Correia, M.S. (2019) Sustainability: an overview of the triple bottom line and Sustainability implementation. *International Journal of Strategic Engineering* 2, 29–38. DOI: 10.4018/IJoSE.2019010103.

Dania, W.A.P., Xing, K., Amer, Y., Jamari, J., Handogo, R. *et al.* (2016) Collaboration and sustainable agri-food suply chain: a literature review. *MATEC Web of Conferences* 58, 02004. DOI: 10.1051/matecconf/20165802004.

Elkington, J. (2004) Enter the triple bottom line. *The Triple Bottom Line: Does It All Add Up* 11(12), 1–16.

FAO (2014) The State of Food and Agriculture: Climate Change, Agriculture and Food Security. FAO, Rome.

FAO (2017) Several Issues. Food and Agriculture Organization of the United Nations, Rome.

Galal, N.M. and El-Kilany, K.S. (2016) Sustainable agri-food supply chain with uncertain demand and lead time. *International Journal of Simulation Modelling* 15(3), 485–496. DOI: 10.2507/IJSIMM15(3)8.350.

Grimsdell, K. (1996) The supply chain for fresh vegetables: what it takes to make it work. *Supply Chain Management: An International Journal* 1(1), 11–14. DOI: 10.1108/13598549610799031.

Hammer, J. and Pivo, G. (2016) The triple bottom line and sustainable economic development theory and practice. *Economic Development Quarterly* 31(1), 25–36. DOI: 10.1177/0891242416674808.

Institute for Human Rights and Business (2015) *State of Play, Business and Sustainable Development Goals; Mind Gap-Challenges for Implementation*. Institute for Human Rights and Business, London.

Jöhr, H. (2004) SY 5-3: Nestlé's efforts towards sustainable Agriculture and SAI, the sustainable agriculture initiative of the International Food Industry. *Journal of Food Science* 69, CRH133–CRH135.

Kamble, S.S., Gunasekaran, A. and Gawankar, S.A. (2020) Achieving sustainable performance in A data-driven agriculture supply chain: a review for research and applications. *International Journal of Production Economics* 2019, 179–194. DOI: 10.1016/j.ijpe.2019.05.022.

Leurs, M.T.W., Mur-Veeman, I.M., van der Sar, R., Schaalma, H.P. and de Vries, N.K. (2008) Diagnosis of sustainable collaboration in health promotion a case study. *BMC Public Health* 8, 1–15. DOI: 10.1186/1471-2458-8-382.

Mark-Herbert, C., Rotter, J. and Pakseresht, A. (2018) *A Triple Bottom Line To Ensure Coprporate Responsibility*. Swedish University of Agricultural Sciences, Uppsala.

Mayer, B. (2022) Building the supply chain of tomorrow. Available at: https://www.foodlogistics.com/warehousing/automation/article/22081135/building-the-supply-chain-of-tomorrow (accessed 10 July 2023).

McKenzie, S. (2004) Social sustainability: towards some definitions. Hawke Research Institute Working Paper Series No 27, University of South Australia, Magill, Australia.

OECD–FAO (2018) The Middle East and North Africa: prospects and challenges. *OECD-FAO Agricultural Outlook* 2017, 67–108.

OECD–FAO (2019) *OECD-FAO Agricultural Outlook 2019–2928: Special Focus: Latin America*. OECD Publishing. DOI: 10.1787/agr_outlook-2018-en.

Onyali, C.I. (2014) Triple bottom line accounting and sustaible corporate performance. *Research Journal of Finance and Accounting* 5, 195–209.

Orr, S. and Jadhav, A. (2018) Creating a sustainable supply chain: the strategic foundation. *Journal of Business Strategy* 39(6), 29–35. DOI: 10.1108/JBS-11-2017-0157.

Pancino, B., Blasi, E., Rappoldt, A., Pascucci, S., Ruini, L. *et al.* (2019) Partnering for sustainability in agri-food supply chains: the case of Barilla sustainable farming in the Po Valley. *Agricultural and Food Economics* 7(1), 1–10. DOI: 10.1186/s40100-019-0133-9.

Peeters, J. (2012) The place of social work in sustainable development: towards ecosocial practice. *International Journal of Social Welfare* 21(3), 287–298. DOI: 10.1111/j.1468-2397.2011.00856.x.

Perera, S., Bell, M.G.H. and Bliemer, M.C.J. (2017) Network science approach to modelling the topology and robustness of supply chain networks: a review and perspective. *Applied Network Science* 2(1), 33. DOI: 10.1007/s41109-017-0053-0.

Rao, R.V. (2007) *Decision Making in the Manufacturing Environment: Using Graph Theory and Fuzzy Multiple Attribute Decision Making Methods*, Vol. 2. Springer, London, p. 294.

Rudloff, B. and Wieck, C. (2020) *Sustainable Supply Chains in the Agricultural Sector: Adding Value Instead of Just Exporting Raw Materials. Corporate Due Diligence Within a Coherent, Overarching and Partnership-Based EU Strategy.* SWP Comment No. 43/2020. German Institute for International and Security Affairs (SWP).

Seuring, S. and Müller, M. (2008) From a literature review to a conceptual framework for sustainable supply chain management. *Journal of Cleaner Production* 16(15), 1699–1710. DOI: 10.1016/j. jclepro.2008.04.020.

Sigrimis, N., Antsaklis, P. and Groumpos, P.P. (2001) Advances in control of agriculture and the environment. *IEEE Control Systems Magazine* 21(5), 8–12. DOI: 10.1109/37.954516.

Sinha, B.R.K. (2010) Working conditions of agricultural workers; A reflection of socio-economic status. *Human Geographies* 4, 35–45.

Slaper, T.F. and Hall, T.J. (2011) The triple bottom line: what is it and how does it work. *Indiana Business Review* 86, 4–8.

Soltanian, S., Kalogirou, S.A., Ranjbari, M., Amiri, H., Mahian, O. *et al.* (2022) Exergetic sustainability analysis of municipal solid waste treatment systems: a systematic critical review. *Renewable and Sustainable Energy Reviews* 156, 111975. DOI: 10.1016/j.rser.2021.111975.

Syahruddin, N. and Kalchschmidt, M. (2011) Towards sustainable supply chain management in agricultural sector. *International Journal of Engineering Management and Economics* 3(3). DOI: 10.1504/IJEME.2012.049894.

Syahruddin, N. and Kalchschmidt, M. (2012) Sustainable supply chain management in the agricultural sector: a literature review. *International Journal of Engineering Management and Economics* 3(3), 237. DOI: 10.1504/IJEME.2012.049894.

Teeuwen, A.S., Meyer, M.A., Dou, Y. and Nelson, A. (2022) A systematic review of the impact of food security governance measures as simulated in modelling studies. *Nature Food* 3(8), 619–630. DOI: 10.1038/s43016-022-00571-2.

Teherani, A., Nicastro, T., Clair, M.S., Nordby, J.C., Nikjoo, A, *et al.* (2023) Faculty development for education for sustainable health care: a university system-wide initiative to transform health professional education. *Academic Medicine* 98(6), 680–687. DOI: 10.1097/ACM.0000000000005137.

UNCTAD (2020) International SDGs investment flows to developing economies down by one third due to COVID-19. Available at: https://unctad.org/system/files/official-document/diaemisc2020d3_en.pdf

Webber, C.M., Chigumira, G. and Nyamadzawo, J. (2013) *Building Agricultural Competitiveness in Zimbabwe: Lessons from the International Perspective.* Zimbabwe Economic Policy Analysis and Research Unit, Harare.

Wiedmann, T. and Lenzen, T. (2006) *Sharing Responsibility a New Life-Cycle Approach and Software Tool for Triple Bottom Line Accounting.* Trinity College Dublin, Dublin, Ireland.

Wu, J. (2013) Landscape sustainability science: ecosystem services and human well-being in changing landscapes. *Landscape Ecology* 28(6), 999–1023. DOI: 10.1007/s10980-013-9894-9.

19 Assessing the Influence of Village Savings and Loan Associations on Climate Resilience and Food Security: A Case of Domboshava, Zimbabwe

Brenda Guruwo[1], Brighton Nyagadza[2]* and Gift Manhimanzi[3]
[1]Department of Agricultural Extension and Technology, Marondera University of Agricultural Sciences and Technology (MUAST); [2]Department of Marketing, Marondera University of Agricultural Sciences and Technology, and Institute for the Future of Knowledge (IFK), University of Johannesburg, South Africa (MUAST); [3]Department of Accounting Zimbabwe Ezekiel Guti University (ZEGU)

Abstract

The research sought to evaluate the impact of village savings and loans associations (VSLAs) on household food security and climate resilience. Underpinned by two theoretical frameworks, namely the VSLA model and the food security approach, the study adopted a positivism research paradigm and quantitative research strategy. The study drew its respondents from smallholder farmers within the Domboshava communal area and distributed a self-administered structured questionnaire to 74 respondents who had been selected using a simple random sampling technique. The said instrument was highly reliable as indicated by Cronbach's Alpha values, which were greater than 0.70 in respect of all variables. Using SPSS version 28, the research analysed data through descriptive and inferential statistics. The key findings of the study were that the VSLAs are significant predictors of climate resilience and household food security. In actual fact, the findings revealed that VSLAs were accounting for 19.3% of the variations in climate resilience. The impact rose to 35.1% when it came to household food security. The debate about village savings and their impact on food security has been complemented by several researches. Adopting a gap-spotting stance, the study measured the impact of VSLAs on food security. It also assessed the impact on climate resilience, which escaped the attention of many authors. The study recommends that the Government of Zimbabwe should create more favourable ground to attract non-governmental organizations that promote VSLAs for enhancing food security.

Keywords: Climate resilience, Food security, Village savings and loan associations (VSLAs)

*Corresponding author: bnyagadza@gmail.com

© CAB International 2023. *Sustainable Agricultural Marketing and Agribusiness Development: An African Perspective* (eds B. Nyagadza and T. Rukasha)
DOI: 10.1079/9781800622548.0019

19.1   Background

19.1.1   Introduction

The threat of climate change remains critical around the world and the impacts are becoming prominent over the last decades. The changing climate around the globe coupled with the economic impacts of the Covid-19 pandemic has forced most people to grapple with food insecurity due to frequency of extreme weather and changes in agricultural productivity (Wright, 2017). Unchecked, global climate change will worsen, pushing 132 billion people into poverty over the next ten years (Adedeji et al., 2014). The Intergovernmental Panel on Climate Change (IPCC) has also identified Africa as one of the most vulnerable continents to climate change. Eriksen (2015) projected that by the year 2030, 75 million people would have been exposed to effects of climate change. Studies have shown that climate change will spark higher agricultural prices and could threaten food security (Nicholas and Wynes, 2016). Hendriks (2011) notes that in sub-Saharan Africa, food insecurity has mainly been sustained by the inability of households and governments to respond to climate change-induced shocks across production. Responding to climate change in Africa has been a major challenge due to low levels of access to technology and reliance on rain-fed agriculture. Evan (2012) notes that between now and 2030 the African continent needs to work on climate policies that enhance adaptive capacity and make it less vulnerable.

Records demonstrate that Zimbabwe is also experiencing the effects of climate change. Increase in frequency of mid-season dry spells has resulted in crop failure. This climate change has seen a decrease in the country's real gross domestic product (GDP) by 36% due to decline in agricultural production as noted by Alcaraz and Zeller (2007). Currently 2.61 million Zimbabweans are food insecure due to climate change variables and the economic impacts of Covid-19. In 2020, the Government of Zimbabwe, through the Ministry of Agriculture, launched a climate resilient program, the Green Climate Fund. This innovative program aimed to raise a total of US$26 million, which aimed to reach 2.3 million farmers to improve food security and strengthen climate resilience. This strengthened the capacities of vulnerable smallholder farmers through farmer field schools and peer-to-peer support to scale up climate resilient agriculture (Zim Vac, 2020). The project however did not reach all the provinces in Zimbabwe including Mashonaland (Barnes et al., 2004).

Following the lack of sustainability in government's effort of funding farmers for climate resilient, local non-governmental organizations (NGOs) such as Care International have engaged an adaptation method known as microfinancing to the rural poor as a way of reducing incidences of food insecurity and strengthen climate change resilience. This microfinance is through use of village savings among the farmers themselves. Yusuf et al. (2016) highlight that this empowers farmers to find their own way out of the poverty trap while avoiding dependency and the hand out shame of conditional aid (Masiyandima et al., 2012). Village savings and loans associations (VSLAs) have played a successful role in women empowerment and household food availability. However, studies still need to make an exploratory approach on their contribution to climate resilience.

19.1.2   Problem statement

Climate change is a key driver of food insecurity in sub-Saharan Africa due to increased vulnerability to droughts (Gregory, 2015). Zimbabwe's smallholder agriculture may be considered the most vulnerable sector to climate change and the consequent droughts due to overreliance on rainfed farming (Mubaya et al., 2012). Most farmers in rural areas over-rely on farming activities or donor handouts for their livelihoods making them more vulnerable to climate-induced shocks. Microfinance has been found out to be the best possible option which will benefit farmers in fostering climate resilience and food security. However, the Government of Zimbabwe has done little to help remote farmers to access finance considering they lack collateral security to access formal banking services. VSLAs, a microfinance type, have therefore been put forward as one of the strategies that these farmers can employ to increase climatic resilience and ensure a food-secure nation. Their effectiveness however still remains contested with conflicting findings in various contexts. This study therefore seeks to make a context analysis of VSLAs and their impact on climate resilience and food security.

19.1.3 Research objectives

The main objective of this study is to explore the impact of VSLA schemes on climate resilience and food security in Domboshava district of Zimbabwe.

The specific objectives for the study are:

- to examine the influence of VSLAs on climate resilience of Domboshava small-holder farmers in Zimbabwe and
- to establish the effect of VSLA on household food security among Domboshava small-holder farmers in Zimbabwe.

19.1.4 Research questions

- How do VSLAs affect the climate resilience of Domboshava smallholder farmers in Zimbabwe?
- What is the effect of VSLAs on household food security among Domboshava small-holder farmers in Zimbabwe?

19.1.5 Hypothesis

In view of the above research objectives, the study hypothesized as follows:

H_1: There is a positive relationship between VSLAs and food security.

H_2: There is a positive relationship between VSLAs and climate resilience.

19.1.6 Theoretical framework

This study was guided by two models which are inextricably interwoven. The first one is called the Village Savings and Loans Association model, which was propounded by Daley-Harris (2009). The model is based on the hallmarks of

microfinance, which involve the provision of a wide range of financial services such as savings facilities, loans and insurance to low-income customers who do not ordinarily have access to formalized banking services. According to Hongo (2013), the beauty of microfinance is that it assists the marginalized groups of society to avoid the dependency syndrome. This means that microfinance has the potential to empower the marginalized groups of society so that they can find their way out of the poverty trap as well as food insecurity.

The second model that is central to this study is the food security approach, the original version of which was provided by FAO (1996) and later modified by Mercy Corps (2009). The revised version places emphasis on the four pillars of food security, namely food availability, access, utilization and food stability.

19.1.7 Village savings and loans associations

The VSLAs model was first conceived by Care International in Niger in 1991. The model has since extended to several parts of the world where it is adapted to suit respective local con-texts (CARE International, 2017). Therefore, it is apparent that VSLA programming is not only a tried but tested model which offers savings-led financial solutions at community level. According to CARE International (2019), the VSLA model is growing from strength to strength and this assertion is supported by its ever-increasing number of subscribers. In terms of membership, the VSLA subscrip-tion base rose exponentially from one million subscribers in 2008 to about 7.6 million scat-tered in 51 countries across the world (CARE International, 2019). The said members represent 357,000 groups composed mainly of rural and poor women cumulatively saving and investing money well above US$500 million per year (CARE International, 2019). It is worth noting that Care International is con-tinually increasing its investment towards the VSLAs, and in doing so it has partnered with several industry players with a view of pulling resources together.

19.1.8   Origin of VSLAs in Zimbabwe

Extant literature is unanimous on the significance of microfinance as a poverty reduction strategy in the development discourse (Rahman *et al.*, 2009; Qureshi *et al.*, 2012). In Zimbabwe, microfinance has similarly been recognized as a viable financing option for both the poor and those financially excluded (Mago, 2013). According to Mago (2013), the microfinance sector in Zimbabwe started in the 1960s when people, mainly the poor, were organized into relatively small groups to form savings clubs. In agreement, Raftopoulos and Lacoste (2001) assert that the origins of the microfinance sector in the country date back to 1963 following the initiative by the Catholic Missionary to form the Savings Development Movement. Largely, the movement drew its membership from rural women. After independence, an upward trajectory was observed in the number of savings clubs in Zimbabwe. In particular, the number of such clubs rose from 5000 in 1983 to 7000 in 1998. This notable increase in the number of microfinance clubs was a clear indication of a rising demand for microfinance services in Zimbabwe (Alise and Teddlie, 2010). The Government of Zimbabwe established, under the then Ministry of Community Development and Women's Affairs, the National Association of Cooperatives Savings and Credit Unions of Zimbabwe (NACSCUZ) in an endeavour to ensure that support in the mobilization of funds by the poor was guaranteed. NACSCUZ was mainly established for purposes of offering technical support services to all registered savings and credit cooperatives.

Most small-scale farmers were excluded from accessing microfinance and this led to the rise of village savings (Yvonne Feilzer, 2010). The savings and credit cooperatives continued to rise and in 2000, the concept of VSLAs was introduced in Zimbabwe courtesy of Care International. The loans scheme was launched after the organization realized that the majority of the rural population lacked access to finance (Zheke, 2017). Locally known as mkando, VSLAs are one of the main activities carried out primarily to address community challenges through the pooling of resources to achieve microfinancing (CARE International, 2011, 2019). Village savings have to date been adopted by all provinces to back up their agricultural production and to have an alternative source of income. Chuma *et al.* (2013) asserts that VSLAs have been successful as a way of cushioning because they can save and borrow money later to cope with unexpected emergencies.

19.1.9   The concept of VSLAs for microfinance and its developmental trajectory

Microfinance has a long history with scholars and policy makers alike. Several initial proponents argued that lack of access to formal microfinance services was the main reason why people remained poor in developing countries (Dahal and Fiala, 2020). Recent developments with regards to poverty alleviation have culminated in the proliferation of studies examining the effectiveness of VSLAs for microfinance on the reduction or elimination of poverty. Microfinance has existed in different forms like VSLA schemes for several decades (Brau and Woller, 2004; Dzisi and Obeng, 2013) and as such, it has steadily gained impetus and global recognition. VSLAs are regarded as a development-oriented tool with a capability of reducing the prevalence of poverty and the magnitude of deprivation among the marginalized and poor people who are usually discriminated against on account of their profiles (Armendariz and Morduch, 2000; Asamoah and Amoah, 2015). Cognizant of the developments taking place in the microfinance sector, the World Bank is of the view that effective and efficient markets are fundamental in creating more sustainable development and, therefore, there is need for creating aswell as increasing opportunities for the less privileged (Sujatha and Malyadri, 2015).

19.1.10   Climate change and food security

There are three areas of global concern that impact food security, namely overpopulation, climate change and urbanization (Havas and Salman, 2011). In light of the foregoing, it is evident that climate change is one of the serious threats to food security. According to Joshi *et al.* (2011), climate change has the potential to affect all the four elements of food security availability, access, utilization and stability. On availability, climate change significantly affects the crucial elements of food production, which include but are not limited to soil and water.

The most affected are the rural communities, who are highly likely to experience rapid and recurrent crop failure and also loss of their livestock, which represents their source of wealth. In addition, rapidly changing temperatures and weather patterns bring with them new variants in terms of diseases and pests, which will in turn affect crops and livestock. As a result, Úbeda *et al.* (2013) posit that this situation creates a significant decline with regards to both the quantity and quality of yields. When that happens, downstream problems such as scarcity and relatively high costs of food will follow. This phenomenon affects both the rural and urban communities as it severely threatens food stability. Numerous countries are already grappling with climate change-induced effects emanating from irregular and changing rainfall patterns (FAO, 2011). In view of the above, it is apparent that financial support is an important component as far as climate resilience is concerned.

19.1.11 The role of VSLAs on food security

There is a significant relationship between low levels of finance in agriculture and food insecurity in the world (FAO *et al.*, 2019). Finance plays a critical role as far as food security is concerned. This stems from the fact that access to reliable sources of finance is associated with increased production and productivity. The importance of the agricultural sector cannot be overemphasized. In Africa, agriculture is a key contributor to the continent's GDP (FAO, 2018) and as such, the majority of the population in African countries rely much on agriculture. However, there are funding challenges especially to the poor rural community. This is the niche that is being fully utilized by VSLAs. Several studies done in different countries have demonstrated that VSLAs have the potential to directly contribute to food security.

In Mozambique, participation in VSLA programs greatly increased the sufficiency of food in the household (Brunie *et al.*, 2014)

while in Zambia, VSLA participants utilized loans to take care of their expenditures, with significant spending on food (Noggie, 2017). In as much as there is evidence to support the positive contribution of VSLAs to food security, there are also dissenting views towards the said relationship. Some scholars (Gash, 2017) argue that enhanced access to savings and loans alone is a necessary but not sufficient condition to consistently lead to improved food security.

19.1.12 The impact of VSLAs on climate resilience

Savings are increasingly being put forward as a means through which both financial and social capital can be accessed, thereby assisting communities and households to cope with climate change induced hazards (VSLA Associates, 2017). In view of the foregoing, it is apparent that VSLAs play a dual role in as far as the enhancement of climate resilience is concerned, namely the provision of financial and social capital. With regard to financial services, it is evident that VSLAs come in the form of loans and savings accounts, which are expected to go a long way in absorbing climate change-related shocks Wydick (2016). In particular, financial services assist households and communities to cope with and adapt to negative impacts which are occasioned by extreme or unexpected weather conditions. This explains why financial services are highly regarded as an effective tool of enhancing resilience (Linnerooth-Bayer and Hochrainer-Stigler, 2015).

Over and above that, Azevedo (2007) notes that savings and loan components of VSLAs act as financial safety nets, which cushion participating members against extreme climatic conditions and disasters, thus enhancing the absorptive capacity of households and communities (Haworth *et al.*, 2016). Further evidence has demonstrated that access to loan facilities and savings accounts greatly improves disaster resilience by enhancing the coping capacity of affected communities and decreasing recovery time after a climate shock (Hudner and Krutz, 2015). In view of the above, it is apparent that literature has revealed beyond any shadow of doubt that VSLAs are significant predictors of climate resilience. VSLAs play a critical role

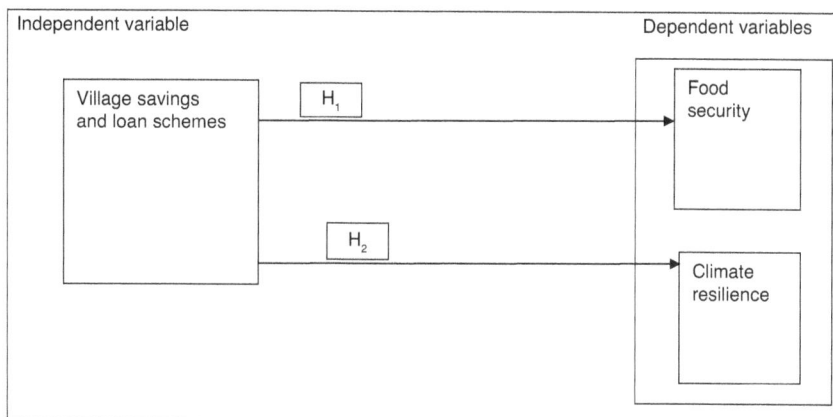

Fig. 19.1. Conceptual framework.

in enhancing climate resilience especially to group members, and by extension, the whole community stands to benefit. It has also been shown that VSLAs enhance the adaptive capability of communities and households to disasters such as desertification and soil salinity (Haworth et al., 2016).

However, the role of VSLAs in climate resilience is not always one-directional. Some community-based savings and loans groups may not be suitable to the kind of covariate risks that are brought about by climate shocks (Jones and Tanner, 2017).

19.1.13 Conceptual framework

The study developed the following research model: village savings are denoted to have a positive impact on climate resilience and food security (Fig. 19.1).

19.2 Research Methodology

19.2.1 Research paradigm

The study adopted a positivist research philosophy on the basis that causality was central to the main aim of the study, which was to determine the impact of VSLAs on climate resilience and food security in Zimbabwe. The study formulated hypotheses and gathered primary quantitative data which ought to be analysed statistically. All variables were measured on Likert scales to enable quantification.

19.2.2 Research design

In this study, an explanatory or causal research design was chosen. The chosen research design enabled the study to collect cross-sectional data for the three variables namely, VSLAs, food security and climate resilience, at once.

19.2.3 Study area

The study was conducted in ward 4 of Domboshava, a peri-urban communal area in Mashonaland East province of Zimbabwe. Domboshava falls under the local authority of Goromonzi Rural District Council and lies approximately 29 kilometres north-east of Zimbabwe's capital city, Harare. Ward four was selected mainly because smallholder farmers in that area largely depend on rainfed agriculture and therefore their susceptibility to climate change-related effects is high. The area is also one where most NGOs have implemented VSLA training to many farmers.

19.2.4 Study population

The target population comprised of all smallholder farmers who are beneficiaries of the

VSLAs schemes in Domboshava district. The farmers should have been actively participating in the scheme for the past 5 years. From the District Agricultural Extension offices, there were 90 such farmers. The number of respondents to consider in a particular study was guided by statistical formulae, an online sample size calculator called Raosoft.

19.2.5   Data analysis procedure

The collected data was populated into the Statistical Package for the Social Sciences (SPSS) version 28 for analysis. The package was used as it eliminates errors, which are associated with manual analysis. The analysis was done using both descriptive and inferential statistics. In addition, descriptive statistics were computed using measures of centrality such as mode, median and mean. Descriptive statistics were mainly used on demographic data. With regards to the remainder of the objectives, the study used inferential statistics.

19.3   Results and Discussion

19.3.1   Descriptive statistics

In this study, the concept of VSLAs, household food security, climate resilience and factors affecting VSLAs were measured on a 5-point Likert scale ranging from 1 (strongly disagree) to 5 (strongly agree). This means that if respondents were in strong disagreement with a given item, they would indicate that with a 1. Similarly, a strong agreement with a particular item would be signified by a 5. The neutral score was 3.0 and as such, it automatically became the cut-off point.

19.3.2   Inferential statistics

The study had formulated the following two alternate hypotheses:

H_1: There is a positive relationship between VSLAs and food security; and

H_2: There is a positive relationship between VSLAs and climate resilience.

To test the above hypotheses, the study conducted a bivariate regression analysis in respect of each hypothesis. As each hypothesis had only one independent variable, multi-collinearity issues did not emerge. Therefore, the study dealt with the assumption of normality in respect of the dependent variables from the two hypotheses. Results in Figs 19.2 and 19.3 show that all the points lie reasonably close to the line of best of fit. Therefore, the data was suitable for regression analysis to be performed.

Having satisfied the above assumption, the study went on to conduct regression analyses as follows:

H1: There is a positive relationship between VSLA and food security

Results in Table 19.1 show that the R square value is 0.351 and this implies that VSLAs account for 35.1% of variations in household food security. From the resultant Analysis of Variance (ANOVA) (Table 19.2), it is clear that the regression model summary achieves a statistically high degree of fit as evidenced by $F(1. 68) = 36.789$; $p<0.001$.

In view of the regression model summary and ANOVA results ($R = 0.593$, R square $= 0.352$, $F(1. 68) = 36.789$; $p<0.001$), the study fails to reject H1 and concludes that at the 5% level of significance, there is enough evidence to show that VSLAs have a positive relationship with household food security.

H2: There is a positive relationship between VSLA and climate resilience

Based on results in Table 19.3 it is clear that the R square value is 0.193 and this means that VSLAs account for 19.3% of variations in climate resilience. In addition, the study used ANOVA results to confirm whether or not the computed model significantly predicts the climate resilience (Table 19.4). From the ANOVA, it is clear that the regression model summary achieves a statistically high degree of fit as evidenced by $F(1. 68) = 15.225$; $p<0.001$.

In view of the regression model summary and ANOVA results ($R = 0.439$, R square $= 0.193$, $F(1. 68) = 15.225$; $p<0.001$), the study fails to reject H2 and concludes that at the 5% level of significance, there is enough evidence to show

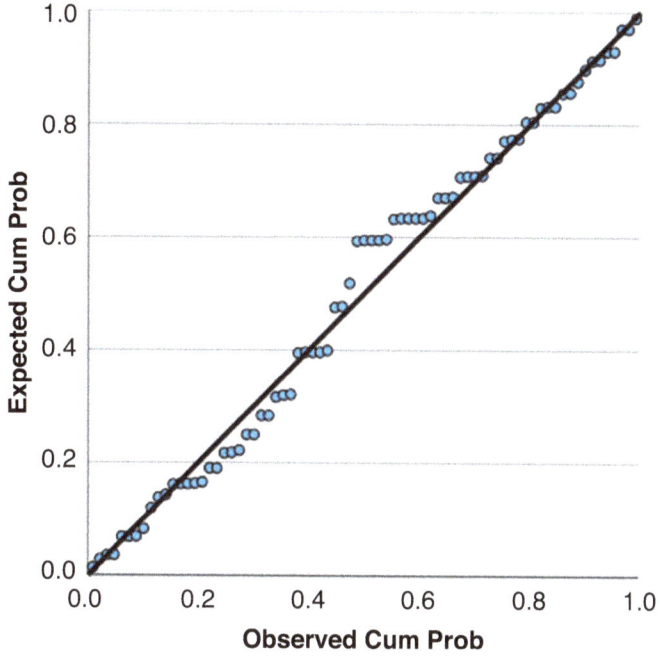

Fig. 19.2. Normal P–P Plot of regression standardized residuals: household food security.

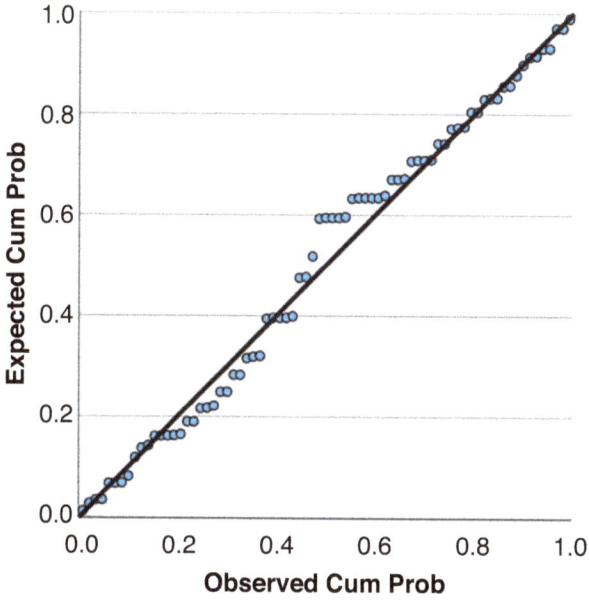

Fig. 19.3. Normal P–P Plot of regression standardized residuals: climate resilience.

Table 19.1. Regression model summary.

Model	R	R square	Adjusted R square	Standard error of the estimate
1	0.593[a]	0.351	0.342	0.56032

[a]Predictors: (Constant), VSLA.

Table 19.2. ANOVA.

Model		Sum of squares	Degrees of freedom	Mean square	F	Sig.
1	Regression	11.550	1	11.550	36.789	<0.001[a]
	Residual	21.349	68	0.314		
	Total	32.899	69			

[a]Predictors: (Constant), VSLA.

Table 19.3. Regression model summary.

Model	R	R square	Adjusted R square	Standard error of the estimate
1	0.439[a]	0.193	0.181	0.52877

[a]Predictor: (Constant), VSLA.

Table 19.4. ANOVA.

Model		Sum of squares	Degrees of freedom	Mean square	F	Sig.
1	Regression	4.536	1	4.536	16.225	< 0.001[a]
	Residual	19.013	68	0.280		
	Total	23.549	69			

[a]Predictors: (Constant), VSLA.

that VSLAs have a positive influence on climate resilience.

19.4   Discussion of Findings

The results above show that VSLAs have a positive and significant influence on both household food security and climate resilience. VSLAs contribute 35% of variations in household food security. However, the influence reduces by almost half when it comes to climate resilience. The multidimensional nature of the concepts of food security and climate resilience also means that they are influenced by several factors.

The results join other studies which have established a positive relationship between VSLA and food security. In Mozambique, participation in VSLA programs greatly increased the sufficiency of food in the household (Brunie *et al.*, 2014) while in Zambia VSLA participants utilized loans to take care of their expenditures, with significant spending on food (Noggie, 2017). Similarly, VSLAs have long been put forward as a means through which both financial and social capital can be accessed, thereby assisting communities and households to cope with climate change-induced hazards (VSLA Associates, 2017). Without access to sources of finance, the vulnerability of communities to extreme climatic conditions cannot be mitigated.

VSLAs act as financial safety nets which cushion participating members against extreme climatic conditions and disasters, thus enhancing the absorptive capacity of households and communities (Haworth *et al.*, 2016). Further evidence has demonstrated that access to loan facilities and savings accounts greatly improve disaster resilience by enhancing the coping capacity of affected communities and decreasing recovery time after a climate shock (Hudner and Krutz, 2015).

19.5 Conclusion

The VSLA model is the brainchild of CARE International (1971) and it is a form of microfinance which targets the marginalized and unbanked groups of society. However, there was a shortage of literature examining the efficacy of VSLAs in enhancing climate resilience and household food security, yet these are some of the principal reasons why the model was introduced.

The key findings revealed that VSLAs were accounting for 19.3% of the variations in climate resilience. The impact rose to 35.1% when it came to household food security.

References

Adedeji, O., Reuben, O. and Olatoye, O. (2014) Global climate change. *Journal of Geoscience and Environment Protection* 02(2), 114–122. DOI: 10.4236/gep.2014.22016.

Alcaraz, G. and Zeller, M. (2007) Use of household food insecurity scales for assessing poverty in Bangladesh and Uganda. *The Journal Nutrition* 5, 45–60.

Alise, M.A. and Teddlie, C. (2010) A continuation of the paradigm wars? Prevalence rates of methodological approaches across the social/behavioral sciences. *Journal of Mixed Methods Research* 4(2), 103–126. DOI: 10.1177/1558689809360805.

Armendariz, B. and Morduch, J. (2000) Microfinance beyond group lending. *The Economics of Transition* 8(2), 401–420. DOI: 10.1111/1468-0351.00049.

Asamoah, M. and Amoah, F.M. (2015) Microcredit schemes: a tool for promoting rural savings capacity among poor farm families: a case study in the Eastern region of Ghana. *Open Journal of Social Sciences* 03(1), 24–30. DOI: 10.4236/jss.2015.31003.

Azevedo, J. P. (2007) Microfinance and poor entrepreneurs. *The Life Cycle of Entrepreneurial Ventures* 4, 301–33.

Barnes, A., Fumagalli, L., Martin, T., Field, S. and Rutherford, D. (2004) Can village savings and loan groups be a potential tool in the malnutrition fight? Mixed method findings from Mozambique. *Children and Youth Services Review* 47, 113–120. DOI: 10.1016/j.childyouth.2014.07.010.

Brau, J.C. and Woller, G.M. (2004) Microfinance: a comprehensive review of the existing literature. *The Journal of Entrepreneurial Finance* 9(1), 1–28. DOI: 10.57229/2373-1761.1074.

Brunie, A., Fumagalli, L., Martin, T., Field, S. and Rutherford, D. (2014) Can village savings and loan groups be a potential tool in the malnutrition fight? Mixed method findings from Mozambique. *Children and Youth Services Review* 47, 113–120. DOI: 10.1016/j.childyouth.2014.07.010.

CARE International (1971) *A Randomized Impact Evaluation of Village Savings and Loan Association and Family-Based Interventions in Burundi*. International Rescue Committee, New York.

CARE International (2011) VSLA reaching the very poor: the need for a new microfinance model.

CARE International (2017) *An overview of the global reach of Care's Village Savings and Loan Association programming*. Care Global VSLA Reach 2017. Available at: https://insights.careinternational.org.uk/media/k2/attachments/CARE_VSLA_Global–Outreach–Report-2017.pdf

CARE International (2019) *Unlocking Access, Unleashing Potential: Empowering 50 Million Women And Girls Through Village Savings And Loan Associations By 2030*. CARE International, Atlanta.

Chuma, P., Sibomana, J.P. and Shukla, J. (2013) Effect of village savings and loan associations on SME growth in Rwanda. *International Journal of Business and Management Review* 4, 57–79.

Dahal, M. and Fiala, N. (2020) What do we know about the impact of microfinance? The problems of statistical power and precision. *World Development* 128, 104773. DOI: 10.1016/j.worlddev.2019.104773.

Daley-Harris, S. (2009) State of the Microcredit Summit Campaign Report. Microcredit Summit Campaign, Washington, DC.

Dzisi, S. and Obeng, F. (2013) Microfinance and the socio-economic wellbeing of women entrepreneurs in Ghana. *International Journal of Business and Social Research* 3, 45–62.

Eriksen, D. (2015) Climate change and soil salinity: the case of coastal Bangladesh. *Ambio* 44(8), 815–826. DOI: 10.1007/s13280-015-0681-5.

Evan, P.J. (2012) Effect of village savings and loan associations on SME growth in Rwanda. *International Journal of Business and Management Review* 4, 57–79.

FAO (1996) *Report of the World Food Summit*. Food and Agriculture Organization of the UN, Rome. Available at: https://www.fao.org/3/w3548e/w3548e00.htm

FAO (2011) *Global Food Losses and Food Waste. Extent, Causes and Prevention*. Food and Agriculture Organization of the UN, Rome.

FAO (2018) *The State of Agricultural Commodity Markets 2018. Agriculture Trade, Climate Change And Food Security*. FAO, Rome.

FAO, IFAD, UNICEF, WFP, and WHO (2019) *The State of Food Security and Nutrition in the World: Safeguarding Against Economic Slowdowns and Downturns*. FAO, Rome.

Gash, M. (2017) The evidence-based story of savings groups: a synthesis of seven randomized control trials. In: *The Bill and Melinda Gates Foundation: The SEEP Network*.

Gregory, H.C. (2015) Contrasting approaches to projecting longrun global food security. *Oxford Review of Economic Policy* 31(4), 26–44.

Havas, K. and Salman, M. (2011) Food security: its components and challenges. *International Journal of Food Safety, Nutrition and Public Health* 4(1), 4. DOI: 10.1504/IJFSNPH.2011.042571.

Haworth, A., Frandon-Martinez, C., Fayolle, V. and Simonet, C. (2016) *Climate Resilience And Financial Services*. BRACED, London.

Hendriks, B. (2011) *Does the Clean Development Mechanism Has A Viable Future?* Discussion Papers. Statistics Norway, Oslo.

Hongo, L.M. (2013) The role of financial services in climate adaptation in developing countries. *Climate Change* 74(2), 196–207. DOI: 10.3790/vjh.74.2.196.

Hudner, D. and Krutz, J. (2015) *Do Financial Services Build Disaster Resilience? Examining the Determinants of Recovery from Typhoon Yolanda in the Philippines*. Mercy Corps, Portland.

Jones, L. and Tanner, T. (2017) "Subjective resilience": using perceptions to quantify household resilience to climate extremes and disasters. *Regional Environmental Change* 17(1), 229–243. DOI: 10.1007/s10113-016-0995-2.

Joshi, N.P., Maharian , K.L. and Piva, L. (2011) Effects of climate variables on yield of major food-crops in Nepal – a time series analysis. *Journal of Contemporary India Studies: Space and Society* 5(2), 19–26.

Linnerooth-Bayer, J. and Hochrainer-Stigler, S. (2015) Financial instruments for disaster risk management and climate change adaptation. *Climatic Change* 133(1), 85–100. DOI: 10.1007/s10584-013-1035-6.

Mago, S. (2013) Microfinance in Zimbabwe: a historical overview. *Mediterranean Journal of Social Sciences* 4, 599–606. DOI: 10.5901/mjss.2013.v4n14p599.

Masiyandima, N., Ngundu, T., Kupeta, K., Moyo, P.S. and Ngwenya, S. (2012) Impact of village savings and loan associations. *Journal of Development Economics* 4, 14–21.

Mercy Corps (2009) Two criteria for good measurement in research: validity and reliability. *Annals of Spiru Haret University* 17(3), 58–82.

Mubaya, C., Tembo, G., Mwamba, Z. and Wamulume, M. (2012) Village savings and loan associations and household welfare. *African Journal of Agricultural and Resource Economics* 12(4), 85–97.

Nicholas, K.A. and Wynes, S. (2016) The climate mitigation gap: education and government recommendations miss the most effective individual actions. *Environmental Research Letters* 12(2017), 1–10. DOI: 10.1088/1748-9326/aa7541.

Noggie, C. (2017) Household Food Insecurity Access Scale: Method made easy hand-out.

Qureshi, M.I., Saleem, M.A., Shah, M., Abbas, Z., Qasuria, A.W. *et al.* (2012) Ensuring the role and impact: reaching the poorest while alleviating the poverty by Microfinance in Dera Ismail Khan, Pakistan. *Developing Country Studies* 2, 38–44.

Raftopoulos, B. and Lacoste, J.P. (2001) Savings mobilization to micro-finance: a historical perspective on the Zimbabwe savings development movement. In: *International Conference on "Livelihood, Savings and Debts in a Changing World"*, Wageningen.

Rahman, S., Junankar, P.N. and Mallik, G. (2009) Factors influencing women's empowerment on micro-credit borrowers: a case study in Bangladesh. *Journal of the Asia Pacific Economy* 14(3), 287–303. DOI: 10.1080/13547860902975648.

Sujatha, G. and Malyadri, P. (2015) Impact of Microfiance on women empowerment: an empirical evidence from Andhra Pradesh. *Journal of Entrepreneurship and Organizational Management* 4, 141–160.

Úbeda, B., Di Giacomo, A.S., Neiff, J.J., Loiselle, S.A., Poi, A.S.G, *et al.* (2013) Potential effects of climate change on the water level, flora and macro-fauna of a large neotropical wetland. *PloS One* 8(7), 112–123. DOI: 10.1371/journal.pone.0067787.

VSLA Associates (2017) What is VSLA (Village savings and loans)? VSLA Associates, reaching the very poor: the need for a new microfiance model. Available at: www.vsla.net/ (accessed 21 October 2021).

Wright, C. (2017) Research methods. *Journal of Business and Economics Research* 5(3), 40–56.

Wydick, B. (2016) Microfinance on the margin: why recent impact studies may understate average treatment effects. *Journal of Development Effectiveness* 8(2), 257–265. DOI: 10.1080/19439342.2015.1121512.

Yusuf, T., Annan, J. and Rutherford, D. (2016) Can village savings and loan groups be a potential tool in poverty alleviation? Mixed method findings from Mozambique. *Children and Youth Services Review* 47(2).

Yvonne Feilzer, M. (2010) Doing mixed methods research pragmatically: implications for the rediscovery of pragmatism as a research paradigm. *Journal of Mixed Methods Research* 4(1), 6–16. DOI: 10.1177/1558689809349691.

Zheke, S. (2017) Microfinance in Zimbabwe: a historical overview. *Mediterranian Journal of Social Sciences* 4(14), 599–606.

Zim Vac (2020) The effects of climate change and variability on food security in Zimbabwe. *International Journal of Humanities and Social Science* 3(6), 270–285.

Index

CABI – who we are and what we do

This book is published by **CABI**, an international not-for-profit organisation that improves people's lives worldwide by providing information and applying scientific expertise to solve problems in agriculture and the environment.

CABI is also a global publisher producing key scientific publications, including world renowned databases, as well as compendia, books, ebooks and full text electronic resources. We publish content in a wide range of subject areas including: agriculture and crop science / animal and veterinary sciences / ecology and conservation / environmental science / horticulture and plant sciences / human health, food science and nutrition / international development / leisure and tourism.

The profits from CABI's publishing activities enable us to work with farming communities around the world, supporting them as they battle with poor soil, invasive species and pests and diseases, to improve their livelihoods and help provide food for an ever growing population.

CABI is an international intergovernmental organisation, and we gratefully acknowledge the core financial support from our member countries (and lead agencies) including:

Ministry of Agriculture People's Republic of China

UKaid
from the British people

Australian Government
Australian Centre for
International Agricultural Research

Agriculture and
Agri-Food Canada

Ministry of Foreign Affairs of the
Netherlands

Schweizerische Eidgenossenschaft
Confédération suisse
Confederazione Svizzera
Confederaziun svizra

Swiss Agency for Development
and Cooperation SDC

Discover more

To read more about CABI's work, please visit: **www.cabi.org**

Browse our books at: **www.cabi.org/bookshop**,
or explore our online products at: **www.cabi.org/publishing-products**

Interested in writing for CABI? Find our author guidelines here:
www.cabi.org/publishing-products/information-for-authors/